ANALYTICAL CHEMISTRY OF URANIUM

Environmental, Forensic, Nuclear, and Toxicological Applications

ANALYTICAL CHEMISTRY OF URANIUM

Environmental, Forensic, Nuclear, and Toxicological Applications

ZEEV KARPAS

CRC Press
Taylor & Francis Group
Boca Raton London New York

CRC Press is an imprint of the
Taylor & Francis Group, an **informa** business

CRC Press
Taylor & Francis Group
6000 Broken Sound Parkway NW, Suite 300
Boca Raton, FL 33487-2742

First issued in paperback 2021

© 2015 by Taylor & Francis Group, LLC
CRC Press is an imprint of Taylor & Francis Group, an Informa business

No claim to original U.S. Government works

Version Date: 20140703

ISBN 13: 978-1-03-224003-9 (pbk)
ISBN 13: 978-1-4822-2058-2 (hbk)

Library of Congress Cataloging-in-Publication Data

Karpas, Zeev.
    Analytical chemistry of uranium : environmental, forensic, nuclear, and toxicological applications / Zeev Karpas.
        pages cm
        "A CRC title."
        Includes bibliographical references and index.
        ISBN 978-1-4822-2058-2 (alk. paper)
        1. Uranium--Analysis. I. Title.

    QD181.U7K27 2014
    546'.4316--dc23   2014025990

Visit the Taylor & Francis Web site at
http://www.taylorandfrancis.com

and the CRC Press Web site at
http://www.crcpress.com

Contents

Preface

When we say uranium, some of the first things that come to the mind are the atomic bomb and Hiroshima, followed by other downbeat associations such as Chernobyl and Fukushima. Indeed, uranium has been at the core of these catastrophic events and is viewed by many as the source of the largest man-made threat to life on our planet. On the other hand, around 15% of global electric power comes from uranium via more than 400 nuclear power plants. In addition, any realistic hope of restraining the release of carbon dioxide into the atmosphere and alleviating the *hothouse effect* and global warming depends on our ability to continue to supply electrical power, and perhaps indirectly build a hydrogen-driven economy, through nuclear reactors. Some may even argue that armed conflicts between nuclear powers have been contained and restrained owing to the repulsive horrors of World War II. However, there is no guarantee that in the future the spread of nuclear weapons would not lead to their use in international conflicts or even in domestic acts of terror. This tome is not judgmental about the pros and cons of production of uranium and the way it is used, but focuses on the analytical chemistry of uranium from the cradle (uranium prospecting and mining) to the grave (disposal of uranium compounds, nuclear fuel, and their radioactive progeny).

Considering the important role that uranium plays in all aspects of our life—supply of electrical energy, depleted uranium (DU) munitions and the controversy concerning their use, the threat to humanity posed by nuclear weapons, and the perils to the environment and life because of radioactive contamination—it is surprising that only a few scientific books are dedicated to uranium in general and analytical chemistry aspects in particular. A search on Amazon.com for "analytical chemistry of uranium" yields just one such title, a 104-page book written by Harry Brearly (born in 1871) and originally published around 1900. The chemistry of uranium was described in several books, and some sections thereof were devoted to the analysis of uranium compounds. On the other hand, the extent of scientific publications concerning uranium is very large. Over 186,000 articles containing the word "uranium" are found in SciFinder®, and searching for the concept "uranium analysis" more than 31,000 of those remain. In Google Scholar, slightly more than one million entries include the word "uranium," and the concept "uranium analysis" encompasses more than 840,000 entries.

As a scientist who has been involved with different analytical aspects of uranium, I felt that it was high time for the publication of a book that covers the fascinating advancements in the field of analytical chemistry of uranium. In view of the breadth of the field and the multitude of articles, I chose to focus on some of the more important facets (in my opinion): industrial processes that involve uranium, its presence in the environment, health and biological implications of exposure to uranium compounds, and nuclear forensics and its role in safeguarding. The approach adopted includes an overview of each topic followed by several examples to demonstrate the analytical procedures. These examples are meant to point out the variety of suitable methods and to whet the curiosity of interested readers, not to provide *cookbook* protocols. Highlights and insights are provided at the end of each section.

Finally, I hope that despite the fact that a broad range of topics are covered, each of which is worthy of a separate book, the readers of this book will gain an understanding of the analytical chemistry approach used today for characterizing the different facets of uranium. At the very least, I hope to provide the readers with a good starting point for investigation into this important element.

Acknowledgments

I thank Dr. Frank Winde for his astute comments on the environmental aspects of uranium determination (Chapter 3). I am grateful to Dr. Ludwik Halicz, who introduced me to the ICPMS technology, and to my colleague, the late Dr. Avi Lorber, who further expanded its analytical performance. I also thank Dr. Eyal Elish and Hagit Sela for their advice and assistance. Last, but certainly not least, special thanks to my wife, Hana, for her support and encouragement throughout my career and in preparing this manuscript.

Acronyms

AAS	Atomic absorption spectrometer (analytical device)
ADU	Ammonium diuranate ($(NH_4)_2U_2O_7$) (used to precipitate uranium)
AFR	Away from reactor (storage of spent fuel)
AGR	Advanced gas-cooled reactor
AMS	Accelerator mass spectrometer (analytical device)
AR	At reactor (storage of spent fuel)
ASTM	American Society for Testing Materials
ATSDR	Toxic Substances and Disease Registry (US Agency)
AUC	Ammonium uranyl carbonate $[(NH_4)_4UO_2(CO_3)_3]$ (U precipitate)
BWR	Boiling water reactor
CANDU	CANada Deuterium Uranium (type of nuclear reactor)
CAP	Canonical analysis of principle coordinates (data processing method)
CDC	Centers for Disease Control and Prevention (US Agency)
CGR	Gas-cooled reactor
CRE	Creatinine (2-amino-1-methyl-1H-imidazol-4-ol) (normalization agent)
CRM	Certified reference material
CTBT	Comprehensive Test Ban Treaty
DART	Direct analysis in real time (ionization method)
DDR	See GDR
DGT	Diffusive gradient in thin-films technique (preconcentration technique)
DM	Dry matter (in plants)
DNAA	Delayed neutron activation analysis (analytical method)
DPP	Differential pulse polarography (electroanalytical method)
DRC	Direct reaction cell (analytical method pertaining to ICPMS)
DU	Depleted uranium ($^{235}U < 0.72\%$, usually $0.2\%-0.4\%$)
EA	Electroanalytical
EDTA	Ethylene diamine tetra acetic acid (complex forming reagent)
EMIS	Electromagnetic isotope separation (calutron)
EMP	Electron microprobe (analytical device)
ENAA	Epithermal neutron activation analysis (analytical method)
EPA	Environmental Protection Agency (US agency)
ESA	Electrostatic analyzer (part of magnetic sector mass spectrometer)
ESI	Electrospray ionization (analytical method)
ETAAS	Electrothermal atomic absorption spectrometer (analytical device)
EXAFS	Extended x-ray absorption fine structure (analytical device)
FAS	Fixed air sampler
FIAS	Flow injection analysis system (analytical device)
FMCT	Fissile Materials Cut-off Treaty
FTA	Fission track analysis (analytical method)

FTIR	Fourier transform infrared spectrometer (analytical device)
FTMS	Fourier transform mass spectrometer (analytical device)
GDR	German Democratic Republic (formerly East Germany)
GFAAS	Graphite furnace atomic absorption spectrometer (analytical device)
GI tract	Gastrointestinal tract
Green Salt	Uranium tetrafluoride (UF_4)
HEU	Highly enriched uranium ($^{235}U > 20\%$)
Hex	Uranium hexafluoride (UF_6)
HPGe	High-purity germanium (gamma detector) (analytical device)
HRN	Human Rights Now (a nongovernmental organization)
IAEA	International Atomic Energy Agency
ICP-AES	Inductively coupled plasma–atomic emission spectrometer (analytical device)
ICPMS	Inductively coupled plasma mass spectrometer (analytical device)
ICP-OES	Inductively coupled plasma–optical emission spectrometer (analytical device)
ICPQMS	Inductively coupled plasma–quadrupole mass spectrometer (analytical device)
ICP-SF-MS	See SF-ICPMS (analytical device)
ICRP	International Commission on Radiological Protection
ID	Isotope dilution (analytical method)
IDMS	Isotope dilution mass spectrometry (analytical method and device)
IM-SIMS	Ion microprobe–secondary ion mass spectrometry
INAA	Instrumental neutron activation analysis (analytical method)
IND	Improvised nuclear device
ISL	In situ leaching—a method for the recovery of uranium from deposits
ITU	Institute for Transuranium Elements (in Karlsruhe, part of JRC)
IUPAC	International Union of Pure and Applied Chemistry
JRC	Joint Research Centre (European agency)
KPA	Kinetic phosphorescence analysis (analytical device)
LA	Laser ablation (pertains to ICPMS or ICPOES) (analytical device)
LB-α/β-CS	Low background alpha/beta counting system (analytical device)
LEG	Low-energy germanium (gamma detector) (analytical device)
LEU	Low enriched uranium ($^{235}U < 20\%$)
LG-SIMS	Large geometry–secondary ion mass spectrometry (analytical device)
LIBS	Laser-induced breakdown spectrometry (analytical device)
LIF	Laser-induced fluorescence (analytical device)
LLNL	Lawrence Livermore National Laboratory
LMFBR	Liquid metal fast breeder reactor
LOD	Limit of detection (see MDL)
MAGIC MERV	Mass, absorption, geometry, interaction, concentration, moderation, enrichment, reflection, and volume (criticality parameters)
MALDI	Matrix-assisted laser desorption ionization (analytical device)
MC	Multicollector (usually with ICPMS or TIMS) (analytical device)
MCN	Micro concentric nebulizer (Cetac, Omaha, NE) (analytical device)

MDA	Minimum detectable activity
MDL	Minimum detectable level (see LOD)
MIC	Multi-ion counting (detector in mass spectrometry)
MOX	Mixed oxide is a nuclear fuel containing UO_2 and PuO_2
MS	Mass spectrometry or mass spectrometer (analytical device)
MSR	Molten salt reactor
NAA	Neutron activation analysis (analytical method)
NBL	New Brunswick Laboratory
NBS	National Bureau of Standards (US agency, now called NIST)
NDA	Nondestructive analysis
NF	Nuclear forensics
NFC	Nuclear fuel cycle
NGO	Nongovernmental organization
NIR	Near-infrared reflectance spectroscopy (analytical device)
NIST	National Institute of Standards and Technology (US agency)
NMR	Nuclear magnetic resonance (analytical device)
NORM	Naturally occurring radioactive material
NPP	Nuclear power plant
NPT	Non-Proliferation Treaty
NRC	Nuclear Regulatory Commission (US agency)
NRCN	Nuclear Research Center, Negev (Israel)
NU	Natural uranium (natural isotopic composition—$^{235}U = 0.720\%$)
NUSIMEP	Nuclear Signatures Interlaboratory Measurement Evaluation Programme (intercomparison organized by the JRC)
ORNL	Oak Ridge National Laboratory
PEG	Polyethylene glycol (chemical reagent)
PERALS	Photon–electron rejecting alpha liquid scintillation (analytical device)
PFA	Perfluoroalkoxy alkanes (polymers) (relatively inert chemical polymer)
PHWR	Pressurized heavy water reactor
PPB	Part-per-billion
PPM	Part-per-million
PROCORAD	An international organization for interlaboratory intercomparisons
PTFE	Polytetrafluoroethylene (Teflon) (relatively inert chemical polymer)
PUREX	Plutonium, uranium extraction
PWR	Pressurized water reactor
QMS	Quadrupole mass spectrometer (see also ICP-QMS) (analytical device)
RBC	Red blood cells
RDD	Radioactive dispersion device (*dirty bomb*)
REE	Rare earth elements
RepU	Uranium obtained from irradiated (reprocessed) nuclear fuel
RMBK	*Reaktor bolshoy moschnosti kanalniy* (high power channel reactor)
RNAA	Radiochemical neutron activation analysis (analytical method)
SAL	Safeguards Analytical Laboratory (part of IAEA)
SDS	Sodium dodecyl sulfate (surfactant)

SEM	Scanning electron microscopy (analytical device)
SEU	Slightly enriched uranium (0.9%–2.0% ^{235}U)
SF	Spent fuel
SF	Spontaneous fission
SF-ICPMS	Sector field–inductively coupled plasma mass spectrometer
SG-SIMS	Small geometry–secondary ion mass spectrometer (analytical device)
SIA	Sequential flow injection (analytical method)
SIMS	Secondary ion mass spectrometer (analytical device)
SNF	Spent nuclear fuel
SNM	Special nuclear materials (^{239}Pu, ^{233}U, and enriched ^{235}U)
SPE	Solid-phase extraction (analytical method for preconcentration)
SRM	Standard reference material
TBP	Tri butyl phosphate (chemical reagent used to extract and purify uranium)
TDS	Total dissolved solids
TE	Total evaporation (method used in TIMS analysis)
TECDOC	Technical document (published by IAEA)
TIMS	Thermal ionization mass spectrometer (analytical device)
TOF-MS	Time of flight mass spectrometer (analytical device)
TOPO	Tri-n-octyl phosphine oxide (chemical used to extract and purify uranium)
TRLFS	Time-resolved laser-induced fluorescence spectrometry (analytical device)
TRU	A chromatographic resin for actinide separation (Eichrom, Darien, CT)
TRXRF	Total reflection x-ray fluorescence (analytical device)
UCF	Uranium conversion facility
ULB	Uranium lung burden
UNL	Uranyl nitrate liquor
UOC	Uranium ore concentrates
UTEVA	A chromatographic resin for actinide separation (Eichrom, Darien, CT)
UV	Ultraviolet (spectroscopy) (analytical device)
WDXRF	Wavelength dispersive x-ray fluorescence (analytical device)
WHO	World Health Organization
WIPP	Waste isolation pilot plant
WLM	Working-level-month (pertains to uranium miners)
WNA	World Nuclear Association
XRD	X-ray diffraction (analytical device)
XRF	X-ray fluorescence (analytical device)
Yellow cake	The term used rather loosely to describe a compound that contains uranium (usually >80% by weight). It has been applied to UOC, ADU, AUC, or U_3O_8.

1 Introduction

Fundamental Properties of Uranium and Its Compounds, the Nuclear Fuel Cycle, and Analytical Methods Used for Characterizing Uranium

[Pitchblende] consists of a peculiar, distinct, metallic substance. Therefore its former denominations, pitch-blende, pitch-iron-ore, &c. are no longer applicable, and must be supplied by another more appropriate name.—I have chosen that of uranite, (Uranium), as a kind of memorial, that the chemical discovery of this new metal happened in the period of the astronomical discovery of the new planet Uranus.

Martin Heinrich Klaproth

1.1 INTRODUCTION

We open with a concise introduction on the sources of uranium in nature and its main physical, chemical, and nuclear properties and then briefly discuss the chemistry of uranium and its compounds with emphasis on those that play an important role in uranium processing, namely, in the uranium fuel cycle. As we are concerned with the modern analytical chemistry of uranium, we present the foremost analytical techniques that are used nowadays to characterize uranium in its different forms.

In Chapter 2, we take a more detailed look at the analytical chemistry pertaining to key commercial activities, that is, uranium mining and its utilization in the nuclear fuel cycle (NFC): first, in the milling process, uranium-containing deposits are processed to form uranium ore concentrates (UOC) that are then shipped to uranium conversion facilities (UCF), where the uranium is transformed into high-purity *nuclear grade* compounds. These can serve as fuel for nuclear power plants or as feed material for isotope enrichment. Then we discuss the analytical aspects of compliance with the strict specifications of the materials used in enrichment plants and in fuel fabrication facilities. Finally, we deal with the analytical procedures to characterize irradiated fuel and waste disposal of spent fuel.

This is followed, in Chapter 3, by the description of the analytical procedures used for monitoring the presence of uranium in the environment: air, water, soil, and plants. As there is a large variety of sample types, there is no universal procedure for all environmental samples. Therefore, we present a myriad of sample treatment procedures and preparation methods as well as some separation and preconcentration techniques. Several examples of analytical procedures, based on the sample preparation methods, are described to demonstrate the diversity required for characterizing environmental samples.

In Chapter 4, we discuss the effects of uranium on human life and well-being with a detailed survey of the methods and means of estimating internal exposure to uranium on the basis of bioassays (urine, feces, blood, hair, nails, and some nonstandard assays). We also present a detailed review of the analytical methods used to assess the amount of uranium in food products and drinking water that are the main pathways of exposure to uranium of the general population.

Finally, in Chapter 5, we delve into the exciting realm of trace analysis and nuclear forensics that can yield information on the source, nature, use (and abuse) of uranium, which in some cases can be based on meticulous analysis of a single grain or particle.

While Chapters 3 through 5 are concerned mainly with the detection of trace levels of uranium in several different matrices and its characterization, in Chapter 2 the focus is on determining the level of contaminants and trace impurities in compounds in which uranium is the main component.

Throughout the book we tried to focus on work that was published in the last decade (after 2005), but older groundbreaking and fundamental studies are also discussed. As this is not a textbook—although it can be used for tutoring purposes—we presented only brief overview explanations of some of the major issues. This is also not a *recipe* book containing the ultimate methods and analytical procedures, dictated by institutional regulations, simply because there is no universally accepted method for assays of uranium. The methodology we use is based on *study by example* approach, and in all cases several examples are presented in order to demonstrate that there are many ways to perform every type of analysis for characterizing the sample. At the end of each section, we attempted to bring forth the highlights of the analytical approach and helpful insights and provide a perspective of the subject.

We are aware that each of the aforementioned topics is worthy of several separate books and hope that readers of this book will gain an understanding of the analytical chemistry approach used today for characterizing the different facets of uranium. At the very least we expect the readers to get a good starting point for further investigation of this important element.

In many cases, it is almost impossible to locate the primary sources of data concerning the properties of uranium (the element, isotopes, and compounds), the flow sheet diagrams of the industrial processes used throughout the NFC (milling, conversion, enrichment, fuel fabrication, spent fuel characterization, and waste disposal), as well as description of the instruments used in analytical chemistry. So, in order to avoid infringement of copyrights, I tried to find publications from which copyright permission could be granted or rely on materials that are in public domain (like Handbooks and

Wikipedia). Unfortunately, due to these restrictions, in a few cases I could not use what I considered the best figures, photos, tables, and diagrams and had to create my own diagrams, settle for substitutes of lower quality, or completely refrain from including them in the book. As an example, consider the phase diagram of UF_6 that is well known, but it is practically impossible to trace its origin (one version in Figure 1.8 later in the chapter).

1.2 FUNDAMENTALS OF URANIUM AND ITS CHEMISTRY

In this chapter, some basic facts about the history of uranium, its occurrence in nature and in minerals, its physical, nuclear, and chemical properties and its isotopes, as well as some of the terminology related to uranium are described in brief. This is followed by an overview of the NFC in which uranium plays a major role. Finally, a short account of the main analytical techniques used to detect and characterize uranium is presented.

1.2.1 HISTORY AND PROPERTIES OF THE ELEMENT

1.2.1.1 Discovery and History

Uranium was discovered in 1789 by the German chemist Martin Klaproth (1743–1817) during his study of the mineral pitchblende. He found that the ore contained a substance that did not behave like iron and zinc and concluded that a new element was present. Klaproth named it after the planet Uranus that was discovered a few years earlier, but after some further tests, he realized that he had found the oxide and not the pure element. Elemental uranium was first isolated in 1841 by French chemist Eugene Melchior Peligot (1811–1890), who converted the oxide into the chloride form and reduced it with elemental potassium (Aczel 2009). The radioactive nature of uranium was discovered by Henri Becquerel quite accidentally in 1896 when he realized that photographic plates that were placed near uranium-containing salts blackened although they were not exposed to light. Some of the physical properties of metallic uranium are summarized in Table 1.1.

After the discovery of radium in the early 1900s, the main interest in mining uranium was for extraction of radium that was considered as a panacea for several diseases. Until World War II, there was only a marginal interest in uranium and it was used in some stains and dyes, colored glass, and some types of specialty steels. The annual global trade consisted of only a few hundred tons (compared with over 50,000 tons today). However, uranium's potential for nuclear weapons (making atomic bombs with enriched uranium and its use in production of plutonium) was realized at the start of the war. In addition, its high density and hardness were utilized for making armor-piercing ammunition. These developments made uranium exploration and processing an important economic commodity (Zoellner 2010). By the 1950s, the construction of nuclear power plants for the production of electricity further increased the interest in mining and extraction of uranium. A detailed treatise of uranium was published as a 450-page chapter (Grenthe 2006) and is part of the monumental work on the chemistry of actinide and transactinide elements.

The world production of uranium in 2009 was 50,572 tons, and about two-thirds of this amount came from mines in three countries: Kazakhstan, Canada, and Australia

TABLE 1.1

Summary of the Properties of Uranium Metal

Property		Property	
Element classification	Radioactive actinide series	Appearance	Silvery white, dense, ductile, malleable
Density	19.05 g mL^{-1}	Oxidation states	6, 5, 4, 3
Melting point	1405.5 K	Boiling point	4018 K
Fusion heat	12.6 kJ mol^{-1}	Evaporation heat	417 kJ mol^{-1}
Pauling negativity	1.38	Specific heat at 20°C	0.115 J gmol^{-1}
Covalent radius	142 pm	Ionic radius	80 (VI); 97 (IV) pm
Lattice structure	Orthorhombic	Lattice constant	2.850 Å
First ionization	686.4 kJ mol^{-1}	Magnetic ordering	Paramagnetic
Electrical resistivity (0°C)	0.280 μΩ m	Thermal conductivity (300 K)	27.5 W m^{-1} K^{-1}
Thermal expansion (25°C)	13.9 μm m^{-1} K^{-1}	Speed of sound (thin rod) (20°C)	3155 m s^{-1}
Young's modulus	208 GPa	Shear modulus	111 GPa
Bulk modulus	100 GPa	Poisson ratio	0.23

Source: Based on data from https://en.wikipedia.org/wiki/Uranium, accessed July 26, 2014.

(see Chapter 2). It should be noted that the uranium market was highly volatile in the past (e.g., prices between 2001 and 2007 ranged from $7 to $137 lb^{-1}) as shown in Chapter 2 (Figure 2.1). Uranium is not freely traded in the commodity markets like other metals, but rather mostly in direct negotiations between buyers and sellers where political considerations, as well as economic concerns, play an important role. For a better understanding of the political issues and the undercurrents of the global uranium market, see Zoellner's *Uranium: War, Energy and the Rock That Shaped the World*, which is highly recommended (Zoellner 2010).

1.2.1.2   Occurrence

Uranium is quite common in our planet. The average concentration of uranium in the Earth's crust is ~2.4 parts per million (ppm) or 2.4 μg U g^{-1}, and in seawater, its level is ~3.1 parts per billion (ppb) or 3.1 μg U L^{-1}. The most common minerals containing uranium are pitchblende, uraninite, carnotite, uranophane, and coffinite, but uranium is present in many other minerals at lower levels (see Section 1.2.2).

1.2.1.3   Physical Properties

Uranium is the heaviest naturally occurring element with atomic number 92. Its appearance as a metal is described as silvery and shiny with a high density of 19.05 g cm^{-3}, a melting point of 1132°C, and a boiling point of 3818°C, and the metallic form is slightly paramagnetic. The metal is ductile and malleable so that it can be formed as wires or thin sheets in addition to rods, ingots, and other bulky configurations. Uranium metal appears in three crystallographic phases: alpha, beta, and gamma. In pure form, it is a bit softer than steel and not hard enough to scratch glass. A summary of the properties of uranium metal is shown in Table 1.1. Alloying with other elements strongly affects these properties.

1.2.1.4   Nuclear Properties and Isotopes

There are 25 known isotopes of uranium, ranging in mass from 217 to 241, but only three occur naturally at significant concentrations—^{238}U, ^{235}U, and ^{234}U—with natural abundance of about 99.274%, 0.7204%, and 0.00548%, respectively. The main nuclear properties—natural abundance, half-life, decomposition pathways, and their energy, as well as the cross section for neutron capture, of the uranium isotopes from ^{232}U to ^{238}U—are summarized in Table 1.2.

When discussing the isotope composition of uranium, the abundance is usually given in units of atom percent (i.e., the relative number of atoms of each isotope in a given mass of uranium) rather than in terms of weight percent (i.e., the relative weight of each isotope in the uranium mass). Thus, the weight percent of each isotope in natural uranium differs slightly from the atom percent as shown in parenthesis in Table 1.2. ^{235}U is the only fissile isotope of natural uranium, but the artificial ^{233}U isotope (produced in thorium-fueled nuclear reactors) is also fissile. All uranium isotopes are radioactive and the half-lives of the isotopes of importance are shown in Table 1.2. Due to the important role of radioactivity in the uranium industry, a short discussion is presented in Frame 1.1.

TABLE 1.2
Some Properties of the Major Isotopes of Uranium

Isotope	Natural Abundance, atom% (wt%)	Half Life (Years)	Decomposition (MeV) Pathway	Energy	Product[a]	Cross Section[b], Barns (σ_f)
^{232}U	Trace	68.9	SF	—	—	73
			α	5.414	^{228}Th	
^{233}U	Trace	$1.595 * 10^5$	SF	197.93		47
			α	4.909	^{229}Th	$\sigma_f = 530$
^{234}U	0.00548 (0.00539)	$2.455 * 10^5$	SF	197.78		96
			α	4.859	^{230}Th	
^{235}U	0.720 (0.711)	$7.04 * 10^8$	SF	202.48		95
			α	4.679	^{231}Th	$\sigma_f = 586$
^{236}U	Trace	$2.342 * 10^7$	SF	201.82		5.1
			α	4.572	^{232}Th	
^{238}U	99.274 (99.284)	$4.468 * 10^9$	SF	205.87		2.7
			α	4.270		
			β^-		^{238}Pu	

Source: Based on data from https://en.wikipedia.org/wiki/Uranium, accessed July 26, 2014.

Note: SF, spontaneous fission; σ_f, the cross section for fission—the cross section of natural uranium is 3.4 barn for capture of thermal neutrons and 4.2 for fission.

[a] Immediate product, in most cases followed by further decomposition.

[b] Cross section for capture of thermal neutrons (barn = 10^{-24} cm²).

FRAME 1.1 RADIOACTIVITY AND URANIUM

The nucleus of an atom contains protons and neutrons as well as other lighter subatomic particles. These are held together by the strong nuclear force that is effective over the short range of 10^{-13} cm within the nucleus. Electrostatic repulsive forces between the positively charged protons and the weak nuclear force also play a role in determining the stability of the nucleus. In some nuclei, the subatomic particles may rearrange spontaneously or due to influence by external forces like penetration of a neutron. This rearrangement may be accompanied by the release of energy in the form of electromagnetic radiation (gamma or X-rays), subatomic particles (electrons, positrons, protons, neutrons, etc.), more massive particles (alpha particles), or even large fragments of the parent nucleus (as in spontaneous or neutron-induced fission). Frequently, several forms of energy are emitted simultaneously due to mass, energy, and momentum conservation laws. In many cases, the daughter nuclei are also unstable and undergo further rearrangement with the release of energetic particles or radiation as in the natural decay series (Figures 1.3 through 1.6, later in the chapter). The emanation of some particles is accompanied by the production of other elements. For example, beta emission leads to formation of an element with a higher atomic number but only a small mass change occurs, while proton emission results in a shift to a lower atomic number with a mass change of one unit (Da). Release of an alpha particle would lead to formation of an element that is lower by two atomic numbers and four mass units. Emission of a neutron or of electromagnetic radiation would not change the atomic number, that is, the parent and daughter nuclei are of the same element, but in a different energetic state and/or a different isotope. A schematic representation of some of these processes is shown in Figure 1.1. As shown here, emission of a beta particle increases the atomic number (Z) of the nuclide and decreases the number of neutrons (N) by one so that the total number of nucleons in the nucleus is retained. Emission of a positron causes the opposite—Z decreases and N increases by one unit. Emission of a neutron decreases N without changing Z while emission of a proton leads to the opposite effect. Emission of an alpha particle (helium-4 nucleus) reduces Z and N by two units.

The term *half-life* ($t_{1/2}$) is defined as the required time for the activity of a radioactive substance to decay to half of its initial value. For some nuclides like ^{238}U, this would take $4.468 * 10^9$ years while in other uranium isotopes like ^{232}U this would occur after 68.9 years (see Table 1.2). Thus, the rate of emission of alpha particles from a given number of ^{232}U nuclei would be about 64.5 million times higher than from the same number of ^{238}U nuclei.

The nuclear transitions are usually between well-defined energy levels within the nucleus so that the gamma rays or alpha particles emitted are characteristic of the parent nucleus. Thus, natural radiation plays an important role in some of the analytical techniques used to characterize samples, as discussed for gamma spectrometry and alpha spectrometry in Section 1.4. In addition, radioactivity

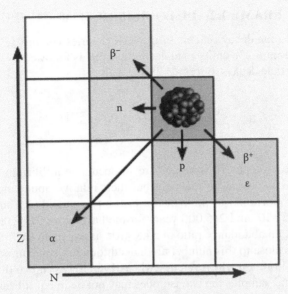

FIGURE 1.1   Schematic of radioactive decay modes. (From http://upload.wikimedia.org/
wikipedia/commons/thumb/7/71/Radioactive_decay_modes.svg/201px-Radioactive_
decay_modes.svg.png, accessed July 26, 2014.)

can be induced when the nucleus of a stable atom captures a neutron (or another
particle) or absorbs enough electromagnetic radiation. The excited atom emits
radiation or particles that can also serve in analytical chemistry for neutron
activation analysis (NAA) and related techniques (see Section 1.4).

It should be noted that the abundance of the naturally occurring isotopes may
vary according to the origin of the uranium ore. Until very accurate measurements
of ^{235}U became available, its abundance was thought to be practically constant in
nature. Variations in the abundance of ^{234}U were recognized several decades ago and
are known as the disequilibrium of ^{234}U (Frame 1.2).

Some of the uranium isotopes undergo spontaneous fission (SF), which is a process
in which the nucleus of the element splits into two or more large fragments (usually
accompanied by emission of neutrons and radiation) without external forces such as
those arising from neutron capture. All the uranium isotopes listed in Table 1.2 and
later in the chapter in Table 1.3 also emit alpha particles, a process that leads initially
to the formation of thorium isotopes that subsequently undergo further dissociation
processes and form other elements like radium (by alpha emission) or protactinium
(by beta emission). Alpha emission is usually accompanied by gamma ray emission.
The energy and probability of the major alpha particles and gamma rays emanating
from uranium are summarized in Table 1.3, later in the chapter.

Note that the values reproduced in these tables for the energy of alpha particles differ slightly from each other as do the values reported in other literature
sources. Uranium-238 may also emit a beta particle (followed by a second beta

FRAME 1.2   DISEQUILIBRIUM OF [238]U–[234]U

According to the decay scheme of the 4n + 2 series (Figure 1.3, later in the chapter), uranium-238 emits an alpha particle that is followed by two consecutive beta particle decays to form U-234:

$$^{238}\mathrm{U} \xrightarrow[\alpha]{} {}^{234}\mathrm{Th} \xrightarrow[\beta]{} {}^{234}\mathrm{Pa} \xrightarrow[\beta]{} {}^{234}\mathrm{U}$$

If these two isotopes of uranium are in secular equilibrium, their alpha particle activity would be identical but their relative abundance would be inversely proportional to the half-life of the two nuclides. Thus, using the values of $4.5 * 10^9$ and 245,000 years for the half-lives of U-238 and U-234, respectively, an abundance ratio of $54.8 * 10^{-6}$ is expected for [234]U/[238]U atom ratio. Values close to this number are indeed found in some mineral deposits that contain uranium. However, as early as 1963 it was shown that in natural environmental samples the two isotopes may not be in equilibrium (Thurber 1963). Since then this phenomenon was widely investigated. Several publications reported that in water samples from various areas of the globe much higher ratios are often found. In some cases, the abundance ratio of [234]U/[238]U may be several times higher than the equilibrium value, with a record high report of samples from hot water springs in Japan where the ratio is 51 times higher (Yamamoto et al. 2003).

The mechanistic basis for this phenomenon was explained in detail (Ivanovich 1994). Simply put, the uranium atoms occupy certain positions in the lattice of the mineral. When an alpha particle is emitted from U-238 (to form U-234), the parent atom recoils and its position in the lattice changes slightly and leads to creation of a fault in the lattice. Water that is in contact with the mineral will preferentially leach uranium atoms that lie close to these faults, leading to an increase in the fraction of U-234 in the leachate. This effect is clearly seen in alpha spectra of natural samples where the [234]U peak is often significantly larger than the [238]U peak or in accurate mass spectral measurements.

The extent of the disequilibrium can serve as a natural isotopic tracer for proving exposure pathways as seen in Figure 1.2 that shows the correlation of the [234]U/[238]U ratio in drinking water and bioassays (hair and nails) of residents of south Finland that consume elevated levels of uranium in their drinking water (Karpas et al. 2006). For some samples, the ratio was more than double the ratio calculated for secular equilibrium.

The [235]U/[238]U ratio is quite constant in natural terrestrial deposits, but slight variations have been found and are discussed in Chapter 3 (Hiess et al. 2012).

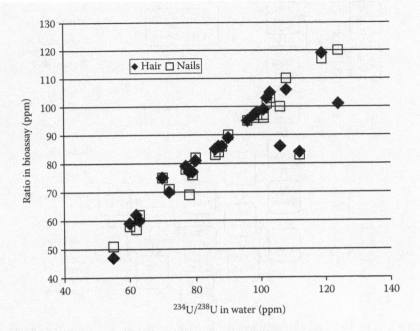

FIGURE 1.2 The $^{234}U/^{238}U$ ratio in drinking water and in bioassays of hair and nails in samples collected from residents of southern Finland that consume elevated uranium levels in their drinking water. (Adapted from Karpas, Z. et al., *Radiat. Prot. Dosimetry*, 118, 106, 2006. With permission.)

particle), leading to the formation of relatively short-lived ^{238}Pu. The isotopes of uranium have considerable cross sections for neutron capture, which explains their importance for construction of nuclear reactors and production of nuclear weapons (Table 1.2).

1.2.1.5 Radioactive Decay Chains

The uranium isotopes U-238 and U-235, after radioactive decay through several steps, eventually form stable lead isotopes (Pb-206 and Pb-207, respectively), as shown in Figures 1.3 and 1.4.

There are two additional radioactive decay chains: another natural decay chain starting with Th-232 ending with stable Pb-208 and an artificial decay chain that begins with Np-237 and goes through U-233 until a stable nuclide of Tl-205 is formed, shown in Figures 1.5 and 1.6, respectively.

The atomic weight of the nuclides in these decay chains changes by four atomic mass units after the emission of each alpha particle so they are commonly labeled as the 4n, 4n + 1, 4n + 2, and 4n + 3 decay chains according to atomic weight of the

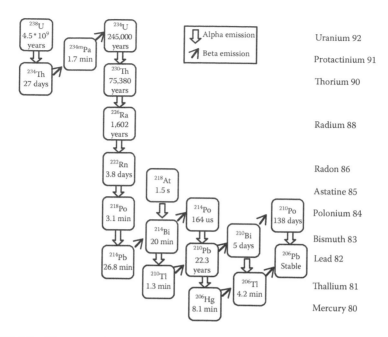

FIGURE 1.3 The natural radioactive decay chain of the 4n + 2 series, beginning with uranium-238 and ending with lead-206.

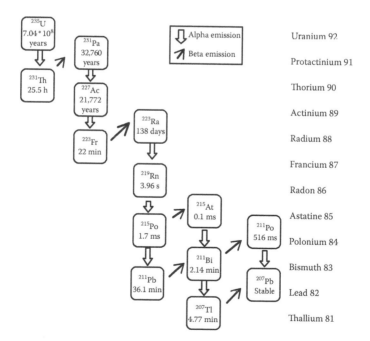

FIGURE 1.4 The natural radioactive decay chain of the 4n + 3 series, beginning with uranium-235 and ending with lead-207.

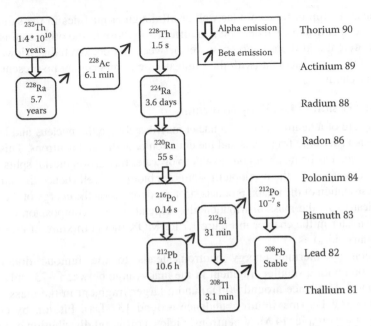

FIGURE 1.5 The natural radioactive decay chain of the 4n series, beginning with thorium-232 and ending with lead-208.

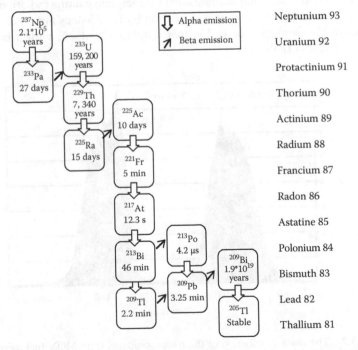

FIGURE 1.6 The artificial radioactive decay chain of the 4n + 1 series, beginning with neptunium-237, decaying through uranium-233 and ending with thallium-205.

nuclide at the top of the chain (as well as all daughter nuclides in the same chain). Thus, the nuclides in each decay chain differ from those in the other chains by their atomic weight and there is no crossover between the decay chains. However, isotopes of the same element (with the same atomic number) may be present in more than one chain.

1.2.1.6 Fission of U-235 by Neutrons

The capture of a neutron by U-235 nuclei may destabilize the nucleus and lead to its fission into two large fragments and the emission of additional neutrons. This process was described as being analogous to a drop of water that grows until it splits into two smaller droplets or to the partition (fission) of a biological cell (hence the name). The mass distribution of the fission products strongly depends on the energy of the neutron. In nuclear fuels, the mass distribution also depends on the composition of the fuel, as shown later in the chapter in Figure 1.7 for MOX fuel (a mixture of uranium and plutonium oxides).

Fission by thermal energy neutrons leads to the famous double-hump distribution with a smaller fragment in the mass range between ~73 and ~110 Da (maximum abundance around 90 Da) and a larger fragment in the mass range of ~125 to ~162 Da (maximum abundance around 137 Da). Fission by energetic neutron, for example, 14 MeV neutrons, yields a different distribution profile with almost no double-hump (a flat top). It should be noted that many of the initially formed fission products are not stable (they usually have an excess of neutrons) and emit beta particles, often accompanied by energetic gamma radiation. In some cases, the capture of a thermal energy neutron by U-235 does not result in fission but leads to the formation of U-236 that is relatively stable with a half-life of ~23 million years (Table 1.3).

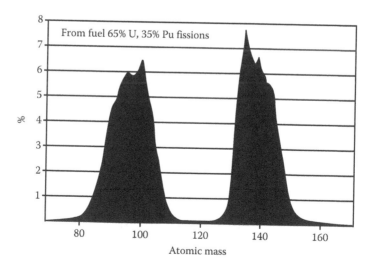

FIGURE 1.7 The mass distribution of the fission products from MOX fuel. (From World Nuclear Association, http://www.world-nuclear.org/uploadedImages/org/info/Nuclear_Fuel_Cycle/distribution_of_fission_products.png, accessed July 26, 2014. With permission.)

TABLE 1.3
Energy (keV) and Probability (%) of Emitted Alpha Particles and Gamma Rays of the Major Isotopes of Uranium

Isotope	Half-Life (Years)	Eγ (keV)	Iγ (%)	Eα (keV)	Iα (%)
U-232	68.9			5263.41	31.7
				5320.17	68.0
U-233	$1.592 * 10^5$			4783.5	13.2
				4824.2	84.4
U-234	$2.455 * 10^5$	53.2	0.123	4722.4	28.42
		120.90	0.0342	4774.6	71.38
U-235	$7.038 * 10^8$	143.76	10.96	4366.1	17
		185.715	57.2	4397.8	55
U-236	$2.342 * 10^7$			4445	25.9
				4494	73.8
U-238	$4.468 * 10^9$	49.55	0.064	4151	20.9
		113.5	0.0120	4198	79.0
U-239	23.45 min	74.664	48.1		

Source: Based on Firestone, R.B. et al., *Table of Isotopes*, book and CD-ROM, 8th edn., John Wiley & Sons, Inc., New York, 1996.

[a] There are slightly different values in different sources.

Natural (NU or Unat), depleted (DU), low-enriched (LEU), and high-enriched (HEU) uranium: the content of the only natural fissile isotope, ^{235}U—is an important feature of uranium applications and value. In natural uranium, the content of this isotope is 0.720 atom % or 0.711 wt% (Table 1.2). LEU is defined as ^{235}U content between 0.720% and just below 20%, while HEU encompasses uranium with ^{235}U content above 20%. The 20% borderline between LEU and HEU is artificial and was based on the assumption that nuclear weapons with 20% or less ^{235}U would not be efficient. The waste, or tails, of the isotope enrichment process contains less ^{235}U than in natural uranium and is defined as depleted uranium (DU). The U-235 content in DU is usually in the range of 0.2%–0.4%. DU is used mainly in armor piecing ammunition, in reactive armor of tanks, in radiation shielding, and is also used as ballast weights in aircraft. In addition, many of the commercially available fine chemicals of uranium compounds are based on the tails of uranium-enrichment facilities and usually labeled as not of natural isotope composition.

Uranium that is used to fuel light water nuclear power reactors is generally enriched to a level of 3%–5% and considered as LEU. Several reactors that are used for research, material testing, and production of isotopes for medical and industrial applications also use LEU fuel, sometimes with as much as 19.75% ^{235}U. Other research reactors and nuclear-powered ships and submarines require a higher content of the fissile isotope in order to reduce the size of the reactor core. Weapon grade uranium typically contains around 90% ^{235}U (HEU). Thus, the value and cost of uranium is strongly dependent on the fraction of ^{235}U present. Some sources also refer to

RepU that is defined as uranium recovered from reprocessed irradiated nuclear fuel (also known as spent fuel) and thus will contain artificial uranium isotopes like ^{236}U and ^{232}U (and also traces of other actinides and fission products) or *SEU* for slightly enriched uranium that contains 0.9%–2.0% of ^{235}U.

1.2.1.7 Alloys

Uranium forms binary alloys with about 70 different elements and a similar number of ternary alloys as well as several quaternary (and more complex) alloys. These may contain different proportions of the components so that overall the number of combinations in which uranium is alloyed is extremely large. Some of the common alloy compositions of industrial importance are discussed in Chapter 2.

1.2.1.8 Chemical Properties

The chemical properties of uranium are derived from its electronic structure: 92 protons and 92 electrons (the number of neutrons varies from 125 to 149). Six electrons are in its outer shell with an electron configuration of $[Rn]7s^2 5f^3 6d^1$ so the two most common valence states are +6 with $[Rn]5f^0$ configuration and +4 with $[Rn]7s^2$ configuration. Trivalent and pentavalent compounds are also known, but their commercial importance is quite negligible. When uranium metal is exposed to air, an oxide layer is formed. Finely divided metal is pyrophoric and can ignite spontaneously in air. The chemistry, production, and reactions of some of the most commercially important uranium compounds, especially those that play a role in the NFC, are discussed later.

1.2.2 MINERALOGY OF URANIUM

One of the interesting features of uranium deposits is their widespread occurrence from a geographical point of view as shown in Table 1.4 that lists the countries with >2% of the world uranium production (there are several other countries that also have uranium production facilities).

There is also a large variability from a geological aspect of the types of minerals and their geological deposits, where as many as 14 types have been described (Grenthe 2006). It has been stated that "The low crustal abundance of uranium belies its mineralogical and geochemical significance: more than five percent of minerals known today contain uranium as an essential constituent." (Burns and Finch 1999).

A detailed discussion of the occurrence of uranium in nature and its mineralogical aspects was presented (Grenthe 2006) so here we shall briefly mention some of the major uranium-bearing minerals. The primary minerals that contain uranium in the oxide form arc uraninite (UO_2), pitchblende (mainly U_3O_8), and coffinite $[U(SiO_4)_{1-x}(OH)_{4x}]$. There are also titanates like brannerite (UTi_2O_6) and davidite $[(REE)(Y,U)(Ti,Fe^{3+})_{20}O_{38}]$, for example, as well as several secondary uranium minerals (some of which are fluorescent or brilliantly colored) like autunite $[Ca(UO_2)_2(PO_4)_2 \cdot 8-12H_2O]$ and carnotite $[K_2(UO_2)_2(VO_4)_2 \cdot 1-3H_2O]$, for example.

TABLE 1.4
List of Countries with More Than 2% of the World Share of Combined Reserves and their Historical Production (in Metric Tons)

Country	Reserves as of 2009	World Share (%)	Historical Production to 2008	World Share (%)
Australia	1,673,000	31.0	156,428	6.5
Brazil	278,700	5.2	2,839	0.1
Canada	485,300	9.0	426,670	17.7
China	171,400	3.2	31,399	1.3
Germany	0	0	219,517	9.1
Kazakhstan	651,800	12.1	126,900	5.3
Namibia	284,200	5.3	95,288	4.0
Niger	272,900	5.0	110,312	4.6
Russia	480,300	8.9	139,735	5.8
South Africa	295,600	5.5	156,312	6.5
Ukraine	105,000	1.9	124,397	5.2
United States	207,400	3.8	363,640	15.1
Uzbekistan	114,600	2.1	34,939	1.4

Source: Adapted from http://en.wikipedia.org/wiki/List_of_countries_by_uranium_reserves, accessed July 26, 2014.

Notes: Historical production for the Czech Republic includes 102,241 tons of uranium produced in former Czechoslovakia from 1946 through the end of 1992. Historical production for Germany includes 213,380 tons produced in the German Democratic Republic from 1946 through the end of 1992. Historical production for the Soviet Union includes the former Soviet Socialist Republics of Estonia, Kyrgyzstan, Tajikistan, and Uzbekistan, but excludes Kazakhstan and the Ukraine. Historical production for the Russian Federation and Uzbekistan is since 1992 only.

Many of the hexavalent compounds of uranium that contain uranyl ions (UO_2^{2+}) are readily soluble in water and can therefore be dissolved and transported by groundwater in the environment (see Chapter 3). Uranium may then be precipitated if oxidation conditions or alkalinity change, thus forming deposits in a wide variety of geological locations. Uranium tends to fractionate when magmas are formed and therefore may accumulate in alkaline granite melts. Exceptionally rich deposits of uranium are found in the Athabasca Basin in Canada where the average grade of U_3O_8 reaches 18%. Among the various types of deposits that contain uranium are major unconformities between quartz-rich sandstone and deformed metamorphic basement rocks (like the Athabasca Basin). Uranium deposits in sandstone may occur when the oxidation conditions change to a reducing environment followed by mineralization, in a river bed, for example. Uranium can often be found in phosphate deposits at levels that reach several hundred mg kg^{-1} or in the waste heaps of gold mines. An interesting anecdotal history of uranium exploration and exploitation in the Congo, Niger, Australia, the United States, and the former Eastern bloc (specifically East Germany and the Czech Republic) among other locations was presented elsewhere and makes sobering reading (Zoellner 2010).

1.2.3 Major Uranium Compounds

As mentioned earlier, the two most common and stable types of uranium compounds are those in which uranium is in the tetravalent and hexavalent states. As far as the NFC is concerned, the binary oxides, binary fluorides, and oxyfluorides are of major importance, although several other compounds (e.g., uranyl nitrate, uranyl sulfate, and ammonium uranates) also play a role in the processing and handling of uranium (Table 1.5). Uranium metal and especially uranium alloys also play a significant role in commercial and military applications of uranium.

Any discussion of uranium and its chemical properties should refer to the classic 1951 book by Katz and Rabinowitch that was the first publication that provided a comprehensive description of the chemical and physical properties of uranium: the element and its binary and related compounds (Katz et al. 1951). More than 60 years later, and despite the considerable developments and extensive research on the chemistry of uranium, this tome is still an excellent primary source. The section on uranium in the series on actinide and trans-actinide chemistry (ATAC) was mentioned earlier as a superb source for understanding the behavior of uranium compounds (Grenthe 2006). A less-known volume that focused on the industrial and technological applications of uranium was translated from Russian in the 1960s and also is useful (although somewhat outdated in parts) for following the production processes of uranium (Galkin 1966).

As this monograph is concerned with the analytical chemistry of uranium, focusing on the processes and compounds that are deployed on industrial and commercial

TABLE 1.5

Some Properties of the Main Uranium Compounds

Compound	Density (g cm^{-3})	Melting Point (°C)	Color	Main Applications	Comments
UO_2	10.97	2827	Black	Nuclear fuel	Insoluble in H_2O
UO_3	5.5–8.7	200–650	Orange-yellow	Intermediate in fuel cycle	Insoluble in H_2O crystallographic forms
U_3O_8	8.38	1300 dec.	Olive green-black	Disposal	Insoluble in H_2O
$UO_4 \cdot nH_2O$			Pale yellow	Intermediate in U mining	$UO_2(O_2) \cdot nH_2O$ Soluble in H_2O
UF_4	6.7	1036	Green	Intermediate in fuel cycle	Insoluble in H_2O
UF_6	5.09	64.05	Colorless	Isotope enrichment	Soluble in H_2O Significant vapor pressure
UO_2F_2			Yellow to orange	Intermediate in fuel cycle	Soluble in H_2O
$UO_2(NO_3)_2 \cdot 6H_2O$	2.81	60	Yellow-green	Intermediate in fuel cycle	Hygroscopic Very soluble in H_2O
UCl_4	4.725	590	Dark green	Calutron enrichment	Decomposes in water Oxidizes in air
UH_3	11.1		Gray-black		

scales, we will not delve into the *exotic* complexes with organic ligands in which uranium may be present in unusual valence states or coordination arrangements or discuss alloys (e.g., ternary alloys with iron and germanium) where several different compositions may be found on a microscopic scale. All these can be found in the literature and proceedings of conferences but are not significant for this treatise on the analytical chemistry of uranium.

1.2.3.1 Binary Oxides of Uranium

The most common oxides are uranium dioxide (UO_2), uranium trioxide (UO_3), triuranium octoxide (U_3O_8), and uranyl peroxide (UO_4 or $UO_2 \cdot O_2$). The main properties of these compounds are summarized in Table 1.5. It should be noted that there are several other oxides in which the oxidation states are not well defined as either tetravalent or hexavalent, like $UO_{2.1}$, U_2O_5, U_3O_7, $U_{12}O_{35}$, etc. Detailed discussions of the uranium–oxygen system and its complex phase diagram were presented elsewhere, for example, by Allen and Tempest (1982) and Grenthe (2006).

Uranium dioxide (UO_2) appears naturally in the minerals uraninite and pitchblende. It is a black powder (the color may vary from brown, through bluish gray to black) with a density of 10.97 g cm^{-3} and is industrially produced in the NFC by the reduction of UO_3 (see Section 1.3). The reaction rate of UO_2 with oxygen strongly depends on the particle size and the temperature. Very fine UO_2 powders (diameter of 0.1 µm) will react with air at room temperature forming $UO_{2.25}$ within a month while larger particles (diameter of 1 µm) would be only slightly oxidized to $UO_{2.02}$ and sintered pellets are practically stable in air at ambient temperatures (Galkin 1966). The most common use of UO_2 is in nuclear fuel, usually after the powder is sintered into pellets. This application utilizes the oxide's high melting point (2865°C), but it should be noted that its low thermal conductivity could lead to hot spots in the fuel element and that in the presence of oxygen above 700°C it may be converted to U_3O_8. Mixed oxide (MOX) nuclear fuel contains a mixture of UO_2 and PuO_2 and may play an important role in future generations of nuclear power plants. Uranium dioxide can be fluorinated by HF, ammonium fluoride, and Freons at elevated temperatures to produce UF_4—the *green salt* that is an intermediate compound in UCF. UO_2 is practically unaffected by dilute acids but is attacked by concentrated nitric acid to form uranyl nitrate and the oxide may also be dissolved by alkaline solutions of hydrogen peroxide to form peruranates (Galkin 1966). There are some other applications of UO_2 that can serve as a catalyst in chemical reactions, as a radiation shield (usually after the uranium-235 has been depleted), and under development are applications in the semiconductor industry and in electric batteries. Uranium oxides were used in the past in the production of colored glass and ceramics, but due to the toxic properties of uranium as a heavy metal and as a radioactive element, this has lost favor over the last decades. A detailed description of the analytical chemistry of this important uranium compound is presented in Chapter 2, where the role of UO_2 in the NFC and its properties and characterization are discussed.

Uranium trioxide (UO_3) is a hexavalent binary compound that appears mainly as an orange-yellow powder with a density of 5.5–8.7 g cm^{-3} and melting point of 200°C–650°C. The broad range of densities and melting points is derived from the strong dependence of the compound on the crystalline form (α, β, γ, and δ) and the production method. In the *Handbook of Chemistry and Physics* (Lide 1999), the density is given as ~7.3 g cm^{-3}. The compound is an important intermediate in the NFC as it is the first well-defined solid uranium compound of nuclear grade purity. In the fuel cycle, it is produced by calcinations of the precipitated ammonium diuranate (ADU), ammonium uranyl carbonate (AUC), or concentrated uranyl nitrate liquor (UNL) (see Section 1.3 and Chapter 2). In the presence of reducing agents, UO_3 is converted to UO_2—an important step in the NFC. The reaction of UO_3 with water leads to the formation of several hydrates that are thermally unstable and lose water when heated. The reaction of UO_3 with HF or HCl leads to formation of uranyl fluoride (UO_2F_2) and uranyl chloride (UO_2Cl_2), respectively, but in the presence of reducing agents,UF_4 and UCl_4 are produced, and in the reaction with fluorine, UF_6 is formed. Heating UO_2 or U_3O_8 in an atmosphere of oxygen will also lead to the formation of UO_3. One of the most interesting chemical properties of UO_3 is its amphoteric nature as an acid and as a base that enables it to form peruranate anions (UO_4^{-2}) and uranyl cations (UO_2^{+2}). Uranium oxide dissolution in strong acids forms uranyl ions that are readily soluble in several types of organic compounds (like diethylether or tributylphosphate—TBP). This property serves in the separation of uranium from other components of irradiated fuel or from other elements in the purification process of uranium ores.

Triuranium octoxide (U_3O_8) occurs naturally in pitchblende and is an olive green to black solid (color depends on the conditions of production) with a particle density of 8.3 g cm^{-3} and melting point of 1150°C. It is one of the most stable uranium compounds and as such has been a candidate for long-term geologic storage and disposal of uranium in repositories. Despite its color, it is sometimes referred to as *yellow cake* when shipped between mills and refineries as uranium ore concentrates (UOC) that contains 65%–85% U_3O_8. U_3O_8 is formed by oxidation of many uranium compounds like UO_2, uranium salts, and metallic uranium when heated in air (dry or moist) above 800°C–900°C or when UO_3 loses oxygen upon heating above 500°C. The reactions of U_3O_8 with dilute sulfuric and hydrochloric acids are slow even upon heating, but addition of an oxidizing agent like nitric acid or hydrogen peroxide accelerates dissolution. Alkaline solutions do not affect U_3O_8 and carbonate solutions selectively leach the hexavalent components (Galkin 1966). Due to its stability at temperatures below 1000°C and the well-defined composition, U_3O_8 serves in the gravimetric determination of uranium. In fact, the ignition of almost all uranium compounds in air would lead to formation of U_3O_8.

Uranium peroxide ($UO_4·nH_2O$) *and uranyl oxyhydroxides*: Uranium peroxide is a pale yellow hygroscopic solid in the form of small needles that is highly soluble in water. It is obtained mainly by reaction of excess hydrogen peroxide on an aqueous solution of uranyl nitrate. It usually appears as a hydrate (n = 0–4) and is found in some minerals (studtite and meta-studtite). Uranium peroxide is an intermediate formed when uranium *yellow cake* is prepared by in situ leaching and resin ion exchange system. When heated to 90°C–195°C, it slowly decomposes to form other oxides, like U_2O_7, and subsequently to UO_3. The oxyhydroxides play an important role in the geochemistry of uranium as they are formed in uranium-rich solutions.

1.2.3.2 Binary Fluorides of Uranium

There are several binary fluorides of uranium in which the formal valence state of uranium ranges from 3 to 6 like UF_3, UF_4, UF_5, UF_6, U_2F_9, U_4F_{17}, and U_4F_{18}. The two most important of these with regard to the NFC are UF_4 and UF_6 and will be briefly discussed here. The other binary fluorides will be mentioned in context in the appropriate sections of this book.

Uranium tetrafluoride (UF_4) plays an important role in the NFC (see Section 1.3) as it is the key compound from which uranium metal and uranium hexafluoride are prepared in UCF. It is a green solid (hence the name *green salt*) with a very low vapor pressure and is practically insoluble in water, like most tetravalent uranium compounds. Its bulk density ranges from 2.0 to 4.5 g cm^{-3}, depending on the preparation process and starting materials. It can be prepared by the action of anhydrous HF (AHF) on UO_2 or by reduction of UF_6 with hydrogen or other reducing agents, like an excess of CCl_4 for example. UF_4 may also be produced by precipitation from aqueous solutions of uranyl compounds with reducing agents, but the precipitated $UF_4 \cdot nH_2O$ contains water molecules (n = 0.5–2.5) that are difficult to remove and it is therefore unsuitable for mettalothermic reduction (with calcium or magnesium) or for fluorination but can be used in the ion exchange conversion electrolytic reduction (Excer) process. One of the potential applications of UF_4 is in *homogenous* molten salt nuclear reactors that may be used commercially in the future. The reaction of UF_4 with water vapor above 200°C leads to an equilibrium reaction (Equation 1.1) that depends on the temperature and concentration of the reactants:

$$UF_4 + 2H_2O \Leftrightarrow UO_2 + 4HF \tag{1.1}$$

If oxygen is also present, then uranyl fluoride (UO_2F_2) may be formed. Heating UF_4 in dry air above 300°C may lead to formation of uranyl fluoride and U_3O_8 (Galkin 1966). An interesting reaction of UF_4 with pure dry oxygen at 800°C may lead to formation of UF_6 without the need to use fluorine (Equation 1.2):

$$2UF_4 + O_2 \rightarrow UF_6 + UO_2F_2 \tag{1.2}$$

The reaction of UF_4 with fluorine at 250°C–400°C results in the formation of UF_6, but it should be noted that nonvolatile intermediate uranium-fluorides (UF_{4+x}, x = 0–2) like those listed earlier may also be formed. Strong fluorinating agents, for example, ClF_3 or BrF_3, may react with UF_4 at lower temperatures to form UF_6.

Uranium tetrafluoride is barely affected by nonoxidizing acids (HCl, H_2SO_4, H_3PO_4) at ambient temperature, but hot concentration phosphoric acid or sulfuric acid may form tetravalent phosphates or sulfates, respectively. On the other hand, UF_4 is fairly soluble in oxidizing acids (HNO_3 and $HClO_4$), especially in the presence of boric acid, leading to the formation of uranyl ions (UO_2^{+2}). Heated alkaline solutions decompose UF_4 and convert it into tetravalent uranium hydroxide (Galkin 1966).

Uranium tetrafluoride forms stable coordination complex fluorides with ammonium fluoride and many alkali and alkali-earth fluorides like NH_4UF_5 or Na_2UF_6 and several other fluoride compounds. The melting point of the complex double salts formed in the NaF-UF_4 system is much lower than that of the components (Galkin 1966).

Uranium hexafluoride (UF$_6$), also called hex, is probably the best known and most widely investigated compound of uranium mainly because it is the only uranium compound with significant vapor pressure at ambient temperatures and therefore an essential raw material for most commercial isotope enrichment processes. UF$_6$ is a white monoclinic crystalline solid that sublimes directly to a gas (reaches atmospheric pressure at 56.5°C), but when heated in a closed vessel will melt at 64.05°C, which is the triple point where the solid, liquid, and gas phases coexist, as shown in Figure 1.8. This is probably one of the most well-recognized phase diagrams in the chemical literature.

UF$_6$ is produced commercially by the reaction of UF$_4$ with fluorine (Equation 1.3), but in practice it can be produced by the action of any strong fluorinating agent (like ClF$_3$, BrF$_5$, etc.) on any uranium compound (metal, oxide, U-containing mineral, etc.) under proper conditions:

$$UF_4 + F_2 \rightarrow UF_6 \tag{1.3}$$

An alternative production pathway for the production of UF$_6$ from UF$_4$ that does not require elemental fluorine or fluorinating agents is shown in Equation 1.2, where oxygen is used.

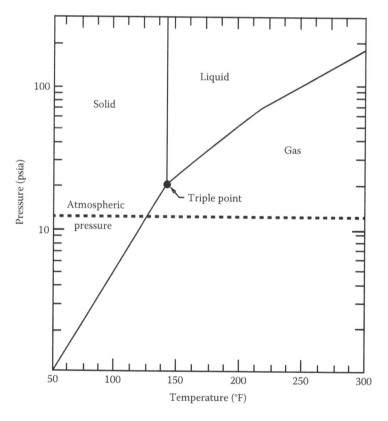

FIGURE 1.8   The phase diagram of UF$_6$. Note the direct sublimation of the solid and the triple point where solid, liquid, and gaseous UF$_6$ coexist. (From Hanna, S.R. et al., *Atmos. Environ.*, 31, 901, 1997. With permission.)

Chemically, perhaps the most important property of UF_6 is its ability to act as an oxidizing agent and is therefore reactive toward many organic compounds, water vapor, and several other chemicals. UF_6 can form stable solutions with some organic solvents, mainly per-fluorinated hydrocarbons as well as anhydrous HF and halogen fluorides, but reacts with common solvents like ethanol, ether, and benzene at ambient temperature. UF_6 is inert in several dry gases like nitrogen, oxygen, CO_2, chlorine, and bromine. Its reaction with reducing agents leads to formation of UF_4 and intermediate uranium fluoride compounds mentioned earlier.

Although dry UF_6 can be stored in glass vessels, its reaction with moisture leads to formation of yellow uranyl fluoride (UO_2F_2) and HF, which is very corrosive and attacks the silica of the glass vessel, as shown next:

$$UF_6 + 2H_2O \rightarrow UO_2F_2 + 4HF \tag{1.4a}$$

$$SiO_2 + 4HF \rightarrow SiF_4 + 2H_2O \tag{1.4b}$$

This is an autocatalytic reaction that can continue in principle until all the silica or the UF_6 is consumed while the HF and H_2O molecules remain in the system.

UF_6 can also react with most metals by the formation of a metal fluoride and is consequently reduced to UF_4 (or other nonvolatile intermediate uranium fluorides). Metals that are suitable for producing, handling, and storing UF_6 are those upon which a continuous, protective, adhesive layer of the fluoride is formed in this reaction (like nickel and some of its alloys, copper and aluminum). Construction materials that form flake-like fluorides that are detached from the surface and expose new surfaces (like iron and several of its alloys) are not recommended for use with UF_6 (as shown in Equation 1.5 for a generic divalent metal, M):

$$M + UF_6 \rightarrow MF_2 + UF_4 \tag{1.5}$$

As a Lewis acid UF_6 forms solid double salts with many solid and gaseous inorganic fluorides, like $NaUF_7$ and Na_2UF_8 with NaF, as shown, for example, in the following reversible reaction:

$$2NaF + UF_6 \Leftrightarrow Na_2UF_8 \tag{1.6}$$

The most interesting feature of UF_6 is its role as the feed material for the two main commercial isotope enrichment processes (gaseous diffusion and gas centrifuges), but it is also the product and tails (or waste) of these processes. Namely, its chemical form does not change during the whole enrichment cycle and only the isotope composition of uranium is altered. The fact that fluorine is mono-isotopic greatly simplifies the enrichment process. Finally, it should be noted that due to the important role UF_6 plays in the nuclear industry there are strict safety and security guidelines and procedures for producing, handling, storing, and transporting the compound. Furthermore, because its nuclear properties strongly depend on the level of the uranium isotope enrichment, these safety guidelines differ according to the [235]U content and the danger of causing an undesired nuclear criticality accident (see Frame 1.3).

FRAME 1.3   CRITICALITY OF URANIUM

One of the most important nuclear properties of uranium is that when a ^{235}U nuclide captures a neutron it can undergo a fission reaction that involves splitting of the uranium nucleus into two smaller nuclei (fission products) and the release of a number of neutrons (Figure 1.29a later in the chapter). The emitted neutrons can cause other ^{235}U nuclei to fission and also release neutrons and so forth, thus creating a chain reaction. Criticality occurs when the number of nuclei undergoing fission from these neutrons is equal to, or larger than, the number of nuclei in the previous generation. In a nuclear reactor, the fission rate is controlled by absorbing excess neutrons with special control rods. However, if the number of fissions increases from generation to generation in an uncontrolled manner, a *runaway* chain reaction may occur, leading to the release of energy (each fission reaction is accompanied by the release of around 200 MeV per fission), neutrons, gamma radiation (from the fission products), and several types of radioactive elements. This unintentional *runaway* chain reaction is termed a criticality accident.

A criticality accident with uranium may occur if ^{235}U nuclei accumulate under certain conditions. The criteria for criticality control are known as MAGIC MERV—the acronym for the parameters that affect the possibility of occurrence of a chain reaction: Mass, Absorption, Geometry, Interaction, Concentration, Moderation, Enrichment, Reflection, and Volume. It should be noted that if uranium is present in a homogeneous solution the probability of reaching criticality is significantly higher than in any other form. The smallest critical mass of ^{235}U for a spherical-shaped solution at optimal concentration and moderation is 0.784 kg for a water-reflected solution (reflection refers to neutrons that are returned to the vessel with the uranium mass instead of escaping from it) and 1.42 kg for an unreflected solution. For the unreflected spherical mass of uranium metal, the required ^{235}U mass is 46.7 kg while with a steel reflector the minimum mass is 16.8 kg (Koelzer 2011).

The first and foremost requirement for reaching criticality is the accumulation of a sufficient mass of U-235; if the mass of fissile nuclei is below a certain value (about 800 g as mentioned earlier), criticality cannot occur. The other factors can be simply described as contributing to the probability of an inadvertent chain reaction like the presence of reflectors and the presence of a moderator that will thermalize the neutron energy increasing the capture probability as well as favorable geometric conditions (spherical symmetry). The concentration of uranium and its level of enrichment determine the density of ^{235}U atoms in the system. On the other hand, neutron absorbers or elements that undergo interaction with the neutrons and unfavorable geometric conditions or concentration (H/U ratio) in solution would lead to an increase in the critical mass and diminish the probability of a criticality incident.

Several tables and graphs showing the calculated dependence of the critical mass of uranium on the level of ^{235}U enrichment and on the H/U ratio can be found in a report from Oak Ridge National Laboratory (Fox et al. 2005).

The calculations included two systems: a configuration of a solution of uranyl nitrate or a mixture of UO_2 and water in spherical symmetry and in slabs of 0.5 in. thickness. For example, the critical mass of a system that consists of a water-reflected sphere of $UO_2 + H_2O$ at 5% and 20% U-235 enrichment with a 30 cm diameter depends on the U/H ratio. The minimum mass drops from 1.8 to 1.0 kg as the enrichment increases from 5% to 20% and the volume fraction of uranium decreases from 0.1 to 0.03. However, if the concentration of uranium is shifted (increased or decreased) from the optimal volume fraction ratio, then the ^{235}U mass needed for criticality will increase.

Additionally, the health hazards posed by UF_6, particularly radiation safety concerns, also depend on the level on enrichment (see Chapter 4). Uranium-containing deposits (called heels) may form in cylinders that are used for storage of UF_6 either through hydrolysis with moisture traces to produce UO_2F_2 or through reduction of UF_6 on the wall surfaces or with impurities. Accumulation of the *heels* gives rise to safety concerns and operational problems.

1.2.3.3   Oxyfluorides of Uranium

Uranyl fluoride (UO_2F_2) is the most common oxyfluoride of uranium while UOF_4 should also be considered as an intermediate in a two-stage mechanism in the hydrolysis of UF_6 (1.7):

$$UF_6 + H_2O \rightarrow UOF_4 + 2HF \qquad (1.7a)$$

$$UOF_4 + H_2O \rightarrow UO_2F_2 + 2HF \qquad (1.7b)$$

This hydrolysis reaction plays an important role in determining the fate of UF_6 that is released into the atmosphere (see Chapters 3 and 5). Accidental release of UF_6 into the atmosphere and its reaction with moisture leads to formation of aerosols or small particles of UO_2F_2 (and clusters with water and HF molecules) that slowly sink to the ground. Uranyl fluoride is hygroscopic, and its color changes from orange to yellow as it absorbs water. If UO_2F_2 is heated in air, it slowly decomposes by the loss of fluorine atoms and U_3O_8 is eventually formed. Controlled hydrolysis of UF_6 and formation of a UO_2F_2 solution is an important stage in the analytical characterization of UF_6 that serves as feed materials for enrichment as described in detail in Chapter 2. This hydrolysis reaction is also the basis for examining the quality of the UF_6 product and tails that are produced in the enrichment process. Uranyl fluoride is also readily soluble in alcohol and acetone but not in ether or higher alcohols like amyl-alcohol. Addition of HF to uranyl fluoride solutions decreases its solubility. The reaction of UO_2F_2 with hydrogen at 500°C–700°C proceeds rapidly to the formation of UO_2 and UF_4.

1.2.3.4   Other Uranyl Compounds

Uranyl ions, UO_2^{+2}, play an important role in the extraction and concentration of uranium-containing minerals and in the purification of uranium compounds.

Strong oxidizing acids, like nitric acid, produce stable hexavalent uranyl ions that can be extracted by suitable complex forming agents like tributylphosphate (TBP). Uranyl compounds are generally very soluble in aqueous solutions and concentrations of several hundred grams of uranium per liter are quite common, like uranyl nitrate with solubility of 660 g L^{-1}. Uranyl acetate, uranyl sulfate, uranyl chloride, and uranyl phosphate are all yellow salts and usually appear as hydrates. Uranyl fluoride, the product of hydrolysis of UF_6, was discussed earlier.

Uranyl nitrate $UO_2(NO_3)_2 \cdot 6H_2O$ is produced by the action of nitric acid on uranium compounds and plays an important role in UCF and in the processing of irradiated nuclear fuel. In crystalline form, it is a yellow-green hexahydrate that is tribo-luminescent (when rubbed, crushed, or shaken, the friction leads to emission of light). It is usually found in solution and has a yellow transparent appearance. It is also soluble in ethanol, acetone, and ether but not in common aromatic solvents or chloroform. Thus, solutions of uranyl nitrate are used for recovery of uranium from various process streams in the industry: high-grade ores, dissolved solid waste, uranium metal shavings, defective fuel elements, etc. When uranyl nitrate is extracted from an aqueous solution into an organic solvent, it is the hydrate of uranyl nitrate containing 2–4 water molecules that is transferred. In the common commercial extraction and purification route, uranyl nitrate forms complexes with tributylphosphate $[UO_2(NO_3)_2 \cdot 2TBP \cdot nH_2O]$ or diethyl ether $[UO_2(NO_3)_2 \cdot (C_2H_5)_2O \cdot 3H_2O]$.

Uranyl carbonate complexes, like sodium uranyl tricarbonate, $Na_4[UO_2(CO_3)_3]$, that is obtained when uranium ore is leached with sodium carbonate solutions and ammonium uranyl carbonate (AUC), $(NH_4)_4[UO_2(CO_3)_3]$, that is used to precipitate the uranium in the UCF, are important in the NFC. These carbonates serve to purify the uranium from several metals (like Fe, Al, Cr, Ni, and other metals) that are precipitated as hydroxides or oxycarbonates, as well as alkaline-earth elements. These purification methods utilize the effect of the ammonium carbonate concentration on the solubility of uranium. Upon heating of AUC to 300°C–500°C, it decomposes to UO_3, ammonia, CO_2, and water and at temperatures of 700°C–800°C, without air, UO_2 may be formed (the ammonia serves as the reducing agent). The solubility of AUC decreases markedly in the presence of ammonium carbonate, for example, from 119.3 g L^{-1} at 50°C without ammonium carbonate to 0.5 g L^{-1} with 35% ammonium carbonate (Galkin 1966). The carbonate complexes also play a role in biological systems and affect clearance by the blood after exposure to uranium compounds.

Uranyl sulfate usually appears as a lemon-yellow trihydrate ($UO_2SO_4 \cdot 3H_2O$) with a density of 3.28 g cm^{-3} and is very soluble in 5 parts of water and 25 parts of alcohol. In geochemistry, oxidation of sulfides would lead to formation of the sulfate, mainly in an acidic environment where carbonates are not present and cause precipitation of uranium. Uranyl sulfate plays a major role in ore processing as it is readily absorbed on anion-exchange resins and may be extracted with amines. As the uranyl sulfate is very stable, the solutions can be heated to elevated temperatures that help dissolve difficult to digest ores.

1.2.3.5   Other Uranium Compounds in the Nuclear Industry

Some additional uranium compounds that play a role in the nuclear industry are discussed in brief here.

Ammonium diuranate $(ADU = (NH_4)_2U_2O_7)$: This compound, together with AUC mentioned earlier, is the compound commonly used in uranium conversion facilities to precipitate uranium from the purified uranyl nitrate solution by bubbling ammonia (and CO_2 for AUC). Although the uranium is in the hexavalent state, it readily deposits from the solution and the precipitate can be filtered mechanically. The advantage of using ADU or AUC is that both compounds produce UO_3 when they are calcined and the ammonium and carbon dioxide released during the calcination can be recycled. In the literature, ADU (and AUC) is sometimes referred to as yellow cake and is often confused with triuranium octoxide (U_3O_8) that is also called by the same name. There are several other carbonates and uranates (Grenthe 2006) that do not play a significant role in the NFC.

Uranium chlorides: Uranium tetrachloride (UCl_4) is the feed material used in the Calutron (or electromagnetic isotope separation (EMIS)) enrichment process. The dark green crystals oxidize in air and decompose in water so should be kept in sealed vials or an inert dry atmosphere. The compound is prepared by reaction of UO_3 with hexachloropropene or by directly reacting chlorine with uranium or uranium hydride. The density of UCl_4 is 4.725 g cm^{-3}, and the melting and boiling points are 590°C and 791°C, respectively. The compound uranium trichloride can be prepared by hydrogen reduction of UCl_4.

Uranium hydride: The action of hydrogen on uranium at 250°C–300°C can produce a fine powder in the form of UH_3 that is a convenient starting material for production of various uranium compounds.

Uranium carbides: UC and UC_2 can be produced by reaction of carbon monoxide with molten uranium. These gray crystals have high melting points of 2790°C and 2350°C, respectively. U_2C_3 also exists but decomposes at ~1700°C. In principle, uranium carbide can replace uranium dioxide in nuclear fuels.

Uranium nitrides: UN and U_2N_3 form cubic crystals and are quite similar to the carbides in their melting point (UN at 2805°C with a density of 14.3 g cm^{-3}). They are formed by reaction of nitrogen with uranium.

1.3   URANIUM IN THE NFC

The following section presents an overview of a generic uranium NFC (Figure 1.9) (IAEA 1613 2009) and a generic flow sheet of the chemical processes in a uranium conversion facility (UCF) where UOC is converted to UO_2 (for nuclear fuel), U metal (for fuel or other metallurgical applications), or UF_6 (feed for enrichment plants) is shown in Figure 1.10.

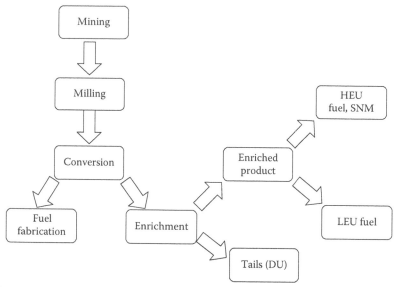

FIGURE 1.9 An overview of a schematic generic nuclear fuel cycle. (Adapted from International Atomic Energy Agency, Nuclear fuel cycle information system, A directory of nuclear fuel cycle facilities, 2009 Edition IAEA TECDOC 1613, IAEA, Vienna, Austria. With permission.)

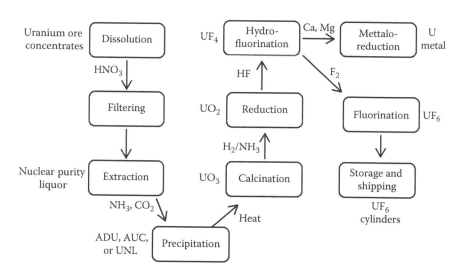

FIGURE 1.10 The main stages of uranium processing carried out at a generic UCF.

It should be noted that there are many variations of the fuel cycle, each consisting of several stages, that we shall not go into. The front end of the cycle starts with an exploration for uranium deposits (see Chapter 2 for details) and continues with mining operations to retrieve the uranium-rich minerals and with milling of the material to augment the fraction with uranium, usually close to the mine site.

The product, called uranium ore concentrate (and sometimes yellow cake), contains 65%–85% U_3O_8, is then shipped to the UCF where the uranium is dissolved and concentrated, and then purified and converted either to the proper form needed for fuel elements (usually uranium oxide for graphite type or heavy water reactors) or to the feed material required for isotope enrichment (usually uranium hexafluoride) (Figure 1.10). Following is either fabrication of fuel elements or enrichment to LEU for fueling light water reactors or to HEU for special reactors or nuclear weapons (special nuclear materials (SNM)). The product of the enrichment process, either LEU or HEU, must then be converted into the suitable form for the application—once again usually an oxide or metal.

After the nuclear fuel is fabricated, placed in a reactor and irradiated the fuel elements become highly radioactive (now called spent fuel) and their removal or replacement requires special protective means. These spent fuel elements need to be placed in cooling ponds until the intensity of radiation abates and then the fuel is either moved into permanent storage or transported to a reprocessing plant. Finally, the used fuel and radioactive products need to be separated according to their chemical and physical properties and then sent to a waste disposal site or transmuted into harmless forms.

The material balance in the NFC for a generic case can be summarized, as detailed later, based on seminar material from the European Nuclear Fuel Management. Starting with 20,000 tons of ore that contains 1% uranium that after milling is reduced to 230 tons of uranium ore concentrates (of which 195 tons consist of uranium). This is converted to 288 tons of UF_6 and after enrichment one part is enriched to 4% ^{235}U (35 tons UF_6 containing 24 tons uranium) and the rest (254 tons UF_6 containing 171 tons uranium) is in the tails. The enriched uranium is converted into 27 tons of UO_2 from which electricity (7000 million kWh) can be produced in the reactor. The spent fuel will contain 23 tons of uranium (~0.8% ^{235}U), ~240 kg of plutonium, ~720 kg of fission products, and some trans-uranium elements.

1.3.1 URANIUM MINING

Mining of uranium can be done by underground excavation, open pits, and by in situ leaching (ISL) techniques in order to recover uranium. Generally, underground excavations are deployed when the uranium bearing ore is 120 m or more below the surface while open mining is used when the deposits are close to the surface. In open pits, a large amount of material must be removed—not just the ore itself but also some of the surrounding surface soil. Underground excavation leaves a smaller mark on the surface but requires more ventilation than other mines (that do not contain uranium) due to the radioactive nature of airborne particles and gases (radon, mainly). Modern techniques, like ISL, are replacing open pit and excavation mining. ISL involves circulation of oxygenated groundwater through the porous ore to dissolve the uranium oxide and bring it to the surface. This can also be done with slightly acid or alkaline solutions to keep the uranium in solution. The uranium oxide is then recovered from the solution. The choice of mining method in a particular deposit is governed by the nature of the ore, safety concerns, and economic considerations.

1.3.2 URANIUM MILLING

In order to minimize shipments of uranium-bearing ores, the milling operations are usually carried out close to the mine. The objective of the milling operation is to produce uranium oxide concentrates (UOC), that generally contain over 80% of uranium while the original ore may contain as little as 0.1% or as much as 18%. The UOC is also referred to as yellow cake—though not yellow (usually grayish black) and certainly not a cake. Milling involves mechanical (grinding), physical (roasting), and chemical (leaching) operations to achieve this objective, as discussed here.

The ore arriving at the mill is first crushed and ground. Physical preconcentration methods are often used to enhance the uranium content in ores that contain low levels of uranium. These may include manual or automatic sorting of lumps, flotation, and/or gravimetric differentiation where particles containing uranium are concentrated in the top phase or sink to the bottom, depending on the mineral and milling process. Sometimes the resultant ore is calcined or roasted at a high temperature in order to dispose of some materials that may interfere with the subsequent stages. This is followed by acid leaching, usually in sulfuric acid, or alternatively with a strong alkaline solution, that separates the uranium from the waste rock, forming a *pregnant liquor* containing uranium, usually in the form of uranyl ions. Acid leaching is effective for hexavalent uranium so that oxidizing reagents may be added if the uranium is in lower valence states. There are several variations of the operating parameters of the actual leaching procedure like temperature, duration, countercurrent leaching, single or multistage processes, and the use of different oxidizing agents. In some cases, carbonate leaching with air oxidation would be preferred, especially if the ore contains high levels of calcium and magnesium. The preferred process depends on the quality and composition of the ore and on economic considerations.

In the past, the recovery of uranium from the *pregnant liquor* solution usually involved precipitation as uranium oxide (U_3O_8) concentrates (UOC). But as the quality of UOC obtained by precipitation from acid leach is not very good ion exchange and solvent extraction are used in most plants nowadays. Precipitation from carbonate leaching produces a reasonably pure uranium concentrate. A detailed description of these processes and the operational and the economic considerations is beyond the scope of this book and can be found elsewhere (Grenthe 2006).

There are two other aspects that should be mentioned here that may directly affect the choice of the milling process. First, the uranium ore often contains other metals that have commercial value, like vanadium or niobium, for example, and their recovery may influence the process selected for uranium recuperation. Second, uranium itself may be a by-product of other processes like gold extraction, niobium, and tantalum production or phosphoric acid manufacture. Thus, recovery of low levels of uranium from phosphates, columbite, or gold-bearing minerals may not be economical in itself, but extracting uranium as a by-product from the waste streams of these operations could be commercially sensible.

The UOC is dried and packed in metal drums and marketed, that is, sent to the UCF. The remainder of the ore, containing most of the radioactivity (from uranium progenies) and nearly all the rock waste material, becomes mine tailings that are placed in storage facilities near the mine. The tailings usually contain low concentrations of long-lived radionuclides and heavy metals. However, the total quantity of radioactive elements is less than in the original ore, and their collective radioactivity will be much shorter-lived.

1.3.3 URANIUM CONVERSION FACILITY

The drums with uranium ore concentrate are shipped to the UCF where the uranium is purified to a nuclear grade and converted into the chemical form needed for its application, usually as uranium dioxide (UO_2), uranium metal, or uranium hexafluoride (UF_6). There are many variations of the methods for conversion of the uranium ore concentrates into commercially valuable compounds. A detailed discussion of each variant is beyond the scope of this book so a generic UCF plant will be described here, shown schematically in Figure 1.10.

First, the UOC is placed in a large stainless steel reactor and nitric acid (concentrated or slightly diluted) is added and heated to dissolve the uranium oxides in the *yellow cake*. Waste materials that are not dissolved sink to the bottom of the reactor and are removed by filtering. The solution containing the uranium in the form of uranyl nitrate [$UO_2(NO_3)_2$] is transported to a series of mixer-settlers where purification takes place by liquid–liquid extraction with an organic phase consisting of kerosene and tributylphosphate (TBP). At this stage, the uranium is considered as nuclear grade (i.e., purified) and called uranyl nitrate liquor. The UNL is then transferred into another reactor where the uranium is precipitated either as ammonium diuranate (ADU), ammonium uranyl carbonate (AUC) by addition of ammonia (and carbon dioxide for AUC), or simply concentrated until solid uranyl nitrate precipitates. The precipitate is filtered and then transferred to a furnace where it is converted to UO_3 by calcinations. The hexavalent UO_3 is subsequently reduced by hydrogen gas (or hydrogen formed by cracking ammonia) to tetravalent UO_2.

Evidently, throughout all these processes the isotopic composition of uranium remains in its natural abundance form, that is, containing 0.72% of [235]U. Uranium of this composition is suitable for use as nuclear fuel in reactors that operate with heavy water (D_2O) such as CANDU reactors or graphite (such as the old Magnox reactors) as the moderator for slowing neutrons. In this case, the UO_2 is ground and sintered to form pellets that will be placed in fuel elements (see Chapter 2 for analytical procedures to characterize these pellets). As an example, Frame 1.4 describes the process used in India for production of UO_2 powder, pellets, and fuel elements as an example of the processes in the UCF.

If the uranium is to be used as a metal or as feed material for isotope enrichment plants, the uranium dioxide is treated with anhydrous hydrogen fluoride (AHF) to form UF_4 (Figure 1.11). The UF_4 can then be reduced metallothermically with magnesium or calcium to produce metallic uranium (still of natural isotope composition), which can also be used to fuel some types of reactors. However, if enrichment of the [235]U isotope is planned, the UF_4 is fluorinated to produce UF_6, which is the feed material for most commercial enrichment plants. The uranium hexafluoride is then transferred into 10–14 ton cylinders where it solidifies. These cylinders with the solid UF_6 are stored until ready to be shipped to the enrichment plant.

FRAME 1.4   NUCLEAR FUEL PRODUCTION IN INDIA

DESCRIPTION OF PROCESS

1. The starting material for natural uranium oxide fuel used in the PHWR type of reactors is the indigenously available uranium concentrate mostly in the mineral belt of Jharkhand that is produced at the uranium mines of Uranium Corporation of India Ltd. (UCIL).

   The fuel assembly for the BWR type of reactors is made of low-enriched uranium (LEU), which is imported in the form of uranium hexafluoride.

2. The conversion of uranium concentrate or UF_6 to nuclear grade UO_2 involves various chemical process steps such as dissolution of uranium ore concentrate in nitric acid to dissolve uranium for the PHWR fuel or hydrolysis of LEU UF_6, in case of BWR fuel production stream.

3. The crude uranium solution is then subjected to purification by solvent extraction in the slurry extraction setup developed indigenously using TBP-uranyl nitrate system of extraction. This is a unique setup, tested on pilot plant scale and then designed, fabricated, installed, and put into operation at the Uranium Oxide Plant of the Nuclear Fuel Complex (NFC) to supply fuel for all PHWR type of reactors in India. Compressed air is used for material transfer, control of flow rates, mixing, settling, etc. A unique feature is the fact that the organic phase is the continuous phase (and is recycled) rather than the normal convention of having aqueous phase as the continuous phase.

4. The purified uranium is precipitated with concentrated NH_4OH solution to form ammonium diuranate (ADU) cake. The precipitation is carried out in batch mode at equilibrium conditions with slow addition of ammonia and mixing. After completion, the ADU slurry is filtered and washed simultaneously to produce the ADU cake. Precipitation of ADU is the most important step in production of reactor grade UO_2 powder and a close control of all the variables is essential for product quality.

5. Pure ADU cake from batch precipitation is dried to obtain dried ADUC/UO_3 in a box furnace (batch operation)/turbo dryer (continuous operation)/spray dryer (semicontinuous operation). All the three types of units are in use in NFC at the three fuel production plants.

6. The dried ADU is collected in leak-proof containers and calcined in the presence of air by passing through electrically heated rotary tubular furnace to get uranium oxide (U_3O_8). This oxide is reduced to UO_2 in a rotary tubular furnace in cracked ammonia atmosphere ($N_2 + H_2$). The off gases comprising water vapor and unused gases ($N_2 + H_2$) are scrubbed in a water spray column before they are released. The UO_2 powder thus produced is subjected to partial oxidation on the surface with limited quantity of air in a rotary tube again, to get a stable uranium dioxide (UO_2) powder.

7. The UO_2 powder is then compacted in a hydraulic press/rotary compactor to increase the bulk density, granulated, and mixed with a binder. The binder-mixed granules are subjected to final compaction to give the cylindrical-shaped pellets of specified dimensions and density (Green Pellets).

8. These Green Pellets are loaded/stacked in molybdenum containers (open)/boats and sintered at high temperature under reducing atmosphere for further densification to achieve at least 96% of the theoretical density of UO_2 pellets. To ensure uniform surface finish and diameter, these pellets are centerless ground and inspected for physical integrity, density, purity, surface defects to meet the specifications.

9. The inspected pellets are loaded in zircaloy tubes after thorough washing and drying that are welded at the both ends by suitable welding techniques. All the required components are then welded on these tubes and the welded elements are assembled in a predetermined matrix to form the fuel assembly. The BWR assembly comprises over 260 components and PHWR assembly comprises 167 components. These fuel assemblies are subjected to a stringent quality control checks using nondestructive techniques before final acceptance and clearance. The accepted assemblies are packed and dispatched to reactor sites.

Source: Sheela et al., *Technology for Production of Uranium Oxide for Nuclear Power Reactor*, Nuclear Fuel Complex, Hyderabad, India, 2000. http://dbtindia.nic.in/women/paper19.htm, accessed July 26, 2014.

An alternative method for production of UF_6 (the fluoride volatility process) for enrichment facilities from uranium ore concentrates was developed by Honeywell and is practiced at its Metropolis plant. Figure 1.11 schematically compares this process with the *classic* process described earlier.

In principle, the fluoride volatility process has fewer stages and is simpler than the conventional process. It consists of reduction of the UOC (yellow cake) to UO_2 with hydrogen (derived from cracking ammonia) followed by hydrofluorination to produce UF_4 (green salt) and then fluorination to produce UF_6 (hex). In principle, uranium ore concentrates may be directly fluorinated to produce UF_6, but this process consumes large amounts of fluorine and is not applied commercially.

1.3.4 ISOTOPE ENRICHMENT

Several methods for increasing the fraction of the fissile isotope, U-235, have been developed, and their principles of operation are briefly discussed in Frame 1.5.

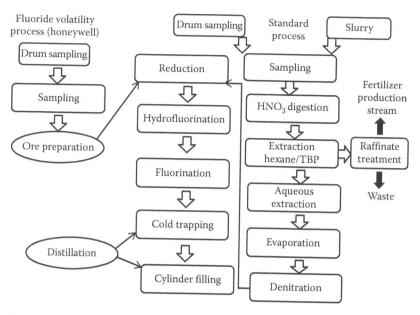

FIGURE 1.11 Comparison of two processes for production of UF_6 from uranium ore concentrates: the classic process and the Honeywell fluoride volatility process. (Adapted from http://www.converdyn.com/product/different.html, accessed July 26, 2014.)

In this section, a generic method where UF_6 is the feed material will be briefly described. The enrichment involves several stages (arranged in a cascade), where the U-235 fraction is gradually increased until it reaches the desired level (3%–5% for LEU nuclear fuel, up to 19.75% for some research, medical isotopes, and material testing or 90% for special nuclear materials). As mentioned earlier, cylinders of the UF_6 produced at the UCF (usually containing about 10 tons of UF_6) are shipped to the enrichment plant and the cylinders are placed in the feed station. The feed cylinder is heated so that the vapor pressure of the UF_6 increases to the appropriate operational level. Once the UF_6 flows into the gas centrifuges or diffusion cells, the enrichment process begins: the lighter molecules with $^{235}UF_6$ are separated from the heavier $^{238}UF_6$ so their fraction increases toward the product end of the cascade and concomitantly the inverse occurs toward the tails end of the plant. While the isotope composition changes during the enrichment process, the chemical form of the feed, product, and tails as UF_6 is retained. The cylinder used to collect the product is normally smaller than the feed cylinder while the tails collection is usually about the same size as the feed cylinder.

The UF_6 enriched product (or depleted tails) needs to be converted to a commercially useful form as uranium dioxide for LEU fuel elements or to metallic uranium for other types of fuel or special nuclear materials. This process is usually called de-conversion or re-conversion. According to one variation, the UF_6 is reduced (by hydrogen or other reagents) to UF_4 that can be further reduced to uranium metal as described earlier. In a different modification, the UF_6 is hydrolyzed to form uranyl fluoride, the uranyl ions are precipitated as ADU or AUC, calcined to form UO_3 that

FRAME 1.5   PRINCIPLES OF OPERATION OF
SOME ISOTOPE ENRICHMENT METHODS

In natural uranium ores, the fraction of the atoms of the fissile isotope (^{235}U) is about 0.72%. For many commercial applications, like production of fuel for light water reactors or several types of research reactors and other nuclear functions, its fraction must be increased, that is, isotope enrichment is carried out. The main isotope separation methods, or isotope enrichment processes, utilize the small differences in between the mass of U-235 and U-238. The two major commercial methods that have supplied most of the enriched uranium to date, gaseous diffusion and gas centrifuges, use the only gaseous compound of uranium, uranium hexafluoride (UF_6), as the feed material. Both methods utilize the difference between the mass of $^{235}UF_6$ (349 Da) and $^{238}UF_6$ (352 Da) where the mass ratio difference that is 0.86%. The product and tails of the enrichment process are also with the same chemical form, but the isotope composition of the material is altered in the enrichment process. Schematic diagrams of the principle of operation of these methods can be found on the web and in many textbooks, so will not be shown here.

Gaseous diffusion: UF_6 molecules are pumped through special barriers (porous membranes) that include small pores to allow the UF_6 molecules to pass through. These membranes are made of an inert material that is compatible with the corrosive feed material. The molecules with the lighter isotope, $^{235}UF_6$, diffuse through the porous membrane at a slightly higher rate than the heavier $^{238}UF_6$ molecules. The molecules are then sent through another membrane where the fraction of $^{235}UF_6$ molecules is further increased, and so on through several consecutive stages (a cascade).

At the end of the process, the enriched UF_6 is withdrawn from the pipelines and condensed into a liquid that is poured into containers in which it solidifies and becomes ready for transportation to a fuel fabrication facility. After about 330 such stages, taking an enrichment factor of ~1.004, the fraction of ^{235}U would theoretically be close to 3.0%, which is typical for the requirements of LEU fuel. Each stage requires heating and compression of the UF_6 so that a large amount of energy is needed to operate the gaseous diffusion facility. This method is now judged as obsolete as is replaced by other methods that require less energy, smaller hold-up of feed material and lower construction costs.

Gas centrifuges process: Here too the feed material is UF_6 but the method relies on the mass difference between $^{235}UF_6$ and $^{238}UF_6$ as the molecules are subjected to a strong gravitational field (centrifugal force) caused by a rapidly spinning cylinder (for more details, see Wood et al. 2008). The heavier molecules move toward the outside walls of the cylinder while the lighter molecules collect closer to the center. A temperature gradient between the warmer bottom and cooler top of the rotating cylinder further increases the isotope separation. Extraction of the molecules from each region is carried out by separate pipes (called scoops)—the stream with lightly enriched UF_6 from the center is passed onward to the next stage, while the stream collected from the wall is passed downward to the previous stage. A typical enrichment factor in gas centrifuges is 1.3, so theoretically enrichment to ~3.5% would be obtained after merely six stages.

Laser isotope separation techniques: Laser-based isotope enrichment techniques deploy selective photo-excitation principles to excite a particular isotope as an atom or molecule (Rao 2003). Each device consists of three parts: the laser system, the optical system, and the separation module. These methods include the atomic vapor laser isotope separation (AVLIS) that uses a fine-tuned laser beam to selectively ionize vapors of atomic [235]U, the molecular laser isotope separation (MLIS), and separation of isotopes by laser excitation (SILEX) that use a laser to selectively dissociate or excite [235]UF_6 molecules.

Other processes: There are also other processes like the aerodynamic nozzle processes that also use centrifugal forces to separate [235]UF_6 molecules from [238]UF_6 molecules but in this case the UF_6 is transported by hydrogen gas through a narrow static nozzle. The historic EMIS (electromagnetic isotope separation) that operates like a large magnetic sector mass spectrometer uses UCl_4 as its feed material has also been used.

Chemical separation methods deployed in solution utilize the very slight difference in the tendency of the two isotopes to undergo oxidation reduction reactions or their affinity toward an ion-exchange resin.

Finally, it should be noted that all the methods that rely on mass difference, mass ratio, or kinetics to separate U-235 from U-238 will also lead to enrichment of U-234 (even to greater relative extent than U-235). Elevated levels of U-234 may complicate the utilization of the enrichment product. Similarly, use of reprocessed uranium that contains U-236 (and perhaps some U-232) may also affect the product quality. On the other hand, laser isotope separation methods will selectively enrich U-235 with only very slight changes in the U-234 and U-236 content.

Source:　Based on Commission, U.N., May 21, 2013. Uranium enrichment, http://www.nrc.gov/materials/fuel-cycle-fac/; http://www.nrc.gov/materials/fuel-cycle-fac/ur-enrichment.html, retrieved February 9, 2014.

is reduced to form UO_2 for fuel elements. On the whole, this procedure in the de-conversion facility is quite similar to the process employed in the UCF.

1.3.5　FABRICATION OF NUCLEAR FUEL AND IRRADIATION

After the uranium compound is produced at the required isotope enrichment level, the fuel elements can be fabricated. Some of the main types of uranium fuel are described later so here it will just be mentioned that the arrangement of the fuel elements in assemblies varies greatly among the different kinds of reactors. The fuel is placed in the reactor and irradiated (burned) for a predetermined length of time that depends on the fuel and the reactor. For nuclear power plants, this would typically be about 3 years before the fuel in the reactor core is replaced, and once again there are many variations, including reactor cores that can be continually loaded without the need to stop the reactor.

1.3.6 TREATMENT AND DISPOSAL OF IRRADIATED FUEL

The back-end of the NFC is concerned with the handling of the irradiated fuel (spent fuel). When the fuel elements are removed from the nuclear reactor, they contain uranium that was not *burned* during the reactor operation, highly radioactive fission products and actinides that were produced when uranium absorbed neutrons. Some of these products may have commercial value while others may pose a serious hazard to the environment and life if not properly treated. Three strategies have been proposed for handling of the irradiated nuclear fuel: long-term storage, *wait and see* interim storage and reprocessing. All approaches first require the removal of the irradiated fuel (spent fuel (SF)) from the reactor vessel and placing them in cooling pools for a period that is long enough for decay of radioactivity to a lower level. The rate of decay and abatement of the heat of the major radionuclides in spent fuel is shown in Figure 1.12 (the abatement of radioactivity is presented in Figure 2.12). This is further discussed in Chapter 2.

Reprocessing is especially controversial due to technical, economic, and strategic (proliferation-related) issues. The technical problems arise mainly from the presence of the unnatural uranium isotopes (^{236}U and ^{232}U) in reprocessed uranium (*RepU*) that may affect the neutron economy in the reactor. ^{232}U in particular is highly radioactive and complicates handling of fuel produced from *RepU*. Re-enrichment of *RepU* would further increase the fraction of ^{232}U aggravating the problem. The economic aspects depend on the prices of uranium and enrichment needed to produce fresh fuel compared to the cost of reprocessing. The biggest concern is that involving the plutonium derived from the reprocessing operation that could in principle be used in

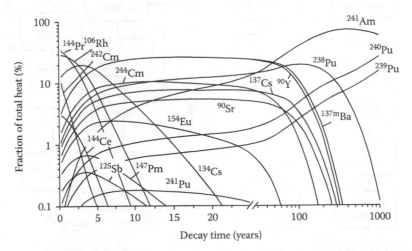

FIGURE 1.12 Decay of heat in irradiated fuel. Note the time scale (linear at first and then logarithmic) and the changes in the contribution of short-lived and long-lived nuclides to the heat generated in the spent fuel.

the production of nuclear weapons or radioactive dispersion devices (RDD or dirty bombs). The plutonium does have an economic value if used in MOX fuels, but it requires special measures to prevent its use in illicit and terrorist acts. A technical problem with plutonium arises from the decay of ^{241}Pu into ^{241}Am that is a gamma emitter. The build-up of ^{241}Am would make handling of MOX much more difficult, so if reprocessed plutonium is to be used in MOX it should be done shortly after the reprocessing when the build-up of ^{241}Am is still low. The question of reprocessing is also discussed in Chapter 2.

1.3.7 TYPES OF NFCs

The three main types of NFCs are depicted in Figure 1.13.

These deal with the way to handle and treat the irradiated fuel elements (spent fuel) that are highly radioactive due to fission products and activation products, including fissile materials that are potential components of a nuclear device. In all cases, the fuel elements are first stored in pools at the reactor site (AR—at reactor) for an initial cooling period. The open fuel cycle involves subsequent removal of the spent fuel rods and placing them in wet or dry storage at a site *away from reactor* (AFR). In the open fuel cycle, the irradiated fuel is not treated at all and is to be left under supervision. This type of fuel cycle is sometimes referred to as *once through* as the uranium components go through the reactor just one single time and are then sent away for disposal. The wait and see option, as the name implies, proposes intermediate storage until some solution for permanent storage and disposal will be developed in the future. The closed fuel cycle presumes that the irradiated fuel elements can be reprocessed after a suitable storage period.

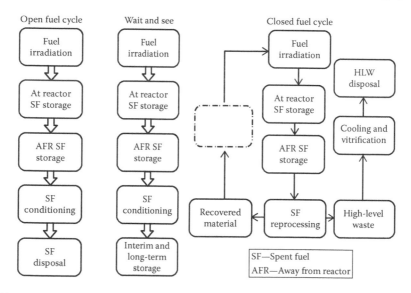

FIGURE 1.13 The main fuel cycle types: the open fuel cycle (once through), the *wait and see*, and the closed fuel cycle (including reprocessing of the spent fuel). (Adapted from International Atomic Energy Agency, Nuclear fuel cycle information system, A directory of nuclear fuel cycle facilities, 2009 Edition IAEA TECDOC 1613, IAEA, Vienna, Austria. With permission.)

The reprocessing involves separating the fission products from the actinides, and then separating the plutonium from the uranium. The best known procedure of this type is the PUREX (Plutonium, URanium EXtraction) process that is used for recovery of uranium and plutonium from irradiated fuel (see details in Chapter 2). The separated plutonium can be used for the production of nuclear weapons or converted into the oxide form, mixed with uranium oxide and can be used as MOX nuclear fuel.

1.3.8 REACTOR FUEL AND FUEL ELEMENTS

There are several different kinds of nuclear reactors that use diverse types of fuel (uranium compounds) that are placed in different sorts of fuel elements and these in turn are arranged in various fuel assemblies. A detailed survey of these is beyond the scope of this book however an elementary understanding of nuclear fuel and its characterization is highly relevant. Figure 1.14 depicts the main nuclear reactor, types of fuel, moderator, and coolants in service.

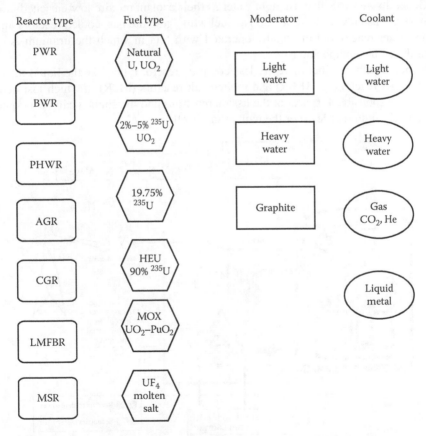

FIGURE 1.14 Classification of reactors according to the reactor type, the fuel type, the moderator, and the coolant (see text). (Based on International Atomic Energy Agency, Nuclear fuel cycle information system, A directory of nuclear fuel cycle facilities, 2009 Edition IAEA TECDOC 1613, IAEA, Vienna, Austria.)

1.3.8.1 Reactor Types, Moderators, and Coolants

Nuclear reactors can be classified according to the kind of fuel they use, the type of moderator and coolant, the operating methods for generating power. The main types of nuclear power plants (NPP) and reactors that are used for production of electricity are the pressurized water reactor (PWR), pressurized heavy water reactor (PHWR), boiling water reactor (BWR), gas cooled reactor (CGR) and advanced gas cooled (AGR) reactors. There is a Soviet design that uses graphite as the moderator that is better known after the Chernobyl accident by its acronym (RMBK) for Reaktor Bolshoy Moschnosti Kanalniy (High Power Channel Reactor). In all these reactors the fission of the fissile ^{235}U in the fuel is mainly by thermal neutrons so the moderator is water (light or heavy water) or graphite. Water also serves as a coolant in PWR and BWR types of reactors while gas cooling (helium or carbon dioxide) is used in graphite moderated reactors. In all power plants, the heat generated by the nuclear fission reactions is used to produce steam at high pressure and temperature that turns turbines for generating electricity. A schematic presentation of a nuclear power plant is shown in Figure 1.15.

Generally, reactors that use light water as their coolant require low-enriched uranium (LEU usually of 2%–5% ^{235}U) as fuel while graphite gas–cooled reactors and heavy water reactors are normally operated with fuel in which the uranium is of natural isotopic composition.

It should be noted that there are less common reactor types like the liquid metal fast breeder reactors (LMFBR) and molten salt reactors (MSR) in which fast neutrons are responsible for most of the fission reactions and the coolant can be a liquid metal (e.g., sodium or lead) or the molten salt itself.

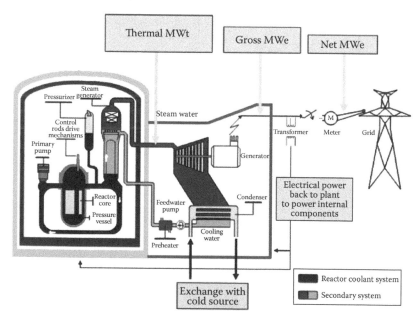

FIGURE 1.15 A schematic of a generic nuclear power plant. (From World Nuclear Association, http://www.world-nuclear.org/info/Nuclear-Fuel-Cycle/Power-Reactors/Nuclear-Power-Reactors/, accessed July 26, 2014. With permission.)

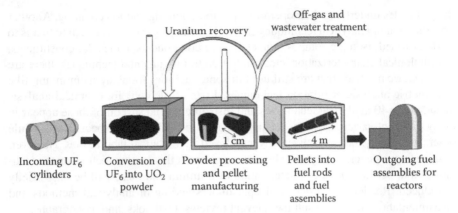

Off-gas and
Uranium recovery   wastewater treatment

1 cm   4 m

Incoming UF₆   Conversion of   Powder processing   Pellets into   Outgoing fuel
cylinders   UF₆ into UO₂   and pellet   fuel rods   assemblies for
                 powder   manufacturing   and fuel   reactors
                                                       assemblies

FIGURE 1.16 The three-step fuel fabrication process: UF_6 is converted to UO_2 powder, which is then processed for pellet manufacturing. The pellets are inserted into fuel rods that are combined in fuel assemblies and placed in the reactor core. (From World Nuclear Association, http://www.world-nuclear.org/info/Nuclear-Fuel-Cycle/Conversion-Enrichment-and-Fabrication/Fuel-Fabrication/, accessed July 26, 2014. With permission.)

1.3.8.2 Type of Fuel

Uranium dioxide, UO_2, is the compound of choice in many nuclear reactors despite its relatively poor heat conduction properties. This is due to its chemical stability, its high melting point, and the ease of production of well-characterized morphological and physical properties. The complete characterization is described in Chapter 2. Uranium metal and particularly uranium alloys like U-Al, U-Zr, U-Si, and U-Mo are also used as fuel. Their heat conduction is superior to that of uranium oxide but the metal and alloys are less stable chemically.

The fuel fabrication process consists of conversion of enriched UF_6 into UO_2 powder that is then pressed into pellets that typically are about 1 cm in diameter and ~1 cm in length. These pellets are inserted into fuel rods that are combined in fuel assemblies and placed in the reactor core as shown in Figure 1.16.

Another promising uranium compound that can be used in nuclear fuels is uranium carbide that has a high melting point and better thermal conductivity than the oxide and in addition does not form oxygen when radiolyzed. Uranium nitride can also be used, but formation of ^{14}C from ^{14}N could be problematic. In addition, other uranium compounds that can be used as a fuel in a nuclear reactor, ranging from aqueous solutions to molten salts that are brought to a high temperature in order to keep them in a molten state. MOX of uranium and plutonium also serve as a nuclear fuel in some reactors.

1.4 MODERN ANALYTICAL TECHNIQUES IN A NUTSHELL

There are several analytical instruments and techniques that are used to characterize uranium compounds and especially the impurities that accompany them. The modern analytical instrumentation can be divided into two main categories according to the principles underlying their operation: photon counting and ion counting. Another classification of analytical techniques is based on the property of the analyte that is to be determined: isotopic composition, elemental components, molecular constituents, morphological characterization, etc. In addition to instrumental techniques, there are some *classic* methods that are still used in some facilities for assaying uranium, like gravimetric analysis or titration techniques. Lists of methods for chemical analysis include over 40 analytical techniques and instruments and several of those appear in many different variations. Some techniques are used only in very special cases while others are widely, or even universally, employed in analytical laboratories. However, as this is not a general analytical chemistry textbook, this section focuses on the techniques that pertain to the characterization of uranium and these will be only briefly introduced. Readers interested in deeper understanding of analytical methods and instrumentation should consult the relevant reviews, textbooks, and monographs.

In this section, we shall first give a simplified description of the principle of operation of the more common analytical techniques that are used to characterize uranium as a major component of a compound or alloy, as a minor component (e.g., in a mineral), as a trace level constituent in an environmental or biological sample and techniques that are used for the determination of other components and impurities in uranium compounds. Frame 1.6 includes quotes from a report that was prepared in 1942 (and declassified in 1947) that compared three analytical techniques for assaying uranium that were considered as state of the art at the time (McInnes and Longsworth 1942).

FRAME 1.6   HISTORICAL NOTE

In a classified report from 1942, which was unclassified and published in 1947, the three main analytical methods that were then used to determine uranium in a sample of hydrated UO_3 were compared. One method used prolonged ignition in air in three stages (700°C, 820°C, and 930°C) to form U_3O_8; the second involved reduction of the product by hydrogen at 900°C to obtain UO_2; and the third included titration with ceric sulfate of uranium sulfate produced by the reaction of UO_3 with H_2SO_4. In their summary, the authors conclude: "The uranium content of a preparation of hydrated UO_3 was determined by three independent procedures... The results of these three methods showed an extreme difference of 0.08%."

A few variations were also examined, mainly based on titration of over-reduced solutions, but these were not as successful as the methods described earlier. The authors state "Extensive application of the differential titration method to over-reduced solutions failed to locate the precise end-point of the U^{+3}–U^{+4} change. The difficulty is believed to reside with the excessively strong reducing power of the U^{+3} ion."

Source: McInnes, D.A. and Longsworth, L.G., *A Comparison of Analytical Methods for the Determination of Uranium*, Rockefeller Institute for Medical Research, Oak Ridge, TN, 1942, MDDC-910 (declassified 1947).

The broad spectrum of analytical methods presented in the following section just emphasizes the great progress of analytical chemistry in general, and uranium assays in particular, during the last seven decades. Techniques that are used mainly for the characterization of microstructural features of alloys and complexes like x-ray and neutron diffraction, electron paramagnetic resonance (EPR), nuclear magnetic resonance (NMR), positron emission tomography (PEL), etc., will be referred to only where relevant to an analytical procedure.

Sampling and sample preparation techniques, which are an integral part of any analytical procedure, are discussed in detail in the context of specific applications in Chapters 2 through 5. In general, liquid samples can usually be measured directly or after removal of interfering substances by liquid extraction, precipitation, ion exchange or chromatographic techniques, with or without preconcentration. Solid samples usually require digestion in order to convert them into liquid form, but there are also methods for direct analysis of solid samples. Gaseous samples or aerosols are usually measured directly but sometimes preconcentration is necessary. Some generic sample preparation and uranium separation techniques are presented in Section 1.4.3.

1.4.1 PHOTON COUNTING TECHNIQUES

Photon counting techniques measure the electromagnetic radiation that is absorbed or emitted from atoms or molecules in a sample. Absorption of photons, or quanta of energy, is usually measured when the substances are in their ground state while emission is measured after the atoms or molecules are in an excited or unstable state and radiate energy in the form of photons. It is quite common to deploy a combination of excitation and emission for analytical purposes. The field of spectroscopy or spectrometry is the term broadly used for these techniques. The range of wavelengths used in these analytical methods covers the electromagnetic spectrum from gamma rays, through x-rays, ultraviolet and visible (UV–Vis) wavelengths, to near infrared and infrared (NIR and IR) and to even longer wavelengths. The shorter wavelengths characterize processes that take place in the nucleus (gamma rays) or the inner shells of atoms (x-rays), the longer wavelengths arise from transitions of valence electrons (UV–Vis) and molecular vibrations and rotations (IR), and the longest wavelengths reflect the motion of the atoms in the compounds as in nuclear magnetic resonance (NMR). In some cases, sharp atomic lines and distinct peaks are observed in the spectrum while in other cases broad molecular peaks are seen. Thus, in some samples, single lines that are usually sharp and strong are measured while in other cases there are only diffuse lines observed in the spectra.

1.4.1.1 Atomic Absorption Spectrometry

This technique is used mainly for determining elemental components in a sample. As the name implies, its principle of operation is based on absorption of photons emitted from a source and measurement of the extent of signal attenuation by the atomized sample that reaches a detector as shown schematically for a generic atomic absorption spectrophotometer (Figure 1.17a). Figure 1.17b shows a representation of an atomic absorption spectrometer (AAS) that was developed in the 1960s for measurement of uranium isotopes (Goleb 1966).

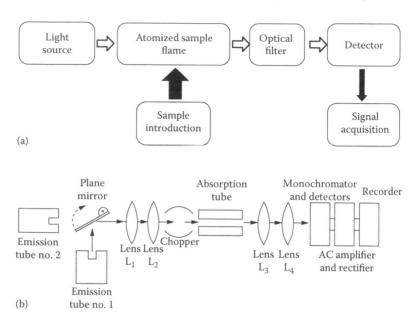

FIGURE 1.17 (a) Schematic diagram of an atomic absorption spectrophotometer (AAS); (b) a schematic representation of an atomic absorption spectrometer (AAS) for measurement of uranium isotopes. (From Goleb, J.A., *Anal. Chim. Acta*, 34, 135, 1966. With permission.)

The absorption of photons by the sample depends on the concentration of the analyte, the length of the optical path of the sample, and the cross section for absorption of photons of that wavelength. Quantification is based on comparison of the signal intensity of standard solutions that contain known (preferably certified) concentrations of the analyte with that of the sample using Beer–Lambert law. The underlying principle is that each element has characteristic wavelengths of absorption and emission that depend on the electronic structure of its atoms that in turn represent the energy levels of the orbital that these electrons occupy.

In the basic configuration, the liquid sample is sprayed continuously into a spray chamber in which fine aerosols are formed (larger droplets are drained from the spray chamber as waste) and mixed with the flame gases and then sprayed into the torch. The flame is produced by burning a fuel (hydrogen, acetylene, propane, etc.) with an oxidizer (air, oxygen, nitrous oxide, etc.) in an elongated torch (burning head is 5–10 cm, typically) that is placed between the light source and the detector (Figure 1.17). The temperature of the acetylene-N_2O flame is typically 2700°C while acetylene-air is about 2200°C. A popular modification includes a hollow graphite tube that is rapidly heated resistively to a high temperature that serves as the atomizer and is particularly useful for samples that are not in pure liquid form like slurries, biological tissues, or other solids that can be volatilized. This approach is called GF-AAS (graphite furnace-AAS) or ET-AAS for electro-thermal-AAS. Some elements, like arsenic, phosphorus, sulfur, etc., can be measured after the sample is strongly reduced by a hydrogen generator (HG-AAS) and converted into volatile hydrides.

In the classic AAS instruments, the radiation source usually consisted of a lamp with a filament (or hollow cathode) made of, or coated with, the same element that is to be determined so that there is a resonant absorption of the emitted photons. The detector used in this system can be a simple photomultiplier and an optical filter is often used to improve specificity. There are other types of photon sources, especially in modern instruments, where several wavelengths can be simultaneously emitted and detected. A wavelength selector is positioned in front of the detector in order to distinguish the absorption of the different elements (wavelengths).

The sensitivity of AAS techniques in liquid samples is usually in the parts-per-million (ppm or mg L^{-1}) range, but for some elements and instruments, parts-per-billion (ppb or μg L^{-1}) limits of detection are attainable. Modern AAS instrumentation deploys advanced background reduction and signal processing methods in order to improve the analytical performance.

1.4.1.2   Atomic Emission Spectrometry or Optical Emission Spectrometry

The principle of operation of atomic emission spectrometry (AES) (or optical emission spectrometry (OES)) is based on the measurement of photons emitted when electrons move from an excited electronic state to a lower state, rather that the inverse when radiation energy is absorbed, as in AAS. Thus, instead of measuring signal attenuation due to photon absorption, the photons emitted from excited atoms and ions are measured as they decay to lower electronic levels. Here, too, each element has its characteristic wavelengths that serve to identify the element and signal intensity is proportional to the concentration of the element in the sample. A comprehensive overview of ICP-AES techniques for measurement of many different types of samples was published (TAMU 2000). The term ICP-OES for optical emission spectrometry is also used sometimes to describe the same technique.

The most widespread type of emission spectroscopy method used to excite the analyte atoms is a plasma that consists of electrons, ions, and excited atoms. In the most frequently deployed system, the plasma is induced in argon gas by a strong radiofrequency (RF) field (or by microwaves), and is thus called inductively coupled plasma (ICP) and the method is known as ICP-AES. The plasma is formed when argon gas flows through a torch made of three concentric glass tubes surrounded by a cooled metal coil with the RF field (Figure 1.18).

Once the argon gas is *ignited* by a spark, the ions and electrons thus created are accelerated by the RF field and additional excited species (argon atoms and ions) are formed. The effective temperature of the plasma is 7,000–10,000 K. An alternate excitation method is to spray the sample into a flame, but the temperature is only around 2000–2500 K leading to weaker signals than those obtained with ICP.

In ICP-AES, as in AAS, the samples are usually in liquid form. In ICP-AES, the liquid is transported by a peristaltic pump and sprayed into a spray chamber where it is transformed into an aerosol (a process called nebulization). The fine droplets are injected into the plasma torch while the larger droplets are drained away from the spray chamber. This is quite similar to the spray chamber used in AAS, but the aerosol here is mixed with argon gas rather than the flame mixture. Under the conditions present in the plasma torch, the solvent rapidly evaporates from fine droplets that are converted into dry aerosol particles that are subsequently

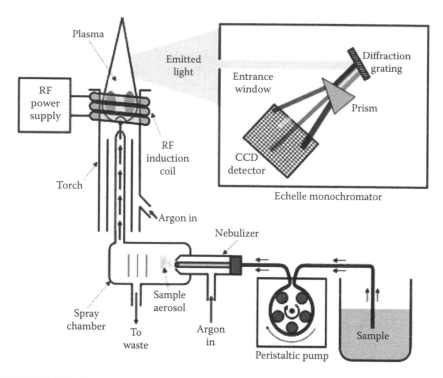

FIGURE 1.18 A schematic of an ICP-AES instrument. The liquid sample is sprayed into a nebulizer in which an aerosol is formed. The finer aerosol particles are injected into a plasma torch and the light emitted from the excited species is reflected by a diffraction grating into a CCD detector array. (Schematic from http://wwwp.cord.edu/dept/chemistry/analyticallabmanual/experiments/icpaes/images/1.jpg, accessed July 26, 2014.)

atomized (molecules break down at the plasma temperature), excited, and ionized. The photons emitted from the excited species (atoms and ions) are separated by a diffraction grating according to their wavelength and reach the photon detection system. Originally, detection was carried out with photographic films, but these were replaced in ICP-AES systems by photon detectors. In one type of system, several detectors were positioned at fixed locations set according to the predetermined wavelengths of the photons emitted from the selected analytes that could be determined simultaneously. In modern instruments, an array of sensitive detectors simultaneously records the whole spectrum that is automatically analyzed by a dedicated software package. In another configuration, popular before the advent of detector arrays, a single detector was positioned at a fixed location and the wavelengths of the photons emitted from the sample were scanned with a diffraction grating component so several elements could be determined sequentially. Other types of samples can also be analyzed, like solid samples that are ablated as dry aerosol particles with laser energy—called laser ablation-ICP-AES (or LA-ICP-AES), or samples with analytes that form hydrides can be selectively introduced into the gas-phase with a hydride generator (HG-ICP-AES) similar to the device used in atomic absorption spectrometry (AAS).

The fact that the liquids or gases can be introduced into the plasma torch, either directly or through the spray chamber, makes it easy to couple chromatographic devices like a gas chromatograph (GC) or a liquid chromatograph (HPLC and its variants like ion chromatography) to ICP-AES and obtain molecular information (from the chromatograph) as well as elemental information. In these configurations, the ICP-AES serves as a sensitive, multielement detector. Thus, for studies of speciation, this combination provides an extremely powerful system.

The advantages of using ICP-AES include its high sensitivity (typically in the ppb range), its broad range of analytes (over 60 different elements), the high throughput of samples (dozens per hour), large linear concentration range (typically six orders of magnitude), quantitative response, and multielement simultaneous measurement capability. Thus, ICP-AES is frequently the method of choice when several elements at different concentration ranges need to be determined.

1.4.1.3 X-Ray Fluorescence Spectrometry

X-ray fluorescence (XRF) is a photon counting technique that is used to characterize the elemental composition of solid samples and in some cases liquid samples as well. The principle of operation is to irradiate the sample with energetic x-ray photons that are capable of knocking out inner shell electrons from atoms in the sample. The vacancy thus created is rapidly filled by an electron from an outer shell concomitant with the emission of a photon whose energy is equal to the difference in energy between the two electronic levels, that is, in the range of x-ray photons (0.25–40 keV), as shown schematically in Figure 1.19. Figure 1.19b shows XRF spectra of two hair samples. One was collected from a person exposed to elevated uranium levels in drinking water in Finland (solid line) and the other sample from a control in Israel (dashed line).

In the laboratory, XRF analysis is usually performed in a vacuum chamber or in an atmosphere of an inert gas (like helium) so that low-energy photons are not absorbed by air. In general, the sensitivity of the method increases with the atomic number of the analyte and in solid samples the limit of detection for many elements is typically in the mg kg^{-1} (ppm) range. One advantage of the technique is the ability to perform a nondestructive assay of several elements simultaneously, although the accuracy for quantitative analysis is quite limited. Due to their simplicity and small size, handheld XRF devices are available and can be carried into the field for onsite analysis as described in detail recently (Bosco 2013).

1.4.1.4 Neutron Activation Analysis

Neutron activation analysis (NAA) is the general term used to describe a nuclear-based technique in which a solid or liquid sample is irradiated with neutrons. Capture or absorption of a neutron excites the nuclide that returns (promptly or after a delay) to ground state by emission of an energetic photon (gamma ray) and/or other particles from the nucleus (Figure 1.20).

There are several configurations of this technique that include instrumental neutron activation analysis (INAA) where the sample is measured without any chemical treatment and radiochemical NAA (RNAA) where post-irradiation separation is done.

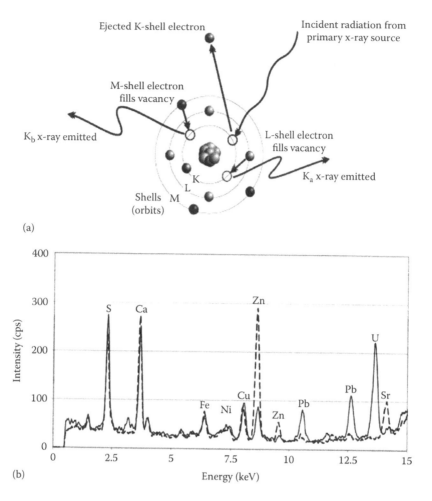

(a)

(b)

FIGURE 1.19 (a) A schematic presentation of the principle of operation of x-ray fluorescence (XRF). (From http://museumbulletin.files.wordpress.com/2012/08/xrf_lg1.jpg, accessed July 26, 2014.) (b) XRF spectra of two hair samples: from a person exposed to elevated uranium levels in drinking water in Finland (solid line) and from a control in Israel (dashed line). (Finnish sample provided courtesy of Laina Salonen from STUK.)

The term PCNAA is used when preconcentration precedes the neutron activation while if epithermal neutrons are used to excite the sample the acronym given is ENAA. The monitoring of the delayed neutrons emitted after excitation is termed DNAA. All these NAA procedures are nondestructive techniques used for characterizing solid (and in some cases also liquid) samples. However, a neutron source and a suitable detector are required and the sample can become quite radioactive after irradiation. The sensitivity of NAA techniques varies widely among different elements and sample preparation and post-irradiation methods employed. Several specific examples of NAA application for analysis of uranium in different matrices will be presented in the appropriate chapters.

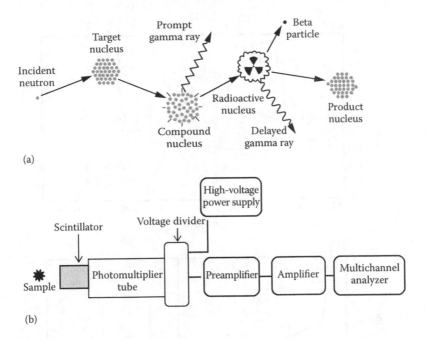

(a)

(b)

FIGURE 1.20 (a) A schematic presentation of the principle of neutron activation analysis (NAA). (Adapted from http://archaeometry.missouri.edu/images/naa_over_fig1.gif, accessed July 26, 2014.) (b) Schematic of a generic gamma spectrometer.

1.4.1.5 Gamma Spectrometry

This is an analytical technique that is used to characterize samples that contain gamma emitting nuclides. Activation by an external source is not required as the radioactive nuclides emit gamma rays (or lower energy x-rays) spontaneously, regardless of the chemical and physical conditions of the sample. In a typical setup, the sample is placed in a shielded chamber (usually lead coated by copper) and includes a liquid nitrogen–cooled HPGe detector and a multichannel analyzer. A simplified schematic of the components of a gamma spectrometer is shown in Figure 1.20b. The high-resolution gamma spectrum of a UF_6 cylinder containing low-enriched uranium (4.4% of ^{235}U) acquired during a period of 3.8 h is shown in Figure 1.21 (Croft et al. 2013). For the sake of clarity, the gamma spectrum was divided into three frames—each showing a different energy range.

In some cases, the presence of a specific radionuclide may be determined only indirectly from the gamma rays emitted by its daughter nuclides (progeny). The sensitivity of gamma spectrometry depends on instrumental parameters (discussed later) but also on the fundamental properties of the analyte nuclides, like its half-life, energy of the gamma photons, and the probability of their emission. The gamma rays emitted from the main isotopes of uranium are shown in Table 1.3 and evidently the most energetic is the 185.7 keV emitted from ^{235}U with a probability of 57%. However, ^{238}U can be detected by the doublet of gamma rays with energies of 63 and 93 keV emitted by its ^{234}Th daughter nuclide, or by the energetic 1001 keV gamma emitted from the short-lived (1.17 min) ^{234m}Pa. Some other examples will be presented in other chapters.

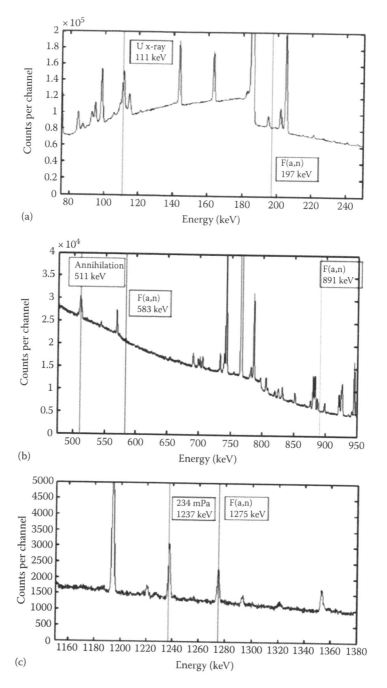

FIGURE 1.21 High-resolution gamma spectrum of a UF_6 cylinder containing low-enriched uranium (4.4% of ^{235}U) acquired during a period of 3.8 h. The photons emitted from the sample at different energy ranges: (a) 80–240 keV, (b) 500–950 keV, and (c) 1160–1380 keV. (From Croft, S. et al., *Nucl. Instrum. Methods Phys. Res. A*, 698, 192, 2013. With permission.)

A detector for energetic photons is required and indeed several types of detectors are available. The main types are solid state detectors like the NaI(Tl) that is relatively inexpensive but has limited resolution and the high-purity germanium (HPGe) that provides excellent resolution. Gas-filled (Ar–methane 90:10) detectors that can serve as an ionization chamber (appropriate for high-energy photons), proportional counter, or Geiger–Müller (unsuitable for quantification) depending mainly on the voltage applied to the anode. There are also scintillation detectors that contain an organic compound that emits photons after absorption of gamma rays. The photons are counted by a photomultiplier detector (Figure 1.20b). Each type of detection system has its advantages and limitations and proper knowledge is required for selection of the right detector for the task at hand. There usually is a trade-off between resolution, efficiency for a given energy range, cost, and counting time. When gamma rays pass through matter, the beam intensity would be attenuated but the photon energy is retained.

Quantification is not straightforward due to self-absorption of gamma photons in the sample material, and there is a strong dependence of the signal intensity on the geometric arrangement of the sample and detector and the shielding of the analytical system from external radiation. These problems can be somewhat overcome by using a standard geometry for the sample arrangement, matrix matching with standards and control of the detector and counting parameters.

1.4.1.6 Electron Microscopy

Electron microscopy (EM) is a general term that includes several analytical techniques used to characterize samples with electron interrogation methods. The sample is placed in a large vacuum chamber and an electron beam is used to illuminate the sample. The image is magnified so features that cannot be seen with a light microscope can be displayed—the resolving power of the EM makes it possible to attain a 50 pm resolution compared to 200 nm of an optical microscope (a factor of 4000). Figure 1.22a shows the tracks irradiating from a UO_2 particle with a diameter of 1 μm (Figure 1.22b) (Esaka et al. 2012). In the EM, the electron beam is focused and controlled by electrostatic and electromagnetic fields that function like the lenses in optical microscopy. The image can be viewed directly or by imaging the backscattered electrons. In addition, other analytical techniques can be introduced into the chamber of the electron microscope and provide elemental composition data (Figure 1.22c shows uranium and oxygen—the carbon is from the stub), information on chemical bonding and valence states, etc. In the realm of nuclear forensics, electron microscopes play an important role in characterizing uranium-containing particles and when combined with other techniques can also provide elemental composition data and valence state information. Several examples will be given in Chapter 5.

There are two main limitations with the common EM apparatus: the sample must be placed in a vacuum chamber (thus imposing limits on the size and volatility) and it must be a conductor of charge (to prevent build-up of static electricity that will distort the impinging electron beam). The advent of e-SEM (environmental SEM) can alleviate the requirement for a vacuum in some cases. Electron microscopy is not strictly a photon counting technique, but as its main application is similar to that of an optical microscope, although on a different scale, we decided to include it in this section.

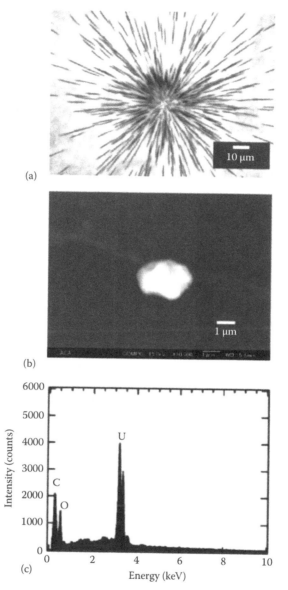

FIGURE 1.22 Electron microscope images of the tracks (a) irradiating from a particle containing uranium (b) and the energy-dispersive x-ray spectrum of the particle after plasma ashing (c). (Adapted from Esaka, F. et al., *Anal. Chim. Acta*, 721, 122, 2012. With permission.)

1.4.1.7 Other Spectroscopic Techniques

There are several other spectroscopic techniques that employ electromagnetic radiation for analytical purposes that can be used to characterize uranium compounds. Among them are UV–Vis (ultraviolet and visible spectroscopy) that reflects transitions of valence electrons in a molecule, infrared and Raman spectroscopy where the absorbed photons arise from vibrational and rotational transitions in the analyte molecules, nuclear magnetic resonance (NMR) that also reflects molecular structure, electron spin resonance (ESR), Mossbauer spectrometry, laser-induced breakdown spectrometry (LIBS), and numerous variations. One of the most popular techniques in the past—colorimetric determination of uranium in a complex with arsenazo-III (e.g., Savvin 1961)—is still used as an example for online preconcentration of uranium in an automatic method for uranium determination in seawater (Kuznetsov et al. 2014). Another example of a spectrophotometric method, as discussed in Chapter 4, used to determine uranium in urine was based on excitation of uranyl ions in a complex with UV radiation and measurement of the visible (green) photons emitted after a short delay (thus called laser-induced fluorescence (LIF)). We shall discuss specific applications of these techniques in the following chapters.

1.4.2 Ion Counting Techniques

1.4.2.1 Mass Spectrometry (MS)

Mass spectrometers are devices that separate and measure ions on the basis of their mass, or more precisely on the basis of their mass-to-charge ratio. Generic mass spectrometers consist of three main components: the ion source where ions are produced from the sample, the ion separation section where ions are differentiated according to their mass to charge ratio, and the detector where the ions are counted. In addition, the device must include a sample introduction system to transfer the analyte molecules from the sample to the ionization chamber and a signal processing element (mainly a dedicated software package) to derive the analytical information from the mass spectrum. Operation of MS requires a high vacuum so that the ions formed in the source region of the device can travel without being affected by collisions to the detector (see Figure 1.23).

There are several types of mass spectrometers that differ from each other in the construction and principle of operation of their ionization source, mass separator, and detector type. Due to important role mass spectrometric techniques play in the

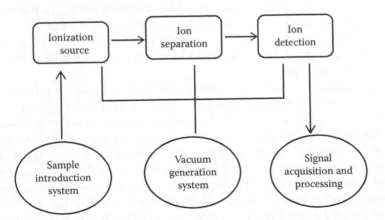

FIGURE 1.23 Schematic of a generic mass spectrometer showing the three main components: ion source, ion separation sector and detector.

characterizing of uranium, as seen throughout this monograph (see also Chapter 5), the main features of some generic MS types are presented later.

1.4.2.1.1   Ionization Sources

Ion sources are usually operated in vacuum, but in some instruments, ionization takes place at elevated or even ambient pressure. The sources suitable for vacuum conditions include filaments made of a fine metal wire (like tungsten, thoriated tungsten, or rhenium) that emit electrons when heated by an electric current and these are very popular especially for characterizing gaseous samples. Thermal ionization sources, where the sample is placed (deposited) on the filament so that analyte ions are directly emitted when the filament is heated are also quite common. In one popular variation, the sample vapors emitted from the coated filament are ionized by electrons emitted from a second filament. The ICP device of the ICP-AES mentioned earlier, operated in argon gas at atmospheric pressure, abundantly produces ions and serves as an ionization source, especially for liquid samples (see ICP-MS later). In other configurations, a beam of energetic ions from one ionization source can be used to produce secondary ions from the sample and those are subsequently characterized by a mass spectrometer as in secondary ion mass spectrometry (SIMS) or a variation where excited atoms are used as a source for production of secondary ions is called direct analysis in real time (DART). Other ionization sources include photo-ionization by energetic photons emitted from a UV-lamp or a short wavelength laser, radioactive materials (alpha or beta emitting nuclides), electric discharge sources (spark-source, glow discharge, corona discharge, etc.), electro-spray ionization (ESI and its derivatives SESI and DESI) used for liquid samples, matrix-assisted laser desorption ionization (MALDI) suitable for heavy biological or complex organic molecules, and other sources. The relevant ionization sources will be discussed in the context of specific analytical and forensic applications of MS for uranium analysis.

1.4.2.1.2   Separation of Ions

Ions are separated according to their mass-to-charge ratio as they travel in an evacuated tube under the influence of electric and/or magnetic fields. The two most common systems for MS devices are the quadrupole and magnetic sector mass separators schematically shown in Figure 1.24 (quadrupole MS in Figure 1.24a and magnetic sector MS in Figure 1.24b). In Figure 1.25, a schematic of a time-of-flight (TOF) mass spectrometer is shown.

In the quadrupole mass spectrometer (QMS), a set of four parallel metal rods (hence the name) creates a uniquely shaped electric field (Figure 1.24a). Variations include a set of eight rods (called octopole-MS) or two caps below and above a circular electrode (commonly known as ion trap-MS). Ions injected from the ionization source into the space between the rods are displaced by a combination of direct and alternating voltages (DC–AC) that create an electric field. At a particular setting of voltages, only ions of a given mass-to-charge ratio can traverse the field and reach the detector. Thus, the device serves as a mass filter allowing only one type of ions to produce a signal at the detector. Scanning the electric voltages on the rods produces the mass spectrum. A mass scan typically lasts several milliseconds (ms). It is common practice to use a *peak hopping* procedure to measure only the analytes

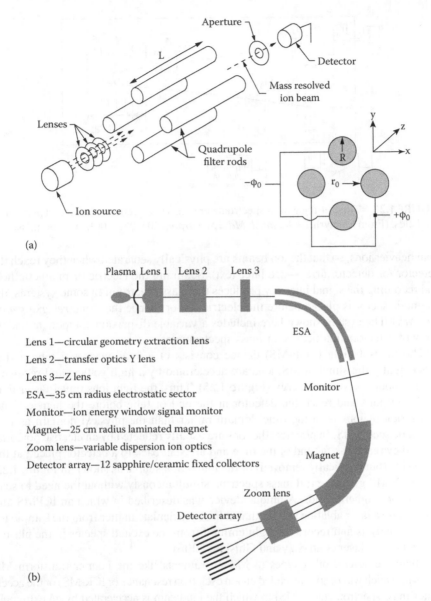

(a)

(b)

Lens 1—circular geometry extraction lens
Lens 2—transfer optics Y lens
Lens 3—Z lens
ESA—35 cm radius electrostatic sector
Monitor—ion energy window signal monitor
Magnet—25 cm radius laminated magnet
Zoom lens—variable dispersion ion optics
Detector array—12 sapphire/ceramic fixed collectors

FIGURE 1.24 (a) Schematic presentation of a quadrupole mass spectrometer. (From Blaum, K. et al., *Int. J. Mass Spectrom.*, 181, 67, 1998. With permission.) (b) Schematic of a double-focusing magnetic sector mass spectrometer with variable dispersion. (From Belshaw, N.S. et al., *Int. J. Mass Spectrom.*, 181, 51, 1998. With permission.)

of interest by, as the name implies, changing the setting of the electric voltages to accommodate the transmittance only of the ions of interest.

In the magnetic sector MS, the ions are accelerated by a strong electric field and the ion beam is then directed into the magnetic field where their trajectory is curved (Figure 1.24b). Ions of lighter mass-to-charge ratio follow a sharper curve

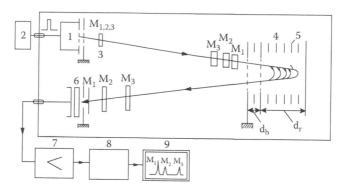

FIGURE 1.25 Time-of-flight mass spectrometer with a two-section reflector and plane electric fields. (From Mamyrin, B.A., *Int. J. Mass Spectrom.*, 206, 251, 2001. With permission.)

than heavier ions, so that the ion beams are physically separated when they reach the detector (or detector array—see later text). Scanning the electric or magnetic field and recording the signal intensity produces the mass spectrum. In some systems, the magnetic sector is placed before the electric sector (hence the name *reverse geometry MS*). The system shown here includes a variable dispersion component and an array of detectors (multicollector mass spectrometer).

The time-of-flight (TOF-MS) device consists of a long evacuated tube and is conceptually the simplest MS: ions are accelerated by a high voltage and receive a given amount of kinetic energy (Figure 1.25). Thus, the light ions travel faster than the heavy ions and reach the detector at the end of tube before the heavier ions. Separation of ions is in the time domain rather than the physical domain as in a magnetic sector MS. In practice, the ions are usually reflected by an electric potential when they reach the far end of the tube and are sent back to a detector placed at the same side (but physically removed) of the ionization source as shown in Figure 1.25. The TOF-MS gives the full mass spectrum simultaneously without the need to scan electric or magnetic fields. A unique device was described in which an ICPMS and LIBS share a laser ablation system that ablates particulate matter from the sample for ICPMS analysis and record the light emitted from the excited species in the plasma created by the laser (Latkoczy and Ghislain 2006).

There are several other types of MS instruments like the Fourier transform MS (FT-MS, which was originally called an ion cyclotron resonance or ICR-MS) or the accelerator mass spectrometer (AMS) in which the ion beam is accelerated by an extremely high electric field and then separated by a strong magnetic field (see next section).

1.4.2.1.3 Detectors

The ions impinging on a detector produce an electric current that is converted into a voltage and amplified. The simplest detector is the Faraday detector that is a flat metal plate that gets charged when ions strike it or the Faraday cup that has a concave configuration and serves a similar function (Figure 1.26a). A more sensitive detector is an electron multiplier made of a series of dynodes. When an ion strikes the surface of the first dynode, a number of electrons are emitted (Figure 1.26b).

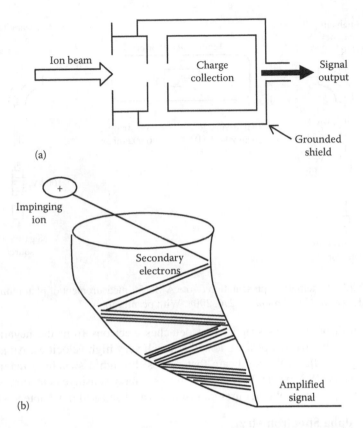

(a)

(b)

FIGURE 1.26 (a) A schematic of a Faraday cup detector and (b) a schematic of an electron multiplier detector.

These are accelerated toward the next dynode where each electron causes the emission of a number of new electrons, and so on until the signal created by a single ion is amplified. The electron current can be registered at one of these dynodes as an analog signal or each pulse can be counted (ion count detector) as a digital signal. If the ion beams are physically separated, as in the magnetic sector MS, then an array of detectors can be used to simultaneously monitor the intensity of several signals (ion currents).

In Chapter 5, the application of mass spectrometric techniques in the field of nuclear forensics, and particularly in the characterization of single particles, is discussed in some detail. These standard types of mass spectrometers are well known and widely deployed in many laboratories and therefore were only described in brief earlier. So here one unique mass spectrometric technique, the AMS, will be discussed.

AMS is a unique type of mass spectrometer due to its principle of operation, size, cost, and sensitivity. A schematic of an AMS for the determination of plutonium and other actinides is shown in Figure 1.27 (Fifield 2008). The sample is placed on a filament that forms negative ions that are accelerated and inflected by a magnetic field. The ions are further accelerated by a very strong electric field (the accelerator)

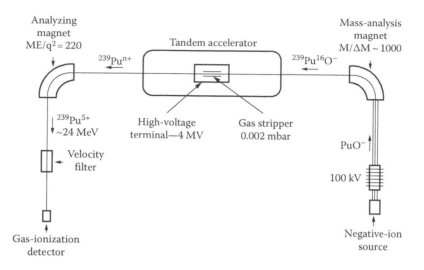

FIGURE 1.27   Schematic representation of the AMS for measurement of plutonium. (From Fifield, L.K., *Quat. Geochronol.*, 3, 276, 2008. With permission.)

and passed through a gas stripper that detaches electrons from the negative ions forming multiply charged positive ions that reach very high velocities. An analyzer strong magnet inflects the ion beam that passes through a switching magnet to a detector. The AMS is particularly suitable for extremely sensitive detection of traces of ^{236}U in the presence of other uranium isotopes as discussed in Chapters 3 and 5.

1.4.2.2   Alpha Spectrometry

The radioactive decay of heavy nuclides commonly involves the ejection of an alpha particle (a helium atom consisting of two proton and two neutrons) and formation of a nuclide that has a lower atomic number by two units and a mass number that is four units lighter than the parent nuclide (Figure 1.28a).

The emitted alpha particles have well-defined energies as they represent a transition from one discrete level of nuclear energy in the parent nuclide to a discrete level in the daughter nuclide. However, this is not always reflected in the alpha spectrum because attenuation of the particles causes tailing toward lower energies and peak broadening (Figure 1.28b). Emission of an alpha particle is usually accompanied by emission of gamma rays, as mentioned earlier. Alpha particles emitted from decay of heavy nuclides have high kinetic energies (between 3 and 7 MeV), but they interact strongly with matter, causing ionization and rapidly losing energy, so that their range in dense matter, and even in air, is very short. Alpha spectrometry can also be viewed as an ion counting technique because each ion is registered by the detector. As mentioned earlier, all uranium isotopes emit alpha particles with different energies and half-life times so that they can be distinguished from each other and from other elements (Table 1.3). One of the advantages of alpha spectrometry is the low background level that makes it possible to detect single events. However, meticulous sample preparation is required as the uranium should be in a pure form and placed as a thin layer in the sample holder (usually a metal plate or planchet). A detailed

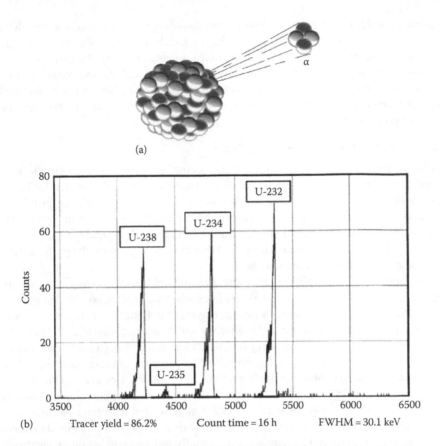

FIGURE 1.28 (a) Schematic representation of alpha particle emission and (b) alpha spectrum of uranium (with a ^{232}U spike) after separation from other actinides in a soil sample (also shown in Figure 3.4). (From Maxwell, 2011. With permission.)

example is given in Chapter 4 in order to demonstrate the tedious procedure of determination of uranium in urine samples by alpha spectrometry. Basically, solid samples must be dissolved, the uranium should be separated and purified, and then concentrated and precipitated or placed on stainless steel plate by electro-deposition. The use of an isotopic spike, ^{232}U in the case of uranium, is frequently deployed to determine the recovery efficiency. Despite the complicated sample preparation and long counting times, alpha spectrometry is often used for the determination of low levels of actinides in general and uranium in particular.

A typical alpha particle spectrum of natural uranium (^{234}U, ^{235}U, and ^{238}U) spiked with the artificial isotope ^{232}U is shown in Figure 1.28b. Unlike the mass spectrometric techniques described earlier, the signal intensity of each isotope does not directly reflect its concentration as it is proportional to the abundance divided by the half-life of each isotope (i.e., the activity of the nuclide in the sample) and the probability of a given transition (see Tables 1.2 and 1.3 for the alpha particles emitted from uranium isotopes). Thus, although ^{234}U is about 18,000 less abundant in natural uranium than ^{238}U, it has approximately the same signal intensity in the alpha spectrum when the two nuclides are in secular equilibrium. Note the tailing toward lower energy of the peaks in the alpha spectrum that is due to attenuation of the alpha particles by the layers of sample material.

Several types of detectors for alpha particles are available, but the solid state ion implanted silicon semiconductors are most commonly used. As the counting time required for measurement of samples with low activity could be quite long (several days) and the cost of an analytical system is not very high, many laboratories operate several alpha spectrometers in parallel.

1.4.2.3 Other Ion Counting Techniques

A photographic film can be used as an ion counting technique as energetic ions would cause damage (black spots) when they impinge on it, as discovered by Henri Becquerel. The modern equivalent of this is a plate upon which charge builds up when struck by alpha particles. A microchannel plate (MCP) is a planar structure that in effect combines several electron multipliers and when struck by ions, electrons, or photons a set of signals is produced that can serve to image the particles that emit radiation or ions from the sample.

Fission track analysis (FTA) is an indirect method of counting heavy ions that are created through nuclear fission of fissile nuclides when a sample is bombarded by neutrons. The sample is placed between two thin layers of polymeric film and placed near a neutron source. The interaction of the neutrons with fissile nuclides, like ^{235}U or ^{239}Pu, induces fission and the recoiling heavy fragments ionize the film, causing craters from which tracks radiate as shown in Figure 1.29. The presence of these tracks and their number reflect the content of the fissile nuclides in the sample (see also in Chapter 5).

Electroanalytical (EA) methods can also be considered as ion counting techniques that are used to selectively measure the potential generated between two electrodes or the currents produced by certain types of ions that are present in a liquid sample. The selectivity is achieved by deploying a specially tailored coating on an electrode that allows only a specific ion type to register a current. In other embodiments, electroanalytical methods can be used to preconcentrate a specific ion and then it can be detected and quantified by stripping the charge from the same electrode or by

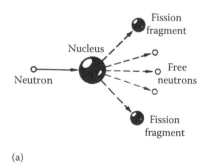

(a)

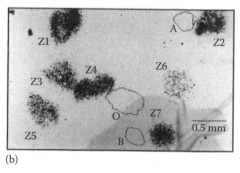

(b)

FIGURE 1.29 The principle of inducing fission in a ^{235}U nuclide by a neutron (a) and the image of radiating tracks from uranium in a mineral (b). (a: From http://www.ratical.org/radiation/CNR/PP/fig3.gif, accessed July 26, 2014; b: From http://www.geotrack.com.au/images/ufission3b.jpg, accessed July 26, 2014.)

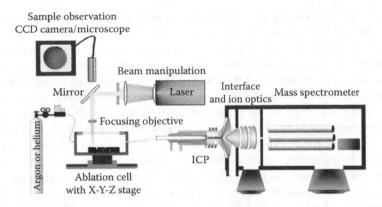

Sample observation
CCD camera/microscope

Beam manipulation

Mirror Laser Interface Mass spectrometer
and ion optics

Focusing objective

Argon or helium

ICP

Ablation cell
with X-Y-Z stage

FIGURE 1.30 A schematic representation of the laser ablation—ICPMS system. (From Gunther, D. and Hattendorf, B., *TrAC*, 255, 2005. With permission.)

other analytical techniques. Electroanalytical methods can also be used to differentiate between ionic species (based on valence state) of the same element by selective reduction or oxidization. In brief, the electroanalytical methods measure the effect of the presence of analyte ions on the potential or current in a cell containing electrodes. The three main types are potentiometry, where the voltage difference between two electrodes is determined, coulometry, which measures the current in the cell over time, and voltammetry, which shows the changes in the cell current when the electric potential is varied (current–voltage diagrams). In a recent review article, 43 different EA methods for measuring uranium were mentioned and that literature survey found 28 voltammetric, 25 potentiometric, 5 capillary electrophoresis, and 3 polarographic methods (Shrivastava et al. 2013). Some specific methods will be discussed in detail in the relevant chapters of this tome.

1.4.2.3.1 Laser Ablation-ICPMS

This was also mentioned earlier in context with mass spectrometry and LIBS. The unique system is suitable for direct analysis of solid samples and combines the sensitivity and isotopic measurement capability of the ICPMS with the spatial resolution of a laser system shown in Figure 1.30 (Gunther and Hattendorf 2005). The system combines a pulsed laser, preferably with short wavelength photon (like 193 nm from an excimer laser) in order to achieve clean ablation without melting of the sample with an ICPMS instrument (any kind is suitable). The sample is placed in a chamber (ablation cell) and a flow of gas (usually argon with some helium) carries the particles that are ablated from the sample by the laser into the torch of the ICPMS.

Thus, imaging of the distribution of uranium on the surface of the sample is possible. A particularly interesting example is the measurement of uranium along a single hair strand that is described in Chapter 4.

1.4.3 GENERIC SAMPLE PREPARATION AND URANIUM SEPARATION TECHNIQUES

Obviously there can be no single universal sample preparation method for the determination of uranium in all types of matrices (biological, environmental, forensic, and industrial) that can be in solid, liquid, or gaseous form. As most of the relevant analytical instruments described in the previous section are best suited for measurement of liquid samples, the common sample preparation methods usually focus on getting the sample into liquid form. In some cases, sample preparation also involves removing and purifying the uranium from potential interfering substances.

Dissolution of solid samples that contain uranium usually deploys acid digestion. Environmental samples (soil and vegetation mainly) are first sieved to remove foreign bodies and then heated to ~105°C in an oven to remove moisture. This is followed sometimes by calcination at an elevated temperature of ~500°C to destroy organic matter or ash vegetation samples. Then the sample may be pulverized or ground to a fine powder to increase the surface area and ensure homogeneity. Digestion is carried out by strong acids, mainly nitric acid, but hydrogen chloride, perchloric acid, sulfuric acid, and hydrogen fluoride are sometimes added. This can be carried out in open vessels or in sealed vessels with microwave-assisted digestion. If further separation and purification are required or desired, then liquid extraction, solid phase extraction, ion exchange columns, or chromatographic resins are deployed. The same approach may be used for assaying uranium in minerals, in industrial materials and even in biological samples. The addition of an isotopic spike, or an internal standard, is often carried out for quantitative assessment of the uranium content in the sample. There are many variations on this generic sample preparation and several examples are presented in Chapters 2 through 5.

Preparation of liquid samples (environmental or biological) is usually simpler and may involve filtering to remove suspended solids and pH adjustment to prevent precipitation of uranium. If separation is required, the same strategies outlined earlier can be deployed.

Air or gaseous samples are usually collected by pumping a large volume of ambient air through a suitable collection device (cold trap, sorbent tube, impinger, etc.). The sample is then prepared for analysis by extracting the uranium by the methods outlined earlier and analyzed with the appropriate device.

1.5 SUMMARY

In this chapter, we tried to present an overview of the properties of uranium (the element, isotopes, and main compounds) that are relevant to the subject matter of this book. An understanding of the industrial aspects of uranium processing, and especially its role in the NFC, is essential in order to appreciate the ways uranium affects our lives. We also tried to briefly discuss the main analytical techniques that are used in the characterization of uranium, whether as a major component, a minor component, or a trace in environmental and biological samples. A detailed description of the analytical techniques will be given in the following chapters.

REFERENCES

Aczel, A.D. (2009). *Uranium Wars: The Scientific Rivalry that Created the Nuclear Age.* New York: Palgrave Macmillan.
Allen, C.G. and Tempest, P.A. (1982). Linear ordering of oxygen clusters in hyperstoichiometric uranium dioxide, *J. Chem. Soc. Dalton Trans.* 39, 2165–2173.

Belshaw, N.S., Freedman, P.A., O'Nions, R.K. et al. (1998). A new variable dispersion double focusing plasma mass spectrometer with performance illustrated for Pb isotopes, *Int. J. Mass Spectrom.* 181, 51–58.

Blaum, K., Geppert, C., Muller, P. et al. (1998). Properties and performance of a quadrupole mass filter used for resonance ionization mass spectrometry, *Int. J. Mass Spectrom.* 181, 67–87.

Bosco, G.L. (2013). Development and application of portable, hand-held X-ray fluorescence spectrometer, *TrAC* 45, 121–134.

Burns, P.C. and Finch, R.J. (1999). *Uranium: Mineralogy, Geochemistry and the Environment.* Chantilly, VA: Mineralogical Society of America.

Commission, U.N. (2013, May 21). Uranium enrichment. Retrieved from http://www.nrc.gov/materials/fuel-cycle-fac/; http://www.nrc.gov/materials/fuel-cycle-fac/ur-enrichment. html (February 9, 2014).

Croft, S., Swinhoe, M.T., and Miller, K.A. (2013). Alpha particle induced gamma yields in uranium hexafluoride, *Nucl. Instrum. Methods Phys. Res. A* 698, 192–195.

Esaka, F., Lee, C.G., Magara, M. et al. (2012). Fission track secondary ion mass spectrometry as a tool for detecting the isotopic signature of individual uranium containing particles, *Anal. Chim. Acta* 721, 122–128.

Fifield, L.K. (2008). Accelerator mass spectrometry of the actinides. *Quat. Geochronol.* 3, 276–290.

Firestone, R.B., Shirley, V.S., Baglin, C.M. et al. (1996). *Table of Isotopes*, book and CD-ROM, 8th edn. New York: John Wiley & Sons, Inc.

Fox, P.B., Petrie, L.M., and Hopper, C.M. (2005). Minimum critical values study, ORNL/TM-2003/211. Oak Ridge, TN: Oak Ridge National Laboratory.

Galkin, S. (1966). *Technology of Uranium* (English Translation). Jerusalem, Israel: Israeli Program Scientific Translation.

Goleb, J.A. (1966). The determination of uranium isotopes by atomic absorption spectrophotometry, *Anal. Chim. Acta* 34, 135–145.

Grenthe, I. (2006). Uranium. In Morss, L.R. (ed.), *The Chemistry of Actinide and Transactinide Elements*, Vol. 1 (pp. 253–698), Chapter 5. Dordrecht, the Netherlands: Springer.

Gunther, D. and Hattendorf, B. (2005). Solid sample analysis using laser ablation inductively coupled plasma mass spectrometry, *TrAC* 24, 255–265.

Hanna, S.R., Chang, J.C., and Zhang, S.J. (1997). Modeling accidental release to the atmosphere of a dense reactive chemical (uranium hexafluoride), *Atmos. Environ.* 31, 901–908.

Hiess, J., Condon, D.J., McLean, N. et al. (2012). 238U/235U systematic in terrestrial uranium bearing minerals, *Science* 335, 1610–1616.

IAEA 1613. (2009). Nuclear fuel cycle information system. A directory of nuclear fuel cycle facilities, 2009 Edition IAEA TECDOC 1613. Vienna, Austria: IAEA.

Ivanovich, M. (1994). Uranium series disequilibrium—Concepts and applications, *Radiochim. Acta* 64, 81–94.

Karpas, Z., Lorber, A., Sela, H. et al. (2006). Determination of 234U/238U ratio: Comparison of multi-collector ICPMS and ICP-QMS for water, hair and nails samples and comparison with alpha spectrometry for water samples, *Radiat. Prot. Dosimetry* 118, 106–110.

Katz, J.J., Katz, B.J.J., and Rabinowitch, E. (1951). *The Chemistry of Uranium. Part I: The Elements, Its Binary and Related Compounds.* New York: McGraw-Hill.

Koelzer, W. (2011). Glossary of nuclear terms—Critical mass. Karlsruhe, Germany: European Nuclear Organization. January, 2011, updated and supplemented.

Kuznetsov, V.V., Zemyatova, S.V., and Kornev, K.A. (2014). Automatic determination of uranium (VI) in seawater using on-line preconcentration by coprecipitation, *J. Anal. Chem.* 69, 116–121.

Latkoczy, C. and Ghislain, T. (2006). Simultaneous LIBS and LA-ICPMS analysis of industrial samples, *J. Anal. Atom. Spectrom.* 21, 1152–1160.

Lide, D. (ed.) (1999). *Handbook of Chemistry and Physics.* Boca Raton, FL: CRC Press.

Mamyrin, B.A. (2001). Time-of-flight mass spectrometry (concepts, achievements and prospects), *Int. J. Mass Spectrom.* 206, 251–266.

McInnes, D.A. and Longsworth, L.G. (1942). *A Comparison of Analytical Methods for the Determination of Uranium.* Oak Ridge, TN: Rockefeller Institute for Medical Research.

Rao, P.R. (2003). Laser isotope separation of uranium—Research account, *Curr. Res.* 85, 615–632.

Savvin, S.B. (1961). Analytical use of arsenazo-III: Determination of thorium, zirconium, uranium and rare earth elements. *Talanta* 8, 673–685.

Sheela, Meena, R., Anuradha, Kartha R.M., Pujar, R.K., Singh, A., and Ganguly, C. (2000). *Technology for Production of Uranium Oxide for Nuclear Power Reactor.* Hyderabad, India: Nuclear Fuel Complex. http://dbtindia.nic.in/women/paper19.htm, accessed July 26, 2014.

Shrivastava, A., Sharma, J., and Soni, V. (2013, June). Various electroanalytical methods for the determination of uranium in different matrices, *Bull. Facul. Pharm.*, Cairo University, 51(1), 113–129.

TAMU. (2000). Guidelines to analytical operating conditions. Retrieved from Ocean Drilling Program. College Station, TX: Texas A&M University. http://www-odp.tamu.edu/publications/tnotes/tnotes/tn29/technot2.htm (March 12, 2014).

Thurber, D.L. (1963). Anomalous 234U/238 ratios in nature. *Nuclear Geophysics* (pp. 176–184). Woods Hole, MA.

Wood, H.G., Glaser, A., and Kemp, R.S. (2008). The gas centrifuge and nuclear weapons proliferation. *Phys. Today*, September 40–45.

Yamamoto, M., Sato, T., Sasaki, K. et al. (2003). Anomalously high U-234/U-238 activity ratios of Tatsunokuchi hot-spring waters, Ishikawa Prefecture, Japan, *J. Radioanal. Nucl. Chem.* 255, 369–373.

Zoellner, T. (2010). *Uranium: War, Energy and the Rock that Shaped the World.* London, U.K.: Penguin Books.

2 Industrial Applications

One pound of uranium is worth about 3 million pounds of coal or oil.

James Lovelock

It may be possible to set up a nuclear reaction in uranium by which vast amounts of power could be released. This new phenomenon would also lead to the construction of extremely powerful bombs of a new type.

Albert Einstein, Letter to President Franklin D. Roosevelt

2.1 INDUSTRIAL CONSIDERATIONS

2.1.1 INTRODUCTION

The main industrial processes that involve uranium are the excavation of uranium-containing ores and its application in the nuclear fuel cycle (NFC). A summary table of the types of facilities involved in the NFC was constructed by the Nuclear Fuel Cycle Information System (NFCIS) published as a technical document by the International Atomic Energy Agency (IAEA-TECDOC-1613 2009) and reproduced as Table 2.1.

The operations and facilities include ore exploration (not included in NFCIS list), mining, ore processing, uranium recovery, chemical conversion to UO_2, UO_3, UF_4, UF_6, and uranium metal, isotope enrichment, reconversion of UF_6 to UO_2 (after enrichment), and fuel fabrication and assembly that are all part of the front end of the NFC. The central part of the NFC is the production of electric power in the nuclear reactor (fuel irradiation). The back end of the NFC includes facilities to deal with the spent nuclear fuel (SNF) after irradiation in a reactor and the disposal of the spent fuel (SF). The spent fuel first has to be stored for some period to allow decay of the short-lived fission products and activation products and then disposed at waste management facilities without, or after, reprocessing to separate the fission products from the useful actinides (uranium and plutonium). Note the relatively large number of facilities in Table 2.1 dedicated to dealing with the spent fuel. Also listed in Table 2.1 are related industrial activities that do not involve uranium, like heavy water (D_2O) production, zirconium alloy manufacturing, and fabrication of fuel assembly components.

Additionally, there are secondary processes, like production of depleted uranium (DU) for commercial purposes or synthesis of fine chemicals that use the waste (or tails) of uranium isotope enrichment plants. When discussing the analytical chemistry of these applications, the emphasis is placed on determining the impurities in materials where uranium is the major component, while in the other chapters of this book, the analytical objective is usually the detection of trace levels of uranium and their characterization.

TABLE 2.1

Types of Facilities in the Nuclear Fuel Cycle

Nuclear Fuel Cycle Stage	NFCIS Facility Subtypes	Description
Uranium production	Uranium mine	Mines from which uranium is extracted
	U ore processing	Facilities in which U ore is processed to produce *yellow cake* includes in situ leach facilities
	U from phosphate	Facilities that retrieve U as a by-product from phosphate
Conversion	Conversion to UO_2	Facilities that convert U_3O_8 to UO_2 for fuel production
	Conversion to UO_3	Facilities that convert U_3O_8 to UO_3 for later conversion to UO_2 fuel or UF_6
	Conversion to UF_4	Facilities that convert U_3O_8 to UF_4 that is later converted to UF_6 for enrichment or U_{metal} for fuel
	Conversion to UF_6	Facilities that convert U_3O_8 or UF_4 to UF_6 for enrichment
	Conversion to U_{metal}	Facilities that convert UF_4 to U-metal for fuel
	Reconversion to U_3O_8 (DU)	Facilities in which depleted UF_6 is converted to U_3O_8 for further storage or processing
Enrichment	Enrichment	Facilities that increase ^{235}U content relative to ^{238}U content
Fresh uranium fuel fabrication	Reconversion to UO_2 powder	Facilities that convert enriched UF_6 to UO_2 powder
	Fuel fabrication (U pellet-pin)	Facilities that produce fuel pellet and/or pins using UO_2 powder
	Fuel fabrication (U assembly)	Facilities that produce fuel assemblies using pellets/pins (powder/pellet/assembly may be combined in a facility)
	Fuel fabrication (research reactors)	Facilities that produce research reactor fuel
	Fuel fabrication (Pebble)	Facilities in which fuel pebbles are produced for pebble bed reactors
Irradiation	Irradiation	Reactors that irradiate the fuel
Spent fuel storage	At reactor (AR) spent fuel storage	Facilities located in the reactor site in which fuel is stored temporarily, usually in reactor pools
	AFR wet spent fuel storage	Facilities located outside the reactor site in which fuel is stored temporarily in pools
	AFR dry spent fuel storage	Facilities located outside the reactor site in which fuel is stored temporarily in dry silos or containers
Spent fuel reprocessing and recycling	Spent fuel reprocessing	Facilities in which spent is reprocessed to retrieve nuclear material
	Reconversion to U_3O_8 (Rep. U)	Facilities in which reprocessed uranium is converted to U_3O_8
	Co-conversion to MOX powder	Facilities in which uranium and plutonium are mixed in the form MOX powder

(Continued)

TABLE 2.1 (*Continued*)
Types of Facilities in the Nuclear Fuel Cycle

Nuclear Fuel Cycle Stage	NFCIS Facility Subtypes	Description
	Fuel fabrication (MOX pellet-pin)	Facilities in which MOX fuel pellets/pins are produced
	Fuel fabrication (MOX assemblies)	Facilities in which MOX fuel assemblies are produced
Spent fuel conditioning	Spent fuel conditioning	Facilities in which spent fuel is conditioned for longer term interim storage or for disposal
Spent fuel disposal	Spent fuel disposal	Facilities in which spent fuel is disposed permanently
Related industrial activities	Heavy water production	Facilities in which heavy water is produced
	Zirconium alloy production	Facilities in which zirconium metal sponge is produced
	Fuel assembly component	Facilities in which other fuel structurals and components are produced
Transportation	Transportation	All transportation related to the nuclear fuel cycle
Waste management	Waste management	Facilities in which all kinds of radioactive wastes are conditioned, processed, stored, or disposed

Source: International Atomic Energy Agency, Nuclear fuel cycle information system: A directory of nuclear fuel cycle facilities, 2009 Edition IAEA TECDOC 1613, IAEA, Vienna, Austria, 2009. With permission.

The price of uranium has seen dramatic changes over the last 30 years, as seen in Figure 2.1, and is at the time of writing (April 2014) about $34 lb^{-1} of U_3O_8, after reaching a maximum of almost $137 lb^{-1} in July 2007.

It should be noted that uranium is not traded like other commodities on open markets but in direct contracts between buyer and supplier. The price is determined by a small number of private business organizations and strongly affects the economic value of extraction and processing of uranium.

As most of the excavated uranium is subsequently used as nuclear fuel, the price of conversion of the ore (yellow cake) to UF_6, the price of enrichment (separative work units—SWU), the cost of deconversion of the enriched UF_6 to uranium oxide (or other chemical forms), and the production of fuel elements determine economic factors. In addition, the cost of electric power production by nuclear power plants in comparison with other plants (gas, coal, and oil) and the overall cost of disposal of the waste from all these processes will influence the worthiness of uranium extraction. In view of the rapid changes in the prices of these processes, it is difficult to assess the threshold concentration of uranium in the ore that will make mining viable economically. Furthermore, nations or organizations that cannot purchase uranium

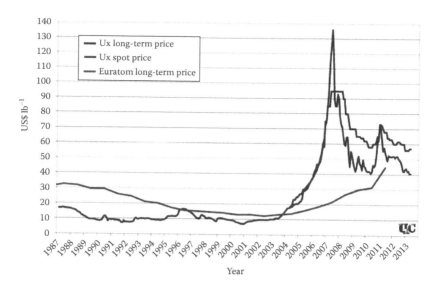

FIGURE 2.1 The price of uranium oxide (US\$ lb^{-1}) since 1987. (Reproduced from the World Nuclear Association, http://www.world-nuclear.org/info/Nuclear-Fuel-Cycle/ Uranium-Resources/Uranium-Markets/, last accessed July 27, 2014. With permission.)

on the open market but have natural resources that contain uranium may develop these ores regardless of the economic aspects. Finally, major accidents at nuclear power plants (Chernobyl 1986; Fukushima 2011), or even minor incidents that did not lead to the release of radioactive material (Three Mile Island in 1979), strongly affect the public's attitude toward nuclear reactors that are used to generate electricity and thus the demand for uranium and its price.

2.1.2 World Resources and Production of Uranium

The OECD Nuclear Energy Agency (NEA) and IAEA publish biannually a survey entitled "Uranium: Resources, production and demand," which is also known in the industry as the "Red Book" (IAEA 2012). This large volume presents a comprehensive summary of the world's known and estimated resources of uranium and thorium, the supply of uranium, and the demand and a nation-by-nation assessment of these parameters. The supply of uranium strongly depends on the price of production and the cost of alternative fuels (like coal, oil, and gas) and the "Red Book" also presents the resources of the major suppliers of uranium on the basis of production price (<\$40, \$40–\$80, \$80–\$130, and \$130–\$260 kg^{-1} U). The main production methods include open mining, underground mining, which is the major source for cheap uranium costing less than \$40 or \$80 kg^{-1}, in situ leaching (ISL) that is gaining popularity, and some other minor sources like heap leaching from open mining operations. The low-cost resources are usually from established rich deposits excavated by underground mining (mainly in Canada) and the by-product category (mainly in Brazil) and the ISL process (mainly in China and Kazakhstan). Australia is the main supplier of the <80\$ kg^{-1} U while at the high end of production costs underground

TABLE 2.2

Major Producers of Uranium and Their Output (in Metric Tons) in the Years 2002, 2006, and 2012

Country	2002	2006	2012
Australia	6,854	7,593	6,991
Canada	11,604	9,862	8,999
China	730	750	1,500
Czech Republic	465	359	228
India	230	230	385
Kazakhstan	2,800	5,279	21,317
Namibia	2,233	3,077	4,495
Niger	3,075	3,434	4,667
Russia	2,900	3,400	2,872
South Africa	824	534	465
Ukraine	800	800	960
United States	883	1,692	1,596
Uzbekistan	1,860	2,270	3,000
World total	**36,036**	**39,670**	**58,394**

Source: Adapted from the World Nuclear Association, http://www.world-nuclear.org/info/Facts-and-Figures/Uranium-production-figures/, last accessed on July 27, 2014. With permission.

Note: The bottom line gives the world total uranium production in those years.

mining is dominant. Table 2.2 shows a breakdown of the major suppliers of uranium in the years 2002, 2006, and 2012 and of the total global supply in those years. Most notable is the great increase in the share of uranium supplied from Kazakhstan that rose from ~7.8% of the global production in 2002 to 36.5% in 2012 concomitant with the decrease of the share of Australia and Canada in the market.

The cost of uranium production varies according to the geological formations where uranium is found and excavated. Unconformity-related deposits of uranium provide the main source of the uranium in the <$40 kg^{-1} category, but in the <$80 kg^{-1} class hematite breccias complex deposits contain about half of the reasonable assured resources while sandstone and unconformity-related deposits each contain about 15% of these resources (IAEA 2012). The "Red Book" also refers to *unconventional uranium resources* that include phosphate rock (dominated by Morocco that holds 85% of this source), as well as much smaller deposits of uranium in nonferrous ores, carbonatite, and black schist/shales lignite (in Sweden, this is a by-product of polymetallic black shales). Perhaps one of the most exciting production methods of uranium that are being developed is extraction from seawater that holds about four billion tons with an average concentration of 3.1 μg L^{-1} (a special session with 25 presentations was devoted to this topic at the 244th American Chemical Society Meeting in 2012). One technique uses long strips coated with an absorbent polymer, and another uses a kelp-type module (Wang et al. 2014) or metalorganic frameworks, which selectively recovered uranium. The cost of uranium thus

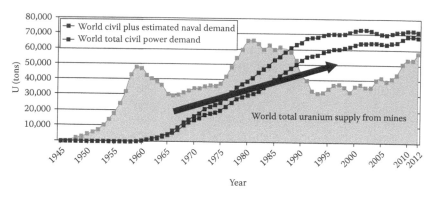

FIGURE 2.2 The annual production of uranium and the annual requirements between 1945 and 2012. (From World Nuclear Association, http://www.world-nuclear.org/info/Nuclear-Fuel-Cycle/Uranium-Resources/Uranium-Markets/, last accessed July 27, 2014. With permission.)

produced is still high >$260 kg^{-1} U, and there are other economical, technical, and legal difficulties, but considering the large amount of uranium in seawater this could be a future inexhaustible source of uranium.

As mentioned earlier, the three major production methods used for the extraction of uranium include open pit mining, underground mining, and ISL. Additionally, in-place leaching, utilizing uranium as a by-product from other processes, heap leaching, and other methods are also deployed. The percentage of uranium extracted by the different processes between 2007 and 2011 was summarized in the "Red Book" (IAEA 2012) and shows that of the three dominant methods, which together include about 90% of the extracted uranium, ISL is gaining popularity while open pit and underground mining are less widely used due to economic and safety considerations. Figure 2.2 shows the global supply and demand for uranium between 1945 and 2012 (from WNA website). After 1991, the demand appears to exceed the production but shortage was avoided due to the fact that there were sufficient reserves of uranium.

2.2 ANALYTICAL CHEMISTRY OF URANIUM ORES

2.2.1 Prospecting Methods

Many approaches have been proposed and adopted for finding uranium-bearing minerals and ores, ranging from high-tech methods that can cover a large area rapidly to methods that can best be described as exotic. Several scientific disciplines have been deployed for uranium prospecting, including mineralogy, classic analytical chemistry, nuclear gamma (Grasty 1979) spectrometry (Barretto 1981), radiometric methods (Sarangi 2000), x-ray fluorescence, radon detection (Charlton and Kotrappa 2006), geobotanical methods (Priyadarshi 2010), examination of animal thyroids (Wogman et al. 1977), ground water analysis, isotopic method (Dickson et al. 1985), geology (Rackley et al. 1968), etc. These also span aerial surveys, air and water sampling, prospective drilling, ground radioactivity inspection, and so on. We shall try to focus on some of the modern prospecting methods as discussed in more detail and reviewed later.

2.2.1.1   Aerial and Ground-Level Gamma-Ray Surveys

The IAEA has published a detailed technical document that summarizes the guidelines for radioelement mapping using gamma ray spectrometry data acquired by aircraft, vehicles, on-foot, in boreholes, on the sea bottom, and in laboratories (IAEA-TECDOC1363 2003). The theoretical design parameters for aerial gamma ray surveys for total gamma count or the more expensive spectrometric data were presented several years ago but are still valid (Pitkin and Duval 1980; Minty 1997). Among the factors that determine the efficiency (and cost) of the survey are the flight line spacing, survey altitude, and performance parameters of the detector. In an earlier publication, an aerial survey for uranium, thorium, and potassium in the top 30 cm of the surface using gamma spectrometry was described (Duval et al. 1971). As the gamma rays emitted from the uranium isotopes have relatively low energies, the gamma rays from Bi-214, which is a daughter of U-238 at 609, 1120, and 1765 keV with yields of 46.1%, 15.1%, and 15.4%, respectively, are monitored. The 1765 keV line is the most energetic and therefore has the highest penetration through soil and is preferred (Figure 2.3). The actual gamma ray airborne spectrum and the conventional energy windows for the determination of uranium (around 1765 keV), thorium, and potassium are shown in Figure 2.4.

Uranium concentration thus determined is usually given in units of *ppm eU* (parts per million of equivalent uranium) as a reminder that this indirect measurement is based on the daughter nuclide that is assumed to be in secular equilibrium with U-238 (IAEA-TECDOC1363 2003). The data from aerial survey may be corrected for effects arising from the background (cosmic radiation and aircraft radioactivity), altitude (changes in pressure, temperature, and air density effects), the presence of

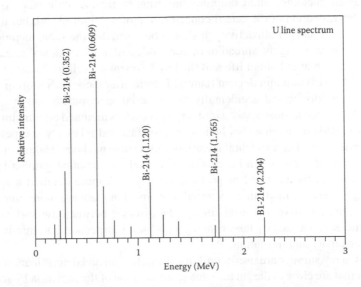

FIGURE 2.3   The line spectrum of the main gamma rays arising from U-238 (the Bi-214 daughter). (Reproduced from International Atomic Energy Agency, Guidelines for radioelement mapping using gamma ray spectrometry data, IAEA TECDOC 1363, IAEA, Vienna, Austria, 2003. With permission.)

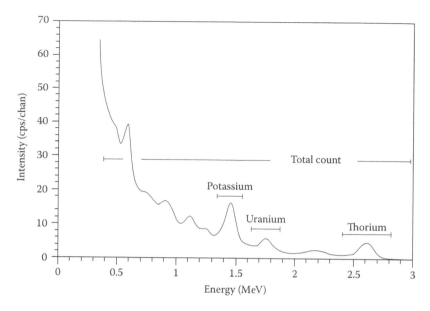

FIGURE 2.4 Typical airborne gamma ray spectrum showing the positions of the conventional energy windows. (Reproduced from International Atomic Energy Agency, Guidelines for radioelement mapping using gamma ray spectrometry data, IAEA TECDOC 1363, IAEA, Vienna, Austria, 2003. With permission.)

airborne Bi-214 (arising from atmospheric Rn-222 gas that escaped from the soil using partial shielding), and Compton scattering. Meticulous calibration is required for the determination of the actual concentration of uranium in soil, but not for the assessment of relative radioactivity. It should be noted that the same methods can be applied for monitoring the spread of radionuclides after an accidental release and to estimate the threat to human life and the environment as well for locating radiation sources. One rich uranium deposit (named Palette after the artist's tool) in Australia was allegedly discovered accidentally by an aerial spectrometer survey (Zoellner 2010). Modern techniques apply gamma spectrometers mounted on unmanned aerial vehicles (UAVs) or autonomous helicopters for fast and relatively inexpensive surveys, particularly after accidental release of radioactive nuclides (Halevy et al. 2014).

Ground-level surveys can be conducted by vehicle-mounted gamma spectrometers or on foot by handheld or portable detectors. A more detailed map may be obtained by these methods than by aerial mapping, but these are more time consuming and labor intensive. In aerial survey, the distance between the surface and the spectrometer is necessarily larger than ground-level inspection and that is reflected by the sensitivity and resolution.

Gamma ray surveys are useful only for the detection and determination of radionuclides that are close to the surface due to attenuation of the radiation by soil, water, vegetation, and air. Ten meters of air or 10 cm of snow will reduce the radiation intensity by ~7% (so from an altitude of 100 m the attenuation in air is ~50%) while 2 cm of soil would attenuate 35% of the gamma rays and a 10% increase in soil moisture would increase the attenuation by about 10% (IAEA-TECDOC1363 2003).

2.2.1.2 Geophysical Methods

A summary of the innovations in geophysical uranium exploration methods was compiled and is shown in Table 2.3 (IAEA-NF-T-15 2013). The gamma ray spectrometric methods were discussed in detail earlier.

An interesting innovation is the measurement of radon gas that escapes from the soil as an indicator of buried uranium deposits. A simple apparatus for measurement of the radon emanating from uranium ore is shown in Figure 2.5 (Sahu et al. 2013).

TABLE 2.3
Summary of Innovations in Geophysical Methods for Uranium Exploration

Method	Application	Innovation
Radiometric	Uranium prospecting	Lightweight, sensitive, field spectrometers, and total count rate meters
	Uranium prospecting and geological mapping	High-resolution airborne systems; spectral smoothing techniques; global baseline
	Borehole logging	Borehole probes that are unaffected by uranium disequilibrium
Radon	Detecting buried U	Instruments differentiating radon and thoron
Electrical, electromagnetic, magnetotellurics	Mapping lithology, structure, alteration and topology	High-resolution and deep penetrating systems; 2-D and 3-D inversion techniques
Gravity	Mapping lithology, structure, alteration and topology	3-D modeling; airborne gradiometry systems permit rapid regional assessment
Magnetic	Mapping lithology, structure and alteration	High-resolution airborne gradiometry systems; 3D modeling
Seismic	Mapping lithology, structure, alteration and topology	Regional seismic, borehole seismic, 3D seismic
Remote sensing	Mapping lithology, structure, alteration and topology	High-resolution, high-precision, multispectral spaceborne and airborne imagery
GPS	Global positioning system	Increased geographical precision in data collection
Software	Geophysical data processing	Development of geological model systems for exploration through geophysical data inversion
GIS and image analysis	Geographic information systems; visualization and interpretation	Integrated interpretation of geological, geophysical and geochemical databases

Source: Adapted from International Atomic Energy Agency, Advances in airborne and ground geophysical methods for uranium exploration, IAEA Nuclear Energy Series NF-T-15, IAEA, Vienna, Austria, 2013 (IAEA-NF-T-15 2013). With permission.

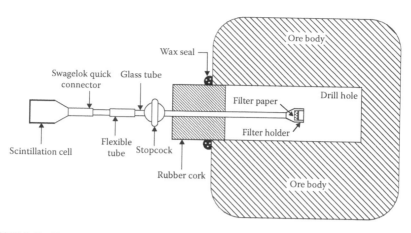

FIGURE 2.5 Experimental setup for in situ measurement of radon emanation rate from uranium ore body. (From Sahu, P. et al., *J. Environ. Radioact.*, 126, 104, 2013. With permission.)

Mapping of rock formations (lithology) and detection of magnetic and gravitational irregularities have also advanced. Developments in photography by aircraft and satellites in space—high resolution, deep penetrating, multispectral, and gradiometry—have enhanced the detection of geophysical and geochemical unconformities and the ability to locate uranium deposits. An example of the combination of several methods to locate geophysical anomalies associated with uranium mineralization in India was published (Mandal et al. 2013). The survey included the use of gravity, magnetic, radiometric, very low frequency electromagnetic (VLF), and gradient resistivity profiling methods. This was carried out around known mineralized zones in order to identify the geophysical signatures on the uranium mineralization bands. The conclusion was that the zones with low-gravity anomaly, low resistivity (or high conductivity) with high magnetic anomaly are indicative of uranium mineralization. In any case, the combination of gravity, resistivity, radiometric, and VLF methods is most suitable for the detection of subsurface hydrothermal alteration zones.

2.2.1.3 Other Prospecting Strategies

An investigation of the uranium content in the plant material covering the surface of a lake area in Labrador, Canada, was reported (Nyade et al. 2013). Several vegetation samples (black spruce, Labrador tea stem and humus) were collected, ashed at 450°C, and analyzed by ICPMS. Uranium content in humus samples varied between 0.05 and 885 µg U g⁻¹ and some soil samples that were also analyzed showed concentrations of 50–405 µg U g⁻¹. The spruce twigs had a better correlation with radiometric measurements than the humus and soil samples.

The interaction between uranium and certain types of microorganisms is interesting, as shown in Frame 2.1, but will not be discussed in detail here.

Highlights: Uranium prospecting has come a long way from the days when prospectors looked for yellow deposits, as so vividly described (Zoellner 2010). Surface or superficial uranium deposits can be located by aerial surveys that are deployed

FRAME 2.1   URANIUM GOBBLING MICROORGANISMS

Although not considered a prospecting method, the mutual effect between uranium and microorganisms has potential in detecting uranium deposits and in bio-remediation of contaminated areas. One of the much cited and interesting processes in which U(VI) may be reduced and deposited as U(IV) in sedimentary environments is the action of sulfur-reducing microorganisms (Lovley and Phillips 1992). The kinetic phosphorescence analyzer (KPA) was used to monitor the hexavalent uranium concentration, and the total uranium concentration was measured by the same technique after oxidizing the sample with concentrated nitric acid under aerobic conditions. A transmission electron microscope (TEM) with an energy-dispersive x-ray analyzer spectrometer (EDS) was used to determine the uranium in the cell suspension while x-ray diffraction (XRD) was used to analyze uranium in samples of filtered cell suspension. It was found that *Desulfovibrio desulfuricans* is capable of causing reductive precipitation of uranium that may be important as a means of immobilizing uranium and preventing it from entering drinking water or the food chain. Many other studies have shown that uranium may be taken up by other types of microorganisms (e.g., Kumar et al. 2011, 2013; Bhattacharjee et al. 2013), but a detailed discussion is beyond the scope of this book (see Section 3.6).

for charting remote and inaccessible regions or large areas in a relatively short time. Their sensitivity strongly depends on the flight pattern (altitude, grid lines, and speed) and on the equipment used. Ground surveys (on foot or by vehicle) are slower but more sensitive. Other methods are preferred for locating uranium deposits that are not close to the surface, like monitoring the radon emanating from the soil, measuring the uranium content in water bodies and streams, looking for uranium in sediments and vegetation. Combining geophysical methods with analysis of geological features can be used to focus on promising areas. Finally, uranium may be readily found without prospecting efforts as a by-product of other mining operations, like phosphate production, gold mining, or columbite extraction.

2.2.2   Assay of Minerals and Uranium Ores

In general, any analytical technique that is used for the determination of uranium can be adapted for assay of uranium in ores after appropriate sample preparation is carried out. Once uranium-bearing minerals are discovered, then an assay of the amount of uranium is necessary in order to assess the economic aspects and value of extracting the uranium from the ore. In most cases, assay of the uranium content can take place only after some sample preparation procedures are carried out. Although many variations exist, a common procedure was described by the IAEA (IAEA341 1992): the ore sample is crushed and dried overnight at 110°C and then pulverized and blended. After that the sample is split and further processed according to the requirements (like dissolution, pressing into pellets, etc.) of the analytical technique used for the assay, as shown later in detail for some examples.

Several analytical methods have been deployed for the determination of uranium in ores. Among the older methods that were used are radiometric methods that were already used over 50 years ago for ore sorting (Mal'tsev 1960), titrimetric methods in which the uranium content in the ore was determined with ferrous ion-phosphoric acid reduction (Hitchen and Zechanowitsch 1980), colorimetric methods where complexes of uranium are formed with standard arsenazo III (Onishi and Sekine 1972), exotic siderophores (Renshaw et al. 2003) reagents and instrumental neutron activation analysis (INAA) based on measurement of ^{239}Np in the ore (Chaudhry et al. 1978) in addition to numerous other approaches. Many modern techniques are now employed for destructive and nondestructive determination of uranium in ores.

A comparison of three methods (ICP-OES, alpha spectrometry, and gamma spectrometry) for the determination of uranium in rock phosphates and columbite was published (Singhal et al. 2011). In that work, the samples were first powdered and sieved with 200 μm mesh. For the ICP-OES analysis, uranium was extracted from the ore by fusion with sodium peroxide followed by leaching with 2 M HNO_3. Selective separation from the leachate was carried out by solvent extraction (tributylphosphate—TBP in CCl_4 and Di(2-ethylhexyl) phosphoric acid [DEHPA] and TBP in petrofin) and subsequently the uranium was stripped from the organic phase with 1 M $(NH_4)_2CO_3$ (ammonium carbonate). After evaporation, the residue was dissolved in 4% nitric acid and the uranium content was determined by ICP-OES. For alpha spectrometry, a standard solution of uranium was added to the powdered samples that were then heated in a muffle furnace to destroy organic matter. The uranium was leached with nitric acid and co-precipitated with ferric hydroxide. The precipitate was dissolved and uranium was separated and electrodeposited on a stainless steel disk for alpha spectrometry. For gamma spectroscopy, a 30 g sample of the powdered ore was placed in a polypropylene container and stored for 30 days to allow for secular equilibrium to be reached between ^{238}U and its daughters. The samples were counted for 10,000 s with a high-purity germanium (HPGe) detector. Method validation was done with reference materials: BCR-032 containing 125 μg U g^{-1} for phosphate rock and IGS-33 containing 260 μg U g^{-1} for columbite. For the phosphate rock, the agreement between the three methods was within 1%–2%, but for the columbite alpha spectrometry results deviated by 20%–30% from the other two methods (Singhal et al. 2011).

Two main types of digestion methods were deployed for the determination of uranium and some other elements in shale and sandstone samples from Poland (Chajduk et al. 2013). Fusion of the sample with sodium peroxide was carried out in zirconium crucibles at 550°C for 1.5 h. The melt was washed with water and then 5 M HNO_3 was added and heated to 80°C to dissolve the melt so a clear solution was obtained. Several microwave acid digestion procedures with different amounts of HNO_3 and HF (some used HCl as well) were also examined. The samples were finally diluted with 0.7% HNO_3 and 13 elements, including uranium, were measured by ICPMS. Method validation was carried out with CRMs with different compositions. The detection limits for uranium in the alkali fusion method and microwave digestion were 46 and 23 pg U mL^{-1}, respectively. The limit of detection of other elements by the two methods varied considerably according to the element and method. The uranium content in the different samples had an average value of 94.1 mg U kg^{-1} for the shale samples and 256 mg U kg^{-1} for the sandstone samples, but in each group the concentration of individual samples varied by a factor of 300–400 between the lowest and the highest values. The authors concluded that the microwave digestion procedure employing HNO_3:HF and removal of fluoride ions by H_3BO_3 was more efficient than the alkaline fusion and conventional acid digestion.

Laser-induced breakdown spectroscopy (LIBS) using a standard addition approach of known amounts of uranium oxide (0.2%, 0.4%, 0.6%, 0.8%, and 1.0% of uranium) to ore samples was used to determine the uranium content in the ore (Kim et al. 2012). The ore was pulverized, the uranium oxide spike was added, and samples were formed by pressing the powder into pellets (10 mm diameter, 10 mm length containing about 4 g of sample) that were sintered in an argon atmosphere. The samples were excited with Q-switched Nd:YAG laser at 532 nm and a calibration curve was generated based on the peak at 356.659 nm. The reported limit of detection was 158 µg U g^{-1}, and the relative standard deviation (RSD) improved with the number of laser pulses, reaching about 4% for 700 pulses. In a later publication, LIBS was used to determine thorium and uranium in oxide powders and uranium in ore samples (Judge et al. 2013). Sample preparation involved grinding the sample to a fine powder and pressing it into a pellet at 5000 psi for 5 min without a binder (the pure oxide powders required the use of stearic acid as a binder). The ore contained Si (22%), Al (6%), Fe (5.9%), Ca (4.0%), Na (3.6%), and uranium (7.09%) in oxide form in addition to lower concentrations of other elements like Cr, Ti, and Pb. All these trace elements were observed in the LIBS spectrum. The authors report that over 1000 emission lines were observed for the uranium ore sample between 200 and 780 nm.

Laser-induced fluorimetry (LIF) using a Sintrex UA-3 uranium analyzer and gamma spectrometry for assay of uranium in ores were compared (Madbouly et al. 2009). Sample preparation involved dry ashing in a muffle furnace for 24 h and acid digestion (8 M HNO_3 and 30% hydrogen peroxide) followed by volume reduction. The sample was reconstituted with 1 M HNO_3 and Fluran (sodium pyrophosphate and sodium dihydrogen phosphate) was added to enhance the fluorescence from uranyl ions in the sample. For a representative measurement, the sample solution must be clear (no color or turbidity). Excitation with a nitrogen laser at 337 nm and monitoring the uranium fluorescence at 540 nm can lead to a limit of detection of 0.005 µg U L^{-1} (5 ng U L^{-1}) in the final sample. Note that quite a similar procedure has been used for determining uranium in urine (Chapter 4). For the gamma spectrometric assay, the sample was placed in a cylindrical container and allowed 30 days for secular equilibrium to be attained, as described earlier. The analysis was based on the emission of the 1001 keV gamma line for [234m]Pa that is free from interferences and a counting time of 65,000 s with a HPGe detector was deployed. The bias of the methods, as determined from a certified standard ore sample (IAEA RGU-1), was 1.55% and 0.9% for the LIF and gamma methods, respectively (Madbouly et al. 2009).

A simple spectrophotometric (or colorimetric) method for the determination of uranium in perchloric acid based on the arsenazo-III complex was described (Khan et al. 2006). While azo-dyes may be oxidized by nitric acid, they are more stable in perchloric acid and organic interferences are effectively eliminated by this acid media. Standards of uranium and ore samples were weighed and decomposed in Teflon beakers with a 1:1 HNO_3:HF mixture. The uranium was then extracted from the nitrate solution by liquid–liquid extraction with TBP in methyl isobutyl ketone (MIBK). The uranium concentration was determined by a spectrophotometer at 651 nm relative to reagent blanks and standard solutions. The optimal conditions were 0.07% weight per volume of arsenazo-III in 3 mol L^{-1} perchloric acid where the recovery was between 96% for addition of a 10 µg U g^{-1} spike and increased to 98.6% for a spike of 50 µg U g^{-1}.

Speciation in uranium in ore samples and contaminated soil was determined by ICPMS after ion chromatography was used to separate hexavalent and tetravalent uranium (Jovanovic and Pan 2013). For the determination of the total amount of uranium, the samples were digested with a mixture of HCl and HNO_3 in a microwave oven at 200°C for 30 min and centrifuged to remove undigested (mainly silica) materials. Matrix effects in the ICPMS measurements were minimized by dilution with a high matrix introduction plasma correction. For the speciation studies, a relatively mild extraction process was deployed to selectively leach the U(VI). This consisted of ultrasonic digestion in a mixture of 4.5 M HCl–0.03 M HF in 2% hydrazine dihydrochloride for 1 h at room temperature. Under these conditions, tetravalent uranium is not oxidized while extraction of U(VI) is effective and unaffected if large amounts of iron (<20% by weight) are not present. Here too the sample was centrifuged to remove undissolved material and the supernatant fluid was analyzed by ICPMS. The ratio between the two digestion methods is indicative of the two uranium fractions [U(VI) and U(IV)] in the sample.

Highlights: The uranium content in ore samples can be assessed by nondestructive methods based on gamma spectrometry or neutron activation or by destructive methods that are usually considered to be more accurate and sensitive. The generic sample preparation procedure (shown schematically in Figure 2.6) involves crushing

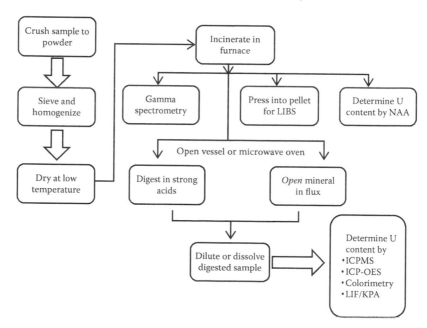

FIGURE 2.6 A generic flowchart of the preparation of ore samples for determination of the uranium content.

into a powder, sieving it to remove large fragments, and placing it in an oven at a low temperature for drying or at a high temperature for destroying organic matter.

The samples are then digested, usually by a combination of strong acids (HNO_3 and HF, and sometimes with addition of HCl or H_2O_2). This can be carried out in open crucibles or closed vessels with microwave-assisted digestion. Fusion digestion is an alternative method in which solid salts are used in an open crucible. If the uranium speciation is studied, then a relatively mild nonoxidative digestion procedure that selectively leaches the U(VI) without affecting the U(IV) can be deployed. In most procedures, after the digestion is completed, the sample is diluted and the uranium content in the solution is determined by an analytical method. The modern methods are based on ICPMS or ICP-AES, but in some cases, the colorimetric arsenazo-III method is still used. If a method such as LIBS is used, the powdered and calcined sample is pressed into pellets that are directly analyzed by LIBS. For gamma spectrometry, the powdered sample must be placed in a closed vessel for at least 30 days to allow for secular equilibrium and then the spectrum is recorded for several hours. It should be noted that all the assay methods require validation that is based on measurement of certified reference materials (CRMs) often with standard addition of known amounts of uranium for derivation of a calibration curve.

2.3 ANALYTICAL CHEMISTRY IN THE NFC

An overview of the NFC was shown in Chapter 1, and in this section we shall describe in more depth the analytical aspects of the major components involved in the NFC. It should be emphasized that the American Society for Testing Materials (ASTM) has published over 200 documents relating to uranium compounds that are used in the NFC. These documents include lists of the specifications of each material as well as detailed sampling, sample preparation, and analytical procedures and also guidelines for calculation of the results and reporting them. Some of these ASTM documents describe very specific analytical methods in great detail (e.g., determination of vanadium in the presence of iron in uranium dioxide) while others express the mechanical or morphological properties of the fuel elements. Some of the methods are used for simultaneous multielement determination while others may only relate to specific isotopic ratio measurements or a single substance.

The main operations that constitute the front end of the uranium NFC are prospecting and assay of uranium in ores that were discussed earlier as well as mining, ore processing, and uranium recovery that were described in brief in Chapter 1. The specifications for uranium ore concentrates (UOC) are presented by the American Society for Testing Materials (ASTM) and outline the following items: chemical composition that includes the uranium content (minimum of 65%), its isotopic composition, the amount of insoluble uranium, extractable organics and impurity content (16 trace elements, carbonate, halogens, and moisture) in addition to some other toxic elements (C967 2013). In addition, the particle size, ability to flow, and foreign matter need to be determined and reported. What can happen when the seller and buyer do not agree on the assay results? A historic episode demonstrates this point as shown in Frame 2.2.

FRAME 2.2   ANALYSIS OF URANIUM CONCENTRATES AT THE NATIONAL BUREAU OF STANDARDS

A historic episode about the problems associated with assaying the uranium content in UOC was presented by the National Bureau of Standards (NBS) (now called NIST—National Institute of Standards and Technology) in 1965 (Richmond 1965). Apparently differences were found in the uranium assay between laboratories of the seller of UOC (in South Africa) and the buyer (US-AEC) that were beyond the *splitting limit* (the accepted difference of 0.3% U_3O_8 in the early 1950s that was later decreased to 0.2%). This required an *umpire* analysis, carried out by NBS, to determine the true uranium content. After extensive research efforts, the reason for the discrepancy was attributed to the different moisture level in the analytical laboratories that affected the absorption of water molecules by UO_3. The solution to the problem was simple: raising the calcination temperature of the UOC from 500°C to 900°C eliminated the UO_3 and produced U_3O_8 that was not sensitive to the moisture level in the laboratory. This demonstrates the importance of using mutually agreed and reliable analytical procedures.

Source: Richmond, M.S., Analysis of uranium concentrates at the National Bureau of Standards 260-8, National Bureau of Standards, Washington, DC, 1965.

A detailed procedure for sampling UOCs is provided by the American Society for Testing Materials (C1075 1997). This 10-page procedure includes primary sampling by falling stream (in a single stage or in two stages) or by Auger sampling procedure, secondary sampling by a straight part cutter or by rotating cutter multisampler. Sample preparation includes drying and pulverizing prior to moisture determination, calcination, and finally sample packaging and sealing (by wax or vacuum) (C1075 1997). This precedes the analysis of uranium and impurities in the UOC sample. Several methods for dissolution of uranium materials (metal and oxides, U–Al alloys, scrap material and ash, as well as refractory U-containing materials) prior to analysis are also described by an ASTM document (C-1347 2008). What happens when an incident occurs in an industrial plant that handles drums with UOC that must be sampled? See Frame 2.3.

As discussed in greater detail in Chapter 5, UOCs bear a *fingerprint* that can be used to identify their origin and production processes, as demonstrated for Australian ore concentrates (Keegan et al. 2008). In this study, three mining facilities that include milling operations that are located in different geological formations were studied: Olympic Dam in breccias complex deposit, Beverly in sandstone deposit, and Ranger in an unconformity-related deposit (see Zoellner 2010 if interested to learn about the tumultuous history of uranium mining in Australia). The ASTM recommended procedures were employed for determination of B and P by spectrophotometry, U and Si by XRF, As and Th by ICPMS and flame-AAS is suggested for the determination of Ca, Fe, Mg, Mo, Ti, V, K, and Na (C1022 2010). Samples of UOC from the three mines were digested by acid dissolution and measured by ICPMS that also provided isotopic composition data for lead. SIMS was used to analyze the particle size in the samples. The

FRAME 2.3   WHAT HAPPENS WHEN THINGS GO WRONG AT A URANIUM CONVERSION FACILITY?

The investigation report of an incident that occurred in June 2012 at the Blind River refinery (BRR) in Canada can serve as an excellent example of the procedure deployed to collect samples from 200 L drums of UOCs (DeGraw 2012). At BRR, the drums are placed on a conveyor where they were weighed and moved to the delidding station where the bolt on the ring holding the lid is loosened and the lid is removed manually by an operator. The open drum is transferred to the Auger sampling room and a core sample is collected. The drum is then moved to the next station where the lid and ring are replaced by another operator and the drum is transported to the storage area. Active dust collection systems are deployed in all these stations to protect the operators. The BRR plant handles 30,000–50,000 drums each year and during a period of about 30 years over one million drums were handled without incident. The incident under discussion involved sudden popping of a lid at the delidding station and release of approximately 26 kg of dust from the uranium concentrates. The operator in the room was exposed to the uranium dust that entered his mouth and nasal passage. The uranium level in several spot urine samples collected over a period of about 6 weeks showed that he was exposed to a level of uranium that exceeded the kidney toxicity guidelines but as far as radiation exposure was concerned it was well below the annual limit for nuclear industry workers. In the final analysis, it was estimated that no permanent damage was caused in the incident.

isotopic composition of uranium was accurately determined by TIMS. Significant differences were found between the three UOC sources for all 39 elements that were examined and principal component analysis (PCA) showed distinct patterns for each mine. It is interesting to note that there were significant differences between the bulk analysis carried out by ICPMS and the particle analysis performed by SIMS for many elements. The lead isotopic analysis showed variability within different samples from the same mine and therefore was deemed as impractical for fingerprinting. The $^{235}U/^{238}U$ ratio was practically indistinguishable between the three mines, but very high precision measurements of the $^{234}U/^{238}U$ ratio did show a distinct difference. The authors concluded that the rare-earth elements may be a rapid means for identifying the source of the UOC material (fingerprinting uranium ores is discussed in detail in Chapter 5).

One manifestation of the arsenazo-III method was deployed for the determination of the uranium concentration in various process streams in a uranium extraction plant in India (Murty et al. 1997). The uranyl-arsenazo-III complex in 4 M HCl was used for determining low uranium concentrations while the complex in dilute phosphoric acid was used for high concentrations.

Highlights: Before shipping UOCs from the mill to the uranium conversion facility (UCF), an extensive series of tests is necessary, starting from proper sampling, continuing with dissolution procedures, and finally a suite of chemical, isotopic, and physical analyses must be carried out in order to produce the certificate that will accompany the shipment. The ASTM procedures are not the only ones that are used, and in general the supplier of the UOC and the purchaser will agree on the specifications and testing procedures. The analytical requirements in these phases of the NFC focus on assaying the uranium content and on determination of select impurities. However, after chemical conversion to UO_2, UF_6, and uranium metal, which are the main commercial products of the UCF, a meticulous analysis of impurities is mandatory. This is so for UF_6 that is shipped to an isotope enrichment facility, for UO_2 that is used for production of nuclear fuel after sintering, or for uranium metal that may be used as fuel or for other nonnuclear applications. These will be discussed in detail in the following sections.

2.3.1 ANALYTICAL CHARACTERIZATION OF UF_6 FOR ENRICHMENT FACILITIES

The methods used for analytical characterization of UF_6 can be divided into two main groups: analytical procedures that are used to assess the quality of the materials and its compliance with specifications and nondestructive analysis (NDA) methods for assay of closed cylinders containing UF_6 or online monitoring.

The analytical requirements and specifications for UF_6 that is intended as the feed material for isotope enrichment facilities are summarized in an extensive series of documents published by the American Society for Testing and Materials (ASTM). A small portion of these, not a representative sample, is listed in Table 2.4.

The key document was last updated in 2011 (C-787 2011), and it should be noted that the analytical procedures are regularly revised and updated so it is always advisable to search the web or ASTM site for the latest version. These documents present the guidelines for control of the purity of the product and of the equipment used in its preparation, storage, and transport as well as detailed procedures for chemical and isotopic analysis. Procedures for sampling liquid and gaseous UF_6, that are an integral part of the analytical process are also given in detail (Table 2.5). The UF_6 may be prepared from natural (unirradiated) uranium or from reprocessed irradiated uranium that require some additional tests for fission products and activation products. A short discussion of the required analyses and procedures is presented later, but for the full procedures readers are referred to the original referenced documents of the ASTM concerning uranium hexafluoride.

As mentioned earlier, ASTM is not the only source and there are several other reports and publications that deal with analytical methods for characterizing UF_6 and its impurities. For example, an Oak Ridge National Laboratory report on characterizing technetium and transuranic elements in depleted UF_6 cylinders (Hightower et al. 2000) or for transport of UF_6 cylinders (IAEA-TECDOC-750 1994).

2.3.1.1 Sampling of Uranium Hexafluoride

ASTM has summarized the practice for sampling UF_6 from the gas phase (C1703 2013) or from the liquid phase (C1052 2007) and also for subsampling of liquid UF_6 (C1689 2008). Liquid phase bulk sampling will usually be preferred as it can offer a representative sample, but this requires special equipment and carefully controlled and stringent safety procedures. The underlying assumption is that the UF_6 is homogenous in liquid form and of a single quality. The sample can then be divided

TABLE 2.4
Some Examples of the ASTM Methods Concerning Characterization of UF$_6$ for Enrichment Facilities

Document Number	Contents of Document
C787-11	Standard specification for uranium hexafluoride for enrichment
C761	Test methods for chemical, mass spectrometric, spectrochemical, nuclear, and radiochemical analysis of uranium hexafluoride
C859	Terminology relating to nuclear materials
C996	Specification for uranium hexafluoride enriched to <5% ^{235}U
C1052	Practice for bulk sampling of liquid uranium hexafluoride
C1287	Standard test method for determination of impurities in nuclear grade uranium compounds by inductively coupled plasma mass spectrometry
C1295	Test method for gamma energy emission from fission products in uranium hexafluoride and uranyl nitrate solution
C-1413	Standard test method for isotopic analysis of hydrolyzed uranium hexafluoride and uranyl nitrate solutions by thermal ionization mass spectrometry
C1428-11	Standard test method for isotopic analysis of uranium hexafluoride by single-standard gas source multiple collector mass spectrometer method
C-1477	Standard test method for isotopic abundance analysis of uranium hexafluoride by multicollector, inductively coupled plasma-mass spectrometry
C-1561	Standard guide for determination of plutonium and neptunium in uranium hexafluoride by alpha spectrometry
C-1539	Standard test method for determination of technetium-99 in uranium hexafluoride by liquid scintillation counting
C1703	Practice for sampling of gaseous uranium hexafluoride
ANSI N14.1	Packaging of uranium hexafluoride for transport

into subsamples for particular analyses to ensure compliance with the specifications listed in C787 (C-787 2011) and C996 (C996 2010). This latter document is concerned with UF$_6$ that is the product of an enrichment plant and with UF$_6$ that was produced by down blending highly enriched uranium (HEU from disassembled nuclear weapons) with other types of uranium (natural, reprocessed, or even depleted) to create nuclear fuel material (UO$_2$ mainly) containing <5% ^{235}U.

An interesting approach to reduce the hazards involved with handling samples of liquid UF$_6$ by trapping the sample in alumina pellets was proposed and is shown in Figure 2.7 (Esteban et al. 2008). The interaction between the moisture contained in the alumina and UF$_6$ leads to formation of uranyl fluoride and HF that is subsequently also trapped (see Equations 2.1 and 2.2). The isotopic ratio of the uranium can then be determined by treatment of the pellets (Esteban et al. 2008):

$$6Al_2O_3 \cdot H_2O + 3UF_6 \rightarrow 3UO_2F_2 + 6Al_2O_3 + 12HF \qquad (2.1)$$

$$12HF + 2Al_2O_3 \rightarrow 4AlF_3 + 6H_2O \qquad (2.2)$$

In some cases, when cylinders have already been filled with UF$_6$ and sampling of liquid UF$_6$ is to be avoided, a representative sample may be obtained during transfer

TABLE 2.5
Test Methods for Subsampling, Chemical, Mass Spectrometric, Spectrochemical, Nuclear, and Radiochemical Analysis of Uranium Hexafluoride UF_6

	Sections
Subsampling of uranium hexafluoride	7–10
Gravimetric determination of uranium	11–19
Titrimetric determination of uranium	20
Preparation of high-purity U_3O_8	21
Isotopic analysis	22
Isotopic analysis by double-standard mass-spectrometer method	23–29
Determination of hydrocarbons, chlorocarbons, and partially substituted halohydrocarbons	29–36
Atomic absorption determination of antimony	36
Spectrophotometric determination of bromine	37
Titrimetric determination of chlorine	38–44
Determination of silicon and phosphorus	45–51
Determination of boron and silicon	52–59
Determination of ruthenium	60
Determination of titanium and vanadium	61
Spectrographic determination of metallic impurities	62
Determination of tungsten	63
Determination of thorium and rare earths	64–69
Determination of molybdenum	70
Atomic absorption determination of metallic impurities	71–76
Impurity determination by spark-source mass spectrography	77
Determination of boron-equivalent neutron cross section	78
Determination of uranium-233 abundance by thermal ionization mass spectrometry	79
Determination of uranium-232 by alpha spectrometry	80–86
Determination of fission product activity	87
Determination of plutonium by ion exchange and alpha counting	88–92
Determination of plutonium by extraction and alpha counting	93–100
Determination of neptunium by extraction and alpha counting	101–108
Atomic absorption determination of chromium soluble in uranium hexafluoride	109–115
Atomic absorption determination of chromium insoluble in uranium hexafluoride	116–122
Determination of technetium-99 in uranium hexafluoride	123–131
Method for the determination of gamma-energy Emission rate from fission products in uranium hexafluoride	132
Determination of metallic impurities by ICP-AES	133–142
Determination of molybdenum, niobium, tantalum, titanium, and tungsten by ICP-AES	143–152

Source: C761, ASTM, Test methods for chemical, mass spectrometric, spectrochemical, nuclear, and radiochemical analysis of uranium hexafluoride, ASTM, West Conshohocken, PA, 2011.

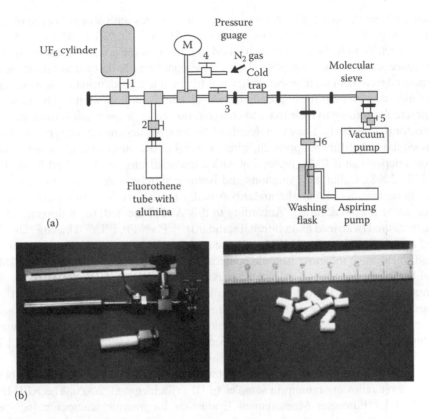

FIGURE 2.7 (a) Schematic diagram of system for sampling liquid UF_6 on alumina pellets and the equipment; (b) the Hoke tube, fluorothene P-10 tube, and alumina pellets. (From Esteban, A. et al., UF_6 sampling method using alumina, in *29th ESARDA Annual Meeting, France, 2007 and 49th INMM Annual Meeting*, Nashville, TN, 2008 [Esteban, UF6 sampling method using alumina 2008]. With permission.)

of gaseous UF_6. It should be noted that this type of sample collection may lead to elevated levels of volatile impurities that are overly abundant in the gas phase, so the results for these impurities should be regarded as conservative (C1703 2013).

2.3.1.2 Isotopic Analysis of Uranium Hexafluoride

Once a sample is collected, the isotopic composition of uranium must be determined as the ^{235}U content is one of the main factors that determine the price of the product. Several mass spectrometric techniques have been developed for direct isotope analysis of gaseous UF_6 and for indirect analysis (usually after hydrolysis) of liquid and gaseous UF_6 samples. The use of a thermal ionization mass spectrometer (TIMS), nowadays equipped with several detectors (i.e., multicollector TIMS), has been the method of choice for many years, but the sample must be hydrolyzed to liquid form (uranyl fluoride or uranyl nitrate solutions) and the uranium must be purified (usually not a problem for UF_6 samples), as mentioned, for example, by ASTM (C1413 2011). The method is used for hydrolyzed samples of UF_6 (UO_2F_2 (uranyl fluoride)) or for uranyl nitrate ($UO_2(NO_3)_2$) solutions and covers samples with a broad range of isotopic compositions (^{235}U from 0.1% to 5.0%, ^{234}U from 0.00555% to 0.05%, and ^{236}U from 0.0003% to 0.5%) but can be applied to other isotopic compositions if suitable reference standards are available. The sample containing 2 μg of uranium must be deposited on a rhenium filament. When this filament is heated, uranium is vaporized and electrons emitted from a second filament ionize uranium atoms. The ions are separated according to their mass to charge ratio and each ion beam is directed to a detector (C1413 2011). The exact details of the analysis depend on the type of TIMS device used. A different approach, where control of the sample purity is not as rigorous, deploys an ICPMS, preferably with a multicollector, as described by ASTM (C1477 2008). Calibration solutions and isotopic standards are used to correct for any mass bias, and internal standards (usually thorium or lead isotopes) are also used for improved accuracy. According to this ASTM method, the following ratios are measured (with lead as an internal standard): $^{207}Pb/^{208}Pb$, $^{234}U/^{238}U$, and $^{235}U/^{238}U$, and again $^{235}U/^{238}U$ and $^{236}U/^{238}U$ and once again $^{207}Pb/^{208}Pb$ and then an algorithm is used to calculate the exact isotopic composition of uranium. Interferences to TIMS and ICPMS measurements arise from mass bias (ions of different masses cause different responses at the detector), abundance sensitivity (a large signal may affect adjacent masses), memory effects (from previous samples), and isobaric interferences from ions of the same mass from other elements. In ICPMS, there is an additional potential interference from polyatomic species that may cause an isobaric effect, like $^{235}UH^+$, which has the same mass as $^{236}U^+$ (C1477 2008).

The European Commission carried out an interlaboratory test for measurement of the isotopic ratios of uranium in samples of UF_6 (Richter et al. 2005). The REIMEP 15 (Regular European Measurement Evaluation Programme) campaign included shipment to the participating laboratories of four samples of low-enriched or slightly depleted UF_6 samples packed in standard Monel capsules. The certified samples were prepared by IRMM (Institute for Reference Materials and Measurements) by mixing uranium reference materials. At IRMM, the $^{235}U/^{238}U$ ratio was measured by a mass spectrometer equipped with an electron impact ionization gas source while the isotopic compositions of $^{234}U/^{238}U$ and $^{236}U/^{238}U$ were determined by TIMS (after hydrolysis). The eight laboratories that yielded results used a variety of methods (gas measurement, hydrolysis followed by TIMS, MC-ICPMS, and quadrupole-ICPMS) for the assay. The results from all laboratories for the $^{235}U/^{238}U$ ratio were in agreement with the expected ratio. The ratio of the minor isotopes was in the range of $1-3*10^{-4}$ and the laboratories performed consistently well. The results submitted by the participants showed that in one of the samples the $^{234}U/^{238}U$ ratio deviated from the expected ratio (based on the mixing ratio of the standards) implying the sample preparation may have been flawed. This was later remedied after IRMM purchased a new TIMS instrument (TRITON, by Thermo Electron) and measured the sample with Faraday cup detectors (avoiding cross-calibration errors between ion counting and Faraday detectors). In their summary of the test, the authors conclude as follows: "The results point to the need for close checking during mass spectrometric measurements. A systematic effect like the one found here can only be isolated and corrected for using independent isotopic reference materials such as IRMM-072" (Richter et al. 2005).

Direct determination of the isotopic composition of uranium in gaseous UF_6 requires no sample preparation and purification, but the accuracy of the method was limited. However, a dedicated UF_6 gas source mass spectrometer (GSMS) was used and a procedure for accurate measurement of the isotopic composition of uranium was developed (Richter et al. 2013). A *memory corrected double standard* (MCDS) procedure was implemented and the results were favorably compared with those obtained with a TIMS device employing *modified total evaporation*.

An interesting prototype hand-portable device—a micro ion trap mass spectrometer for measurement of gaseous UF_6—was described (Whitten et al. 2004). An ion trap mass spectrometer was scaled down in size, resulting in diminished pumping requirements and power consumption so that it could be battery operated. Ionization was based on an electron gun source, three electrodes formed a 1 mm diameter cylindrical ion trap and focusing electrodes and grids transferred the ions to a multiplier ion detector. The device was operated in a special chamber that contained an atmosphere of highly reactive UF_6 vapors. The measurement of the ion current as a function of exposure time to UF_6 vapors showed a marked decrease of about two-thirds within the first 2 h and after that a steady signal was achieved (probably due to passivation of the surfaces). It was stated that the isotope ratio could be measured within 10–100 s, depending on the desired precision, but results of such a measurement were not presented. The conclusion was that UF_6 could be measured by the mass spectrometer and responded to electron impact ionization like other gases.

Determination of the $^{235}U/^{238}U$ isotopic ratio inside UF_6 cylinders without the need to withdraw a sample was demonstrated (Jordan et al. 2012). The work presented a novel concept for automated enrichment assay of UF_6 cylinder by nondestructive methods. The hybrid approach combined the traditional ^{235}U passive gamma line at 186 keV with the signature manifested by the high-energy gamma continuum (3–8 MeV) called HEVA (hybrid enrichment verification array). The neutrons are generated by (α,n) reactions originating from alpha particles derived mainly from ^{234}U interacting with fluorine atoms in UF_6. Different sensor arrays were constructed and tested on cylinders with various UF_6 contents (LEU of 2%–5% ^{235}U, natural uranium and DU) and data analysis algorithms were developed. The automated enrichment assay system is intended to replace handheld measurements in verification and safeguard scenarios. When combined with data from the weighing station, the system may also provide full mass-balance calculations of an entire facility, thus contributing to the effectiveness of safeguards. The reported results, tested so far on 30B cylinders (30 in. or 75 cm diameter) but not on the larger type 48 in. (120 cm diameter) cylinders, suggest that HEVA has the potential for unattended cylinder enrichment verification. As mentioned earlier, one great advantage of this nondestructive assay approach is that drawing UF_6 samples from the cylinders is not needed.

The potential of an online uranium enrichment monitor for assaying the isotopic ratio in UF_6 gas at a centrifuge enrichment plant, based on calculations and Monte Carlo modeling, was presented (Smith and Lebrun 2011). The unattended monitor would be located on the high-pressure header pipes that receive the gas flow from multiple cascades. The online monitor is to be equipped with a NaI gamma detector for measurement of the 186 keV line from ^{235}U. The major uncertainties are due to the thickness of uranium containing deposits on the walls of the pipes and to fluctuations in the operating parameters (pressure, enrichment).

2.3.1.3   Determination of Impurities in Uranium Hexafluoride

The specifications for UF_6 list the impurities that must be determined (C-787 2011), and 34 detailed procedures including subsampling, chemical analysis, mass spectrometric, spectrochemical, nuclear and radiochemical methods are presented (C761 2011). Many of these methods also appear separately as independent procedures (as listed within C761) that are regularly revised and updated. As mentioned earlier, some methods are dedicated to the determination of a single element in UF_6 (e.g., see the method for ^{99}Tc; C1539 2002) while others may be suitable for multielemental analysis like the methods used for simultaneous determination of 67 elements (C1287 2010) or Section 62 (Spectrographic Determination of Metallic Impurities) and Sections 133–142 (Determination of Metallic Impurities by ICP-AES) in Table 2.5.

The main specifications concern the uranium content in UF_6 by gravimetric and titrimetric methods, isotopic composition (as listed earlier), elemental impurities by atomic absorption, ICP-AES, spectrographic, colorimetric, titrimetric, spark source mass spectroscopy and radiochemical (alpha and gamma spectrometry and neutron absorption) methods, and organic impurities like hydrocarbons and their derivatives. Modern methods also deploy ICPMS techniques for the determination of impurities and isotopic composition (C1287 2010). It is beyond the scope of this book to present all these methods in detail so a number of examples are shown later.

Special attention is dedicated to elements that pose a health hazard, mainly radioactive fission products and activation products that may be present in UF_6 from reprocessed or recycled uranium. For example, technetium-99, which can be determined by liquid scintillation counting (C1539 2002). According to this procedure, uranium is precipitated by ammonium hydroxide from a solution of hydrolyzed UF_6. After centrifugation and acidification of the supernatant by sulfuric acid and extraction by TBP, an aliquot is placed in a vial with $SnCl_2$ and HCl and a scintillation *cocktail*. The ^{99}Tc beta activity is determined by counting and a detection limit of 0.4 ng Tc g U^{-1} is reported. Technetium impurities are important because TcF_6 is volatile and may accompany the UF_6 in the enrichment product, so that even a low concentration of technetium in the feed material may lead to significant levels (and radioactivity) in the product. The same rationale applies to the presence of traces of plutonium and neptunium in reprocessed UF_6 that is used as feed material. They also form volatile hexafluorides, so although they are heavier than UF_6 they may pose a health hazard even at low concentrations due to their enhanced radioactivity and therefore should be eliminated. A procedure for the determination of Pu and Np in UF_6 based on alpha spectrometry was developed (C1561 2003). The method converts the hydrolyzed UF_6 solution to oxalic acid–nitric acid and the uranium is removed by solid phase extraction (SPE). Plutonium and neptunium are further purified by SPE and co-precipitated with NdF_3 and counted by alpha spectrometry. Isotopic tracers are used to determine the yield of the process that is typically 75%–90%. Incomplete removal of uranium may lead to an interference from ^{234}U, which is especially problematic if UF_6 contains enriched uranium (>5%) (C1561 2003).

Other specifications concern elements that form volatile fluorides that may concentrate in the enriched uranium product (like molybdenum, niobium, tantalum, titanium, and tungsten listed in Sections 143–152 in Table 2.5). Among those volatile fluorides, boron is of special interest as it has a large cross section for neutron absorption and may affect the quality of nuclear fuel (Sections 52–79 in Table 2.5). Other concerns arise from volatile organic compounds, like the Freon-114 refrigerant, that may increase the pressure in the enrichment facility. The latter group of compounds is determined by FTIR that can also be used to determine nonmetallic gaseous fluorides like SiF_4, PF_5, BF_3, HF, and MoF_6 (C1441 2004).

The specific gamma activity must also be determined according to the test procedure in uranium-containing materials, in this case UF_6 (C1295 2013). The measurement is carried out with a high-resolution HPGe detector to ensure that health and safety specifications (C787, C788, and C996) are met. The gamma radiation is emitted from fission and decay products. The limit of detection of 5000 MeV Bq kg^{-1} of uranium depends on the detector efficiency and the background level and was determined in pure, aged (i.e., in secular equilibrium) natural uranium solutions. The nuclides that are measured include ^{106}Ru/^{106}Rh, ^{103}Ru, ^{137}Cs, ^{144}Ce, ^{144}Pr, ^{141}Ce, ^{95}Zr, ^{95}Nb, and ^{125}Sb. If other gamma-emitting nuclides are present at detectable levels, they should be identified and quantified.

Highlights: Uranium hexafluoride is one of the most important commercial uranium products due to its role as the feed material in the large isotope enrichment facilities that use gas centrifuges or gaseous diffusion techniques. Sampling of UF_6 requires special equipment and handling procedures, especially if liquid samples are to be drawn from large cylinders that may contain tons of UF_6. A representative sample should be drawn after the contents of the cylinder are liquefied and homogenized. The sample should then be subdivided into several smaller vessels in order to perform the various tests to ensure compliance with specifications. The quantitative determination of the uranium content and the isotopic composition are of utmost commercial importance as the cost of the material is determined by their results. As the enrichment process greatly affects the content of impurities, increasing the fraction of volatile impurities in the product so that even a small concentration in the feed material may lead to significant concentrations in the enriched product and detrimentally influence the performance of the nuclear fuel derived from the product. Therefore, a large suite of analytical procedures were devised to measure all the impurities required by the rigorous specifications. It stands to reason that the extensive testing required has an effect on the price of UF_6 that is used as a feed material in enrichment plants. The enriched UF_6 that is used to form nuclear fuel must also be tested meticulously, but this can be done after it is converted into a solid uranium compound like UO_2, UF_4, or metallic U.

2.3.2 Analytical Characterization of UO_2 Fuel

The processing of UO_2 for manufacturing nuclear fuel involves two stages: powder production followed by fabrication of ceramic UO_2 pellets. The processing of the powder includes homogenization, that is, blending to ensure uniform particle size distribution and surface area, and adding reagents to improve the mechanical and nuclear properties of the pellets. The additives may comprise U_3O_8 and lubricants to control the microstructure, density, and pore size of the pellets and burnable absorbers (e.g., gadolinium) that affect the performance of the pellets in the reactor. The concentration of the burnable absorber decreases as the ^{235}U in the fuel is also burned so fuel with initial higher reactivity may be used thus extending its utility in the reactor.

Green pellets (not to be confused with the UF_4 *green salt*) are formed by feeding UO_2 powder into a press where cylindrical pellets are produced by biaxial pressing of several hundred MPa. The pellets are sintered by heating in a furnace to ~1750°C under a controlled atmosphere of argon–hydrogen, and this naturally leads to a decrease in their volume. Finally, the pellets are machined to exact dimensions and rigorous quality control is applied to ensure pellet integrity and precise dimensions (usually just under 1 cm diameter and about 1 cm in length).

The standard methods for characterization of nuclear grade uranium dioxide powders and pellets are outlined in an ASTM document (C696 2011) that is equivalent to the uranium hexafluoride document mentioned in the previous section (C761 2011). In order to ascertain that powdered UO_2 is suitable for use as nuclear fuel, it must comply with the specifications listed in C753 (Standard Specification for Nuclear-Grade, Sinterable Uranium Dioxide Powder). Document C776 (Standard Specification for Sintered Uranium Dioxide Pellets) lists the specifications of the pellets that include mainly the uranium content and its isotopic composition as well as the O:U ratio (stoichiometry), moisture content, and impurity content. In addition, physical characteristics of the pellets must also be tested, including pellet dimensions and density, grain size, pore morphology, pellet integrity (surface cracks and chips), cleanliness, and workmanship. The stability of fuel pellets upon irradiation shall be estimated. Another parameter that must be assessed is the *equivalent boron content* (EBC) that is an overall measure of the neutron absorption properties of the uranium oxide. The EBC is calculated by summing the contribution (concentration multiplied by the cross section) of all the impurities that absorb neutrons and is an important feature of nuclear fuel.

The O:U ratio is needed for predicting the sintering behavior of the powder in pellet production. The uranium content (of the dry sample after moisture determination) and isotopic composition are required for estimation of the performance of the fuel. The impurity content is needed for calculating the EBC. The methods outlined in the document (C696 2011) cover a broad range of analytical techniques some of which are listed in Table 2.6.

These include chemical methods (reduction by ferrous sulfate and titrimetric determination), gravimetric methods for O:U ratio, moisture analysis by coulometric techniques, and determination of H, C, N, Cl, and F with a specific method for each of these elements. In addition, the isotopic composition is determined by mass spectrometric methods and several metallic and nonmetallic impurity elements are determined by spectrochemical methods.

Some detailed examples of ASTM protocols for the characterization of UO_2 are presented later in order to demonstrate the procedures. Determination of uranium, oxygen, and the O:U atomic ratio by the ignition (gravimetric) impurity correction method is part of the general test methods for UO_2 (C696 2011). The sample (5–10 g of powdered

TABLE 2.6
Analytical Procedures for Characterizing UO_2

Uranium by ferrous sulfate reduction in phosphoric acid and dichromate titration method

C1267: Test method for uranium by iron (II) reduction in phosphoric acid followed by chromium (VI) titration in the presence of vanadium

Uranium and oxygen uranium atomic ratio by the ignition (gravimetric) impurity correction method

C1453 standard test method for the determination of uranium by ignition and oxygen to uranium ratio (O/U) atomic ratio of nuclear grade uranium dioxide powders and pellets

Carbon (total) by direct combustion-thermal conductivity method

C1408 test method for carbon (total) in uranium oxide powders and pellets by direct combustion-infrared detection method

Total chlorine and fluorine by pyrohydrolysis ion-selective electrode method

C1502 standard test method for the determination of total chlorine and fluorine in uranium dioxide and gadolinium oxide

Moisture by the coulometric, electrolytic moisture analyzer method

Nitrogen by the Kjeldahl method

Isotopic uranium composition by multiple-filament surface ionization mass spectrometric method

Spectrochemical determination of trace elements in high-purity uranium dioxide

Silver, spectrochemical determination of, by gallium OxideCarrier D-C arc technique

Rare earths by Copper Spark-Spectrochemical method

Impurity elements by a Spark-Source mass spectrographic method

C761 test method for chemical, mass spectrometric, spectrochemical, nuclear, and radiochemical analysis of uranium hexafluoride

C1287 test method for determination of impurities in uranium dioxide by inductively coupled plasma mass spectrometry

Surface area by nitrogen absorption method

Total gas in reactor-grade uranium dioxide pellets

Thorium and rare earth elements by spectroscopy

Hydrogen by inert gas fusion

C1457 standard test method for determination of total hydrogen content of uranium oxide powders and pellets by carrier gas extraction

Uranium isotopic analysis by mass spectrometry

C1413 test method for isotopic analysis of hydrolyzed uranium hexafluoride and uranyl nitrate solutions by thermal ionization mass spectrometry

Source: Adapted from C696, ASTM, Standard methods for chemical, mass spectrometric and spectrochemical analysis of nuclear grade uranium dioxide powders and pellets, ASTM, West Conshohocken, PA, 2011.

UO_2 or up to 50 g pellets) is placed in a preweighed platinum crucible and weighed to within 0.1 mg. The sample is then heated for 4 h (powder at 45°C, pellets at 160°C) in a vacuum oven after flushing with nitrogen several times, placed in a dessicator and weighed immediately after cooling to room temperature. The sample is then transferred to a muffle furnace and ignited at 900°C for 3 h and weighed again after cooling in a dessicator. This treatment converts the UO_2 to U_3O_8 and the stable oxide is analyzed for impurities by spectrographic methods. The atomic O:U ratio is calculated from both the weight on vacuum drying and the weight gain from the UO_2–U_3O_8 conversion. The standard deviation of the method for pellets in which the O:U ratio is between 2.0 and 2.1 is reported as 0.007 absolute with a confidence level of 95% (C696 2011).

Due to the importance of methods that use ICPMS for the determination of impurities and isotope composition in nuclear grade uranium compounds, including UO_2, we will present it in some detail (C1287 2010). The method covers the determination of 67 impurities without requirement for separation of the uranium matrix and is suitable for any form of nuclear grade uranium that can be converted into liquid form by acid dissolution. Some elements require special preparation methods (B, Si, K, and Ca) and are not included in the following procedure. A sample of UO_2, containing ~2.5 g of uranium, is weighed into a platinum dish (solutions of UO_2F_2 or $UO_2(NO_3)_2$ can also be determined). Water (10 mL) and concentrated nitric acid (15 mL) are added and the sample is heated. Hydrofluoric acid (2.5 mL of 40% solution) is added and warmed to 80°C for 5 min. The solution is cooled and transferred to a 50 mL plastic volumetric flask and water is added to fill the volume. The flask now contains 50 g L^{-1} of uranium in 25% HNO_3 and 5% HF. An aliquot of 4 mL is transferred to 100 mL polypropylene volumetric flask, 1 mL of an internal standard solution (Rh 10 μg mL^{-1}) is added and the volume is made up with water so that the uranium concentration is now 2 g L^{-1} (0.2%), Rh 0.1 μg mL^{-1}, 1% HNO_3, and 0.2% HF. A uranium-free blank is prepared by the same procedure and several standard solutions (that will not be elaborated here) are also prepared. The ICPMS operating parameters are set according to specific instructions supplied by the manufacturer and depend on the particular instrument. The blank, standards, and samples are analyzed by peak-hopping according to the mass selected for each analyte, including the mass of the rhodium internal standard. Uranium-matched calibration solutions are run at the start and end of each batch, and if instrument drift is found, then recalibration is carried out. The analyte counts are normalized using the internal standard to correct for matrix effects and changes in sensitivity of the ICPMS. The four elements mentioned earlier (B, Si, K, and Ca) go through a different sample preparation procedure (dissolution is in a capped polypropylene tube and scandium is used as an internal standard) the details of which can be found in the protocol (C696 2011). The protocol also presents the precision and bias of the method as well as the lower and upper reporting limits for each element. All these values should be examined periodically as the performance of analytical devices keeps improving as far as detection limits, reproducibility, and precision are concerned.

In advanced nuclear fuel pellets, the uranium oxide is mixed with burnable neutron poisons, mainly gadolinium oxide (Gd_2O_3 gadolinia) to extend the functional life of the fuel pellets. As the fraction of ^{235}U is reduced during reactor operation by fissions, the gadolinium content also decreases allowing the fuel to maintain its activity level. The specifications for sintered Gd_2O_3–UO_2 pellets are summarized in an ASTM document (C922 2008), for the pure gadolinium oxide (C888 2008) and the test methods for the pellets in another document (C968 2012). The specifications outline the impurity content for twelve elements, the stoichiometry, moisture content, gadolinia concentration, isotopic composition (of U and Gd), and physical characteristics like pellet dimensions and density, grain size, pore morphology, and pellets integrity (surface cracks and chips, pellet ends, and irradiation stability) (C922 2008).

The tests include determination of the carbon content (by combustion to CO_2 and IR detection), analysis of chlorine and fluorine (by pyrohydrolysis and specific ion electrodes), gadolinia and uranium oxide content (by XRF), the hydrogen concentration (by inert gas extraction), the isotopic composition of uranium (by TIMS), nitrogen analysis (by photometric measurement with Nessler reagent), the oxygen-to-metal ratio (by ignition and impurity determination described earlier), total gas content (by hot vacuum oven extraction), and some physical tests for homogeneity and average grain size (C968 2012). Several of these tests for UO_2–Gd_2O_3 pellets have a specific detailed ASTM procedure—for example, hydrogen determination (ASTM C1457) or XRF method for gadolinia and UO_2 (ASTM C1456).

Other analytical techniques are also available for characterizing UO_2 powders and pellets, so adherence to the ASTM procedures is not the only way to ascertain compliance with the specifications. An overview of the spectrometric techniques and sample preparation procedures (including separation of the uranium matrix) for the determination of impurities in nuclear fuel grade materials summarizing several methods was published (Souza et al. 2013). Among the spectrometric techniques surveyed are FAAS, GFAAS, ICP-OES, and ICP-MS as well as strategies for matrix separation and preconcentration steps.

Some specific examples include an ICP-TOF-MS (ICPMS with a time-of-flight [TOF] mass spectrometer) that was used to determine 27 elements that appear as impurities in matrix-matched uranium solutions (Burger et al. 2007). This method was claimed to have overall precision and accuracy of 5% and 14%, respectively, and to be particularly suitable for rapid evaluation of seized smuggled nuclear materials (that is the realm of nuclear forensics as discussed in Chapter 5). Another example is the use of ICP-OES for the determination of some trace impurities (REEs, Y, Cd, Co, V, Mg, B, Ca, Cr, Mn, Ni, Cu, Zn, and Al) in nuclear grade uranium oxide (Satyanarayana and Durani 2010). According to this method, the uranium matrix must first be separated by solvent extraction and relative standard deviations of 1.5%–6% were reported for trace levels of 0.5 µg mL^{-1} and 5.5%–12% for 0.2 µg mL^{-1}. A different approach to uranium matrix removal was carried out by ion exchange with a Dowex 1 × 8 sulfate medium (Aziz et al. 2010). The concentration of 10 impurities (Cr, Co, Cu, Fe. Mn, Cd, Gd, Dy, Ni, and Ca) was determined by ICP-OES with the objective of measuring the UO_2–Gd_2O_3 ratio in ceramic nuclear fuel and comparing the ion exchange procedure with the solvent extraction method. The best results for recovery of the impurities were obtained when the sulfate concentration on the column was 0.1 M at pH = 3 (Aziz et al. 2010). LIBS was applied for the determination of uranium in thorium–uranium mixed oxide fuel pellets in which boric acid was used as a binder (Sarkar et al. 2009). Calibration curves were prepared by two synthetic mixed oxide samples and the uranium content was determined. The advantage of the LIBS system is that dissolution, which is especially difficult for sintered oxides, is not required. ICPMS techniques have been deployed for the determination of 25 impurities, including iron, with a reaction collision cell (Quemet et al. 2012). The analysis of iron by ICPMS is complicated by the interference from polyatomic ArO^+ ions at mass 56 Da that overlap with ^{56}Fe that is the dominant isotope of iron. Measurement of the other iron isotopes also encounters isobaric interferences (e.g., $ArOH^+$ is isobaric with ^{57}Fe). The use of a reaction collision cell, under cold plasma conditions, with reactant gases (He, NH_3, and CH_4) breaks up the polyatomic ions and removes the isobaric interferences. In their study, U_3O_8 was dissolved by a mixture of HCl and HNO_3 and a UTEVA chromatographic column was used to remove the uranium matrix prior to measurement of the 25 impurities. The removal of the uranium matrix improved the limit of detection for most impurities, for example, for Ti and Mn by a factor of 2, for Zn by a factor of 30, but for Mo and Pb the matrix removal slightly degraded the limit of detection (Quemet et al. 2012).

One important aspect, often overlooked, is the availability of suitable reference standards for testing the analytical methods. The recertification of a standard UO_2 fuel pellet for uranium isotopic composition is helpful for validation of analytical methods (Kraiem et al. 2013). Exact isotope ratios were measured by TIMS and the $^{235}U/^{238}U$ ratio was recertified in six randomly selected CRM 125-A pellets. The minor isotope ratios $^{234}U/^{238}U$ and $^{236}U/^{238}U$ were also recertified and the absence of ^{233}U was confirmed by meticulous measurements. The new values are regarded as considerable refinement of the older values, and in particular the uncertainties of the reference material were reduced (Kraiem et al. 2013).

One of the tests that are occasionally carried out in gadolinia-doped UO_2 fuel pellets is the air-oxidation behavior at high temperature (225°C–450°C) (Ollila 2003). Thermogravimetric analysis (TGA) was used to study the temperature effects and dissolution of the $(Gd,U)O_{2+x}$ oxide in 0.1 M HCl and 10 mM $NaHCO_3$ under anoxic conditions.

In some cases, uranium scrap materials from various processes, like UO_3 or U_3O_8, that are enriched to <5% ^{235}U are recovered for conversion to nuclear grade UO_2 and standard specifications for materials of this type were issued (C1334 2010). The test methods for these types of uranium compounds are identical to those described for UO_2 and UF_6. First and foremost, the isotopic composition of the uranium compounds, which is between natural composition (0.72% ^{235}U) and 5%, must be determined. This was followed by tests of the uranium content (as outlined in C696 and C799) and then the level of several impurities must be determined and examined for compliance with the specifications listed in C1334, particularly for the equivalent boron content (EBC) that reflects the neutron absorption properties (C1233). The moisture content shall not exceed 1%, and physical parameters like the particle size and ability of the recovered UO_2 to flow are also measured. The ability to dissolve the powder is also examined with regard to temperature, time, and nitric acid concentration, and the amount of insoluble matter and foaming during dissolution is also determined (C1334 2010).

Highlights: The characterization of UO_2 powders and pellets to examine compliance with specifications involves the use of several physical and analytical test methods. The chemical analyses include determination of the uranium content and isotopic composition, the O:U ratio, and the measurement of the content of several elemental impurities. Of special importance are elements that may affect the neutron absorption properties of the fuel pellets. Each of these elements is determined and the total neutron absorption of all these impurities is summed up as EBC. Modern nuclear fuel may include *burnable neutron poisons* that are used to increase the operational lifespan of the fuel. The intentional addition of these *poisons*, like gadolinia, must be carefully controlled to avoid fluctuations of the neutron density in the reactor. Basically, after dissolution of the uranium oxide samples the analytical methods that are used for analysis of hydrolyzed UF_6 are suitable for these analyses. For the solid UO_2 samples, some unique chemical and physical tests are required. The content of H, C, N, F, and Cl and the O:U ratio, not carried out as such for UF_6, is to be determined in the uranium oxide samples. In addition, the physical properties of the powder and pellets (density, pore size and particle size distribution, surface area and added lubricants, etc.) must be determined to ensure conformity with the specifications. Overall, the specifications for UO_2 and UF_6 differ in several aspects due to the different applications: in UF_6 impurities of elements that form volatile fluorides are of major concern while for UO_2 the focus of impurity analysis is on neutron absorbers, the U:O ratio and the physical properties of the pellets.

2.3.3 ANALYTICAL CHARACTERIZATION OF URANIUM AND URANIUM ALLOYS

Several types of nuclear fuels based on uranium or uranium alloys have been fabricated and tested. Among these, uranium–aluminum alloys are probably the most popular, but alloys of uranium with molybdenum (U–Mo), U–Mo–Pt, U–Nb, and U–Mo doped to 1% with several other elements have been tested (Hofman et al. 1998). There are also alloys with U–Zr and U–Si, not to mention mixed oxide fuels like MOX (with plutonium oxide) or uranium with other actinides.

Impurities in metallic uranium or uranium alloys can be determined by the same methods that are used for UO_2, UF_6, or other compounds, as long as the sample can be dissolved. The procedures for dissolution of uranium-containing materials for analysis were detailed in an ASTM publication (C-1347 2008). In brief, uranium metal, scraps, and residues are dissolved by nitric acid while uranium–aluminum alloys are dissolved by hydrochloric acid and the residue may be treated with nitric acid or carbonate fusion.

Dissolution may be avoided if other analytical methods for determining impurities are employed like the use of DC-Arc emission spectroscopy (C1517 2009). This procedure involves conversion of uranium metal to U_3O_8 in a muffle furnace, addition of a weighed amount of the oxide to a spectrographic carrier (like AgCl/AgF, AgCl/LiF, AgCl/SrF$_2$, Ga$_2$O$_3$/LiF, or Ga$_2$O$_3$) and loading into a graphite electrode. The sample is excited by a DC-Arc and the emitted light is dispersed and measured electronically by a detector array (replacing the old photographic film). Line intensities are compared with standards with a known amount of impurities and the concentration in the sample is calculated. Detection limits depend on the instrumentation used and on the carrier, and are generally in the 0.1–100 parts-per-million (ppm) range, depending on the analyte. For example, sub-ppm detection limits are achieved for Cd, Be, and B, while for phosphorus and tungsten the LOD is 100 ppm by this method (C1517 2009).

Low enriched uranium (LEU) that contains between 15% and 20% [235]U can be produced by blending of high enriched uranium (HEU) with natural uranium or uranium from other sources. The product can be used to fuel research reactors. The standard specifications for this type of material were outlined and include the isotopic composition and the level of 23 select impurities (C1462 2013). The uranium content in the product must exceed 99.85% by weight and the metal should be in small pieces, free of loose or excessive oxides, with dimensions between 10 and 40 mm weighing between 130 and 300 g. The EBC shall not exceed 4 $\mu g\ g^{-1}$ (see ASTM C1233). The maximum level allowed for three minor isotopes are $^{232}U < 0.002\ \mu g\ g^{-1}$, $^{234}U < 0.01\ g\ g^{-1}$, and ^{236}U <0.04 g g^{-1}, and the allowable level for ^{235}U is below 20% by weight. In addition, due to health physics requirements, the gamma activity shall not exceed 600 Bq g^{-1} U as measured by the standard test method (C1295 2013). However, in some cases, as agreed upon between buyer and seller, the gamma activity may be as high as 6000 Bq g^{-1} U (C1462 2013).

Highlights: Practically all uranium compounds that can be dissolved can be analyzed by the same techniques for the uranium content, isotopic composition, and impurities concentrations that are used to characterize UO_2 or UF_6. This includes metallic uranium and uranium alloys. An alternative approach is to oxidize the metal to form U_3O_8 that may then be characterized directly as a solid by DC-Arc.

2.3.4 ANALYTICAL CHARACTERIZATION OF FUEL ELEMENTS

After the UO_2 pellets are produced and characterized, the fuel elements need to be fabricated. In most reactor designs, the pellets are inserted into rods that in turn are inserted into sealed tubes that are arranged in fuel assemblies. These are then placed in the core of the reactor in a designed lattice configuration that also provides specific gaps for positioning control rods. Within the fuel assemblies, there are spaces to allow the coolant (usually water or gases like He or CO_2) and moderator (usually water, heavy water, or graphite) to prevent overheating of the fuel elements and to slow down the neutrons to thermal energies. The structures forming the fuel assemblies must be precision engineered. All the elements must be constructed with materials that are corrosion resistant, can withstand high temperatures and physical forces (mechanical impact), and in addition have low cross sections for neutron absorption. Thus, zirconium and zirconium alloys (referred to as zircalloy), stainless steel, magnesium, and aluminum alloys are the materials of choice for cladding the fuel rods. Each tube must be leak-tested under extreme conditions of pressure (usually tested with helium) and after insertion of the rods that are held in place by a spring, leaving some free space for thermal expansion, the ends are sealed with precision welding. The World Nuclear Association (WNA) website provides details about fuel assemblies for different type of reactors (WNA 2013).

A variety of fuel elements are used for different types of reactors, but there are some common features. In most commercial nuclear power plants (BWR [boiling water reactors] and PWR [pressurized water reactors] that are called in Russian VVER), the pellets are inserted into rods or tubes (usually zirconium alloys) that provide a barrier to prevent escape of fission products, the tubes or rods are arranged in bundles that are loaded into the reactor core. Usually a number of short rods are inserted into the sealed tube and held in place by a spring as described earlier. In some cases, like advanced gas cooled reactors (AGR), pellets are inserted into short narrow steel pins. Magnox reactors use magnesium alloys (usually with aluminum) rather the zirconium alloys. The fuel in some advanced reactors (TRISO) is in the form of microfuel particles with a UO_2 (or UC (uranium carbide)) core surrounded by layers of pyrolytic carbon and silicon carbide that prevent escape of fission products from the fuel pellets, therefore eliminating the need for sealed tubes and enabling continuous refueling. A similar type of fuel includes burnable neutron poisons (e.g., gadolinia mentioned earlier or europium oxide) in the core of the microfuel particles (QUADRISO). In some reactor types, the fuel consists of metallic uranium or an alloy of uranium that are placed in tubing and not UO_2, and the mechanical, physical, and chemical properties of the uranium and tubing material need to be examined.

The main tests that are required for the sealed tubes containing the fuel elements are mechanical and physical like the quality of pellet insertion and spring insertion, inspection of the welding system for the fuel rods, x-ray inspection, and rod inspection.

The tubes themselves must be fully characterized in order to ensure that the zirconium alloys, or magnesium alloys, do not contain impurities that may affect the performance of the fuel. For example, the presence of traces of neutron absorbers, like hafnium that always accompanies zirconium in nature, or elements that modify the corrosion resistance of zirconium, must be determined. The ASTM has outlined the specifications for seamless wrought zirconium alloy tubes that are used for nuclear fuel cladding (B811 2013). The exact technical details and analytical test procedures do not directly involve uranium and are beyond the scope of this book.

Highlights: This section did not deal directly with the analytical chemistry or characterization of uranium, but is highly relevant to ensure the safety of the main commercial application of uranium—production of electric power in nuclear power plants. Therefore, the analytical procedures were only briefly described. The main safety function of the fuel elements is to prevent the escape of fission products (mainly gaseous elements or volatile compounds) into the coolant and atmosphere. Therefore, very rigorous quality control measures are employed to ensure the physical integrity and chemical composition of these elements.

2.3.5 ANALYSIS AT THE BACK END OF THE NFC

According to the WNA website (WNA 2013a,b), used fuel (spent fuel) from light water reactors (at normal US burn-up levels) contains approximately

- 95.6% uranium, of which over 98.5% is U-238 (the remainder consists of trace amounts of U-232 and U-233; <0.02% U-234; 0.5%–1.0% U-235; around 0.5% U-236; and around 0.001% U-237—which accounts for nearly all of the activity)
- 2.9% stable fission products
- 0.9% plutonium
- 0.3% cesium and strontium (fission products)
- 0.1% iodine and technetium (fission products)
- 0.1% other long-lived fission products
- 0.1% minor actinides (americium, curium, neptunium)

Characterization of the elemental and isotopic composition of SNF is like throwing stones at a moving target due to the fact that the composition depends on the initial inventory of uranium in the fuel, the position of a particular fuel element in the reactor core, the irradiation time (or burn-up), and the cooling period. These fine points could lead to considerable differences in the analysis of spent fuel (SF). After irradiation in the reactor, the SNF elements are removed from the core and placed in water pools (called spent fuel pools) to cool for a period that will allow the heat and reactivity that are due to the short lived fission products to decrease. The heat decay as a function of the burn-up ranging from 33 to 63 GWd/tHM was calculated (Feiveson et al. 2010). During the first 40 days, the heat decay drops by factor of four and during the first year by more than 1 order of magnitude. Another feature of irradiated spent fuel is the fact that the actinide and fission product inventory changes as a function of the burn-up. Figure 2.8 shows the reduction in the ^{235}U content and the concomitant buildup of ^{236}U calculated for a VVER-1000 type reactor (Kovbasenko et al. 2003).

The buildup of other actinides (239, 240, 241, and 242 isotopes of plutonium and ^{241}Am), some fission products (^{149}Sm and ^{151}Sm) as well as the decrease in ^{238}U were also calculated. A partial representation of the actinides that are formed during

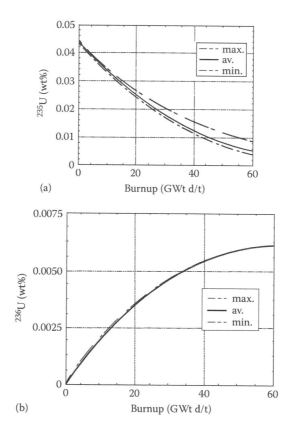

(a)

(b)

FIGURE 2.8 The decrease in ^{235}U content in fuel (a) and buildup of ^{236}U (b) as a function of the burn up. (From Kovbasenko, P. et al., Comparative analysis of isotope composition of VVER-1000 spent fuel depending on their manufactory and operation conditions, in *Seventh International Conference on Nuclear Criticality Safety*, ICNC, Tokai-Mura, Japan, 2003, p. 13. With permission.)

irradiation of LEU fuel can be seen elsewhere (Kilger 2010). Although this is not the complete scheme of the processes taking place in the reactor, the complexity of the actinide inventory is evident here.

On-site interim storage, mentioned earlier, must be supervised to ensure that radioactive nuclides are not released into the environment. The spent fuel elements may be eventually moved from the pool and placed in dry cask storage. An overview of the effects of water quality in the cooling pools for spent fuel in zircaloy cladding (usually from power plants) and aluminum cladding (usually from research reactors) was presented (Sindelar 2010). The conditions in the cooling pool, particularly the pH, conductivity, dissolved impurities species, undissolved solids, colloids, organic substances, biological organisms, and temperature, all affect the integrity of the cladding and therefore need to be monitored. In cases where the spent fuel is reprocessed, it is transported to the reprocessing plant and treated according to the standard procedures, as described later.

2.3.5.1 Characterizing SNF

Basically, there are two approaches to characterizing SNF: analysis of the intact fuel elements by active or passive remote sensing methods (i.e., measurement of the electromagnetic radiation and neutron emission) and analysis of fuel elements after dissolution.

If no reprocessing is planned, the spent fuel elements are moved to long-term storage facilities (water pools or dry storage) or transported to repository disposal sites where they are to be left (for eternity or until an acceptable technical solution is developed). In the United States, where reprocessing is not carried out, the ASTM has published guidelines and procedures for characterizing SNF in support of geological repository disposal (C1682 2009) and to ensure compliance with the Federal Regulations (Title 10, Part 63). Among the required tests is inspection of the cladding to ensure that it is not damaged to an extent that fission products may escape or that water may leach radionuclides from the spent fuel matrix—considered as *failed fuel* or *breached fuel*. Small pinholes or cracks may develop during storage into larger defects—termed as cladding *unzipping*. The main concern associated with geological disposal is leakage of radionuclides into the environment, particularly the water basin. The pyrophoric behavior of uranium and uranium hydride increase the chemical reactivity and may cause enhanced dissolution and leakage that may go beyond the repository boundary. Thus, thermal analysis of the spent fuel package is required in order to assess the potential chemical heat source. In addition, the inventory of radionuclides must be characterized. In summary, the tests include physical properties (appearance, weight, density, shape, microstructure, and cladding inspection), chemical attributes (radionuclide content, corrosion products), and environmental characteristics (drying and oxidation rates, ignition temperature, and dissolution rates) (C1682 2009). Specifications for SNF in aluminum cladding that is to be sent to a repository disposal site were also listed (C1431 2010). Due to the fact that aluminum-based cladding behaves differently than zircaloy-based cladding upon storage, some different tests are required. For example, uranium-rich particles are often found in microstructures in aluminum cladding. So tests of solubility, leaching, oxidation/reduction, and corrosion, as well as physical and environmental tests like those outlined earlier are required for storage of aluminum-clad spent fuel (C1431 2010).

One of the important parameters that need to be determined in SNF is the burn-up, that is, the portion of ^{235}U nuclides that were consumed during the reactor operation. Although this seems like a simple measurement, after all the isotopic composition of uranium in the front end of the NFC can be easily determined by a myriad of mass spectrometric techniques as mentioned earlier, the fact that the SNF contains fission products and activation products and is highly radioactive makes it difficult to handle, so nondestructive analytical techniques that do not require sample dissolution are sought. In addition, selection of fuel elements to obtain a representative sample is complicated by the fact that in different positions in the reactor core the burn-up could vary considerably. In principle, the gamma radiation and neutron emission from spent fuel are correlated with the burn-up, but the presence of fission products, activation products from structural materials, and actinide elements complicate the measurement (Bevard et al. 2009). Bombarding the spent fuel sample with neutrons to induce fission in the fissile ^{235}U, that is, neutron activation, and measuring the resultant delayed neutrons and gamma radiation can also be applied for burn-up assessment, but here too complications may arise from the background of neutrons produced by transuranics like ^{244}Cm in the spent fuel from power plants where high burn-up rates are common. However, this technique may be suitable for research reactors where the level of ^{244}Cm in the fuel is low. Another option is measurement of gamma photons at 661.7 keV emitted from ^{137}Cs, which after 5 years of cooling becomes the major gamma ray emitter. However, a 5-year delay of the assessment could be too long and in any case there are other gamma emitters that increase the background requiring longer counting times. ^{244}Cm mentioned earlier is formed by successive neutron captures of ^{238}U and is a spontaneous neutron emitter with a half-life of 18.1 years so that after 20 years of cooling it becomes responsible for 92% of the source neutrons (Bevard et al. 2009). Some external factors affect the precision of the measurement like positioning of the monitoring equipment and presence of boron (a neutron absorber) in the pool water. In addition, there are other methods to estimate the burn-up of SNF, as further described, but none are in principle simple and straightforward like mass spectrometric isotope analysis.

If we look at the burnt fuel pellets on a microscopic level, we can see that analysis of the material is complicated. The formation of fission products in the pellet causes defects that arise from gaseous fission products like xenon that result in microscopic bubble-like pores. The radioactive decay of short-lived xenon isotopes leads to formation of cesium and other fission products that may be inside these pores. Furthermore, the temperature gradient inside the pellet causes the more volatile fission products to migrate to the cooler parts at the edge of the pellet, while less volatile elements migrate to the center. Total dissolution of a pellet (or better yet, several pellets) can yield a representative sample that can be characterized analytically, but care must be practiced to prevent the loss of gaseous products.

A report that summarizes the state-of-the-art analytical methods for assay of SNF was published by the Expert Group on Assay Data of Spent Nuclear Fuel (EGADSNF 2011). Their stated objective was to view the techniques that serve for "destructive post-irradiation examination (PIE) for analysis of the isotopic composition and concentrations in spent nuclear fuel sample." First, the sampling procedures and sample dissolution methods are considered, followed by techniques for separating the radionuclides and the measurement procedures. However, it should be emphasized that these procedures can only be carried out in laboratories equipped for handling the highly radioactive samples of SNF, namely *hot cells* and the peripheral equipment. As expected, the analytical methods of choice for the determination of the isotopic composition of the actinides are mass spectrometric techniques (TIMS and ICPMS), while for the fission products (those that are not short-lived) mass spectrometry and gamma spectrometry are used (EGADSNF 2011).

Sampling of SNF must take into consideration the fact the burn-up and distribution of the fission products and actinides is not uniform axially or radially within pellets or rods, as determined by ^{137}Cs spectrometry. If possible, the sample should be large enough to include pellets and inter-pellet zones, and if large samples are not available then the inhomogeneous nature of the rods should be considered. Cladding should also be sampled because some fission products may be contained there, but mechanical decladding (as opposed to chemical dissolution) is not desired, especially with low burn-up samples.

Dissolution of UO_2 and MOX fuels is usually carried out with hot (near boiling) concentrated (8–10 M) nitric acid, preferably without dissolving the cladding but dissolving fuel remnants that may be on the cladding (EGADSNF 2011). The acid is filtered and the residue is treated with a mixture of 8 M HNO_3 and 0.1 M HF to ensure plutonium oxide dissolution. The dissolution allows the volatile fission products (xenon, krypton, iodine) to escape, so for determination of ^{129}I the off-gasses are passed through an alkaline solution to trap the iodine. Carrier iodides (NaI or KI) can be added before dissolution to ensure volatilization of iodine. This degree of dissolution brings the actinides and major fission products into solution and is sufficient for burn-up determination. However, if extensive characterization is required, the residues need to be fully dissolved and this is carried out by treatment of the residue in a closed vessel (i.e., bomb digestion) with suitable salts (fusion). After complete dissolution, the sample is usually diluted in the *hot cell* to reduce the activity level to an extent that aliquots may be transferred to the laboratory for further processing. Naturally, all these operations in the *hot cell* contribute to the uncertainty of the results: weighing errors, dilution uncertainties, loss of volatile constituents, cross-contamination, reagent purity, etc. The normal method to assess the uncertainty—the use of reference standards—is not applicable for spent fuel because such standards are lacking at present (EGADSNF 2011).

Chromatographic separation is needed for accurate analysis of concentrations and isotope composition. ICPMS can be used for online measurements without extensive purification and separation, but for TIMS analysis the sample must be purified and this has to be done offline. Due to their high concentrations, uranium and plutonium should be separated from the other actinides and fission products, and this is readily carried out with chromatographic resins (e.g., see Horwitz et al. 1992, 1993; Fajardo et al. 2008) or ion-exchange columns. The eluent without the uranium and plutonium may be processed further by chromatographic techniques (HPLC, ion chromatography, capillary electrophoresis, etc.) to determine the fission products. This can be combined with ICPMS for online analysis. Radiometric analytical techniques are employed for measurement of radionuclides by alpha, beta or gamma spectrometry in dissolved SNF, by the same methodology described earlier that is used for characterizing radionuclides.

The analytical literature includes several methods that have been developed for online or offline straightforward assay of SNF after dissolution. Among those are

ICPMS (Alonso et al. 1994) or ICPMS after ion-chromatography (Moreno et al. 1996) and analysis of high burn-up fuel (Wolf et al. 2005) to mention just a few examples. Methods based on gamma spectrometry for multi-isotope process (MIP) monitor (Orton et al. 2011, 2012), high-energy delayed gamma spectroscopy (Campbell et al. 2011), neutron resonance transmission analysis (Chichester and Sterbentz 2011), lead slowing-down spectroscopy (Smith et al. 2010), and a combination of passive neutron measurement and collimated total gamma measurements with an online depletion code (Lebrun and Bignan 2001) were also tested.

A detailed investigation of the analytical characterization of SNF by mass spectrometric techniques combined with chromatographic separation was published (Gunther-Leopold et al. 2008). Handling highly radioactive spent fuel requires special equipment for sample preparation and analysis that must be carried out in a shielded glove box. Dissolution of irradiated fuel pellets (cut out of the fuel rods) was by reflux in 8 M HNO_3 at ~170°C for 4 h. Most of the gases released during the dissolution were retained in two traps with alkaline 4 M NaOH solution. The liquid sample was then filtered to remove the residues (mainly refractory elements like Mo, Ru, and Rh) and diluted to 1 M HNO_3 with a final concentration of about 0.5 mg fuel g^{-1} solution. Isotopic spikes (^{233}U, ^{242}Pu, and ^{150}Nd) were added for quantification and an unspiked sample was also retained for analysis. Chromatographic separation of fission products in the sample (70 µg fuel g^{-1} solution) was carried out with an HPLC device: lanthanides (Sm, Pm, Nd, and Ce) were separated on a C18 column while Sr and Rb were separated on a CS12A column. The analysis of the fission products and actinides in the sample was carried out with an ICPMS device so that the isotopes (142, 143, 144, 147, 148, and 150) and the 148/150 isotope ratio in samarium and neodymium were determined as shown in Figure 2.9 (Gunther-Leopold et al. 2003).

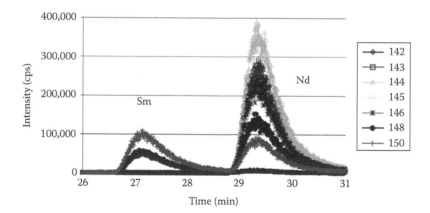

FIGURE 2.9   Separation on Sm and Nd in irradiated mixed oxide sample after injection of 1 mL containing 20 µg fuel mL^{-1} measured with the HPLC-Quadrupole ICPMS system. (From Gunther-Leopold, I. et al., Characterization of spent nuclear fuel by an online combination of chromatographic and mass spectrometric techniques, in *Proceedings of the Seventh International Conference on Nuclear Criticality Safety*, ICNC, Tokai, Japan, October 2003, JAERI-Conf 2003-019, 2003. With permission.)

Preliminary measurements were made with a quadrupole-based system and reported (Gunther-Leopold et al. 2003), but after a multicollector ICPMS was installed in the hot laboratory, more precise isotope ratio measurements were made. The results were adjusted by use of internal correction (known isotope ratio of the same element) or by a bracketing procedure with certified reference materials (external correction). Additional measurements were made with a specially designed laser-ablation system that was coupled to the ICPMS instrument. The nuclear fuel burn-up assessment was based on measurement of the ratio between ^{148}Nd isotope and the four main fissionable nuclides (^{235}U, ^{238}U, ^{239}Pu, and ^{241}Pu) for UO_2 or MOX fuel. ^{148}Nd is formed through neutron capture by ^{147}Nd that is practically independent of the burn-up, and it is destroyed by another neutron capture that increases linearly with the burn-up. As there is an isobaric interference from ^{148}Sm, separation is needed for the ^{148}Nd determination. The precise determination of the isotope ratio of the actinides also requires separation due to isobaric interferences. HPLC-MC-ICPMS was used for the separation and measurement of the uranium and plutonium isotopes for a sample concentration of 15 µg fuel g^{-1} solution as shown in Figure 2.10 (Gunther-Leopold et al. 2003).

Laser ablation can be used for direct analysis of solid samples and that is particularly advantageous for eliminating or reducing the sample preparation steps. However, as irradiated pellets are not homogenous the precision and reproducibility of laser-ablation signals is inferior to analysis of the dissolved samples. The variation of the $^{235}U/^{238}(U + Pu)$ ratio along the cross section of a UO_2 fuel pellet (i.e., the distance from the cladding) at four different angles (30°, 90°, 180°, and 270°) was compared with the ratio measured by HPLC-MC-ICPMS in dissolved pellets (Gunther-Leopold et al. 2008). The variability with the distance from the cladding was evidently large (>30%), and the variation of the ratio at the different angles was

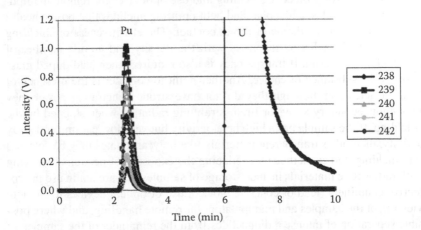

FIGURE 2.10 Separation of plutonium and uranium in an irradiated UO_2 sample measured by HPLC-MC-ICPMS after injection of 25 µL of a solution containing 15 µg fuel mL^{-1}. (From Gunther-Leopold, I. et al., Characterization of spent nuclear fuel by an online combination of chromatographic and mass spectrometric techniques, *Proceedings of the Seventh International Conference on Nuclear Criticality Safety*, ICNC, Tokai, Japan, October 2003, JAERI-Conf 2003-019, 2003. With permission.)

also quite significant (about 10%). In should be noted that in addition to the variability within the fuel pellet there were significant differences in the burn-up between different pellets and within the fuel rod and among bundles of rods. Finally, the main advantage of an online coupled HPLC-MC-ICPMS system over analysis by TIMS was the fact that separation and purification of the actinides and fission products was rapid and the isotope ratios of the different elements could be carried out sequentially without the need to separate, purify, and deposit each element on a dedicated filament. The sample throughput with this system was incomparably higher while the accuracy was similar. Another advantage was that handling of radioactive samples was minimized (Gunther-Leopold et al. 2008). A short overview of handling radioactive samples is presented in Frame 2.4.

In an earlier work by the same group (Gunther-Leopold et al. 2003), sample dissolution was done by digestion in a high-pressure close vessel at 150°C for 3 h with HNO_3/HF and the sample was then similarly diluted to 1 M HNO_3 with 0.5 mg fuel g^{-1} solution. In this work, the performance of the quadrupole-based ICPMS was compared with the MC-ICPMS and with TIMS measurements, and unsurprisingly the quadrupole was less precise than the other two instruments, but as mentioned earlier this was remedied in their later work.

FRAME 2.4   HANDLING RADIOACTIVE SAMPLES

Handling of radioactive samples, and particularly SNF elements, even after an extended cooling period, requires special measures to ensure the health and safety of the operators (for details, see Perkin-Elmer n.d.). The radioactivity of the fission products and actinides can be divided into two groups: gamma radiation that requires protective shielding and use of devices for remote manipulation of the samples, and alpha- and beta-emitting nuclides that pose a health hazard only after inhalation or direct contact. The effectiveness of shielding electromagnetic radiation increases with the density and atomic number of the shielding material. If transparency is also required, then lead-doped glass is often the substance of choice. Another point to consider is the presence of beta particles that although they do not have strong penetration power they may emit secondary x-rays or Bremsstrahlung radiation when stopped by the shielding. If the sample has a high beta-activity but very low gamma and x-ray activity, then other transparent materials like polycarbonates may be selected for shielding. Other methods for reducing the exposure of personnel working with radioactive materials include storage of samples that are not in use in protective containers (lead or concrete blocks), use of optical devices for remote viewing of the samples and manipulators for remote handling, and where possible separation of intense radionuclides from the remainder of the sample.

All these safety measures must be taken in addition to the standard protective measures deployed in any chemical laboratory: lab coats or overalls, gloves and footwear, eye and face protection and respiratory protection. Contamination of surfaces must be avoided and control of airborne contaminants must be practiced (Perkin-Elmer n.d.).

Highlights: The characterization and analysis of spent nuclear is greatly complicated by its high radioactivity. Assessment of the burn-up of the fuel is further problematical due to the variability of the spent fuel composition between fuel elements from different parts of the reactor core as well as variation within each individual fuel pellet. Furthermore, the elemental and isotopic composition, the radiation and heat output of the fuel keeps changing with storage time. Nondestructive assays of spent fuel, based on gamma and neutron spectroscopy, are of limited value due to the mutual interferences from the myriad of radionuclides within the fuel. Dissolution and analysis, mainly by mass spectrometric and gamma spectrometric techniques, provides a more comprehensive picture of the burn-up and composition of the spent fuel. Procedures have been developed for dissolution and separation of the uranium (and plutonium) matrix in order to obtain better analytical assessment of the fission and activation products. However, the high radioactivity necessitates special handling procedures, laboratory equipment, and safety measures that encumber the sample preparation and analysis. Waiting for the fuel to cool (physically and radioactively) for a few years is usually unacceptable.

2.3.6   REPROCESSING SNF

The main methods used for reprocessing of SNF flowsheet were reviewed in a 120 pages report by the Nuclear Energy Agency (NEA 2012). The three main processes are the so-called hydrometallurgy processes (PUREX and UREX), pyrometallurgy processes and its variations, and the fluoride volatility process (quite like the method used at the uranium conversion facilities discussed in Chapter 1). The report reviewed in detail several of these processes that are deployed in different facilities for various types of spent fuel (NEA 2012). In this section, we shall try to briefly present an overview of the main points and the analytical aspects.

As mentioned earlier, spent fuel from a reactor using LEU fuel contains many components that have a commercial value. Among these, uranium is the major component with an isotopic composition that depends on the original level of the fuel enrichment and on the extent of burn-up (typically ~0.5%–1.0% ^{235}U, ~98.5% ^{238}U, ^{236}U ~0.5%, and some ^{234}U, ^{233}U, and ^{232}U). Actinides, mainly plutonium, are also present in the spent fuel as well stable and radioactive fission products. Reprocessing is used to generate three main product streams: uranium, plutonium, and fission products (that may include some activation products).

As mentioned in Chapter 1, the PUREX (Plutonium, URanium EXtraction) process is most widely used for recovery of uranium and plutonium from irradiated fuel. A schematic of a generic PUREX process is shown in Figure 2.11.

The SNF (after a cooling period to allow for decay of short-lived radionuclides) is chopped up and dissolved in nitric acid. The gasses emitted in the process are treated to avoid their release to the environment. The solution is filtered to separate the insoluble residues and sent to the solvent extraction stage in which the uranium and plutonium are extracted into the organic phase (usually TBP in a hydrocarbon solvent) and the fission products and minor actinides remain in the aqueous phase. The radioactive fission products may then be treated as high-level-waste while the uranium and plutonium are then separated from each other by selective back-extraction.

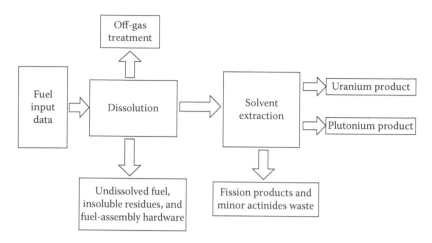

FIGURE 2.11 Generic description of the PUREX process. (Based on NEA, Spent nuclear fuel reprocessing flowsheet, OECD Nuclear Energy Agency, Paris, France, NEA/NSC/WPFC/DOC(2012)15, 2012.)

Each of these elements may be used for production of nuclear fuel or other purposes. The recovery efficiency for uranium is reported as ~99.87% and for plutonium 99.36%–99.51% (NEA 2012). The extended PUREX includes separation of neptunium and technetium as well as recovery of americium and curium that are also separated from each other by additional extraction stages as given in detail in the flowsheet (NEA 2012). The advanced UREX+3 process generates six streams after separation: uranium for re-enrichment; Pu–U–Np for mixed oxide fuel; ^{99}Tc for managed disposal; Am-Cm to be used as burnable poisons and for transmutation; high-heat-generating products (Cs and Sr) and a composite vitrified waste with all other fission products. Some fuel types may require preliminary steps like grinding to enable their dissolution.

Pyrometallurgy processes are less known and generically consist of the following steps: voloxidation; electrowinning of uranium oxide; electrochemical reduction of residual actinide oxides; electro-refining and removal of salt or cadmium from cathode deposits (NEA 2012). There are many variations of this generic process depending on the facility and fuel that are practiced mainly in Japan and South Korea. In the United States, a direct electro-refining process for metallic fuel is being developed at Argonne National Laboratory for recovery of actinides (see details in NEA 2012). Finally, a so-called PyroGreen process to reduce the radiotoxicity level of spent fuel to low and intermediate level wastes. Apparently this process adds three steps to the Pyrometallurgy process described earlier: zircaloy hull cleaning; salt waste purification, and ceramic waste fabrication (see details in NEA 2012).

The third process that can be deployed for reprocessing of SNF is the fluoride volatility process that has an advantage of not being affected by radiation and is considered to be especially relevant to fuel from fast-breeder reactors (NEA 2012). This process utilizes the fact that uranium and plutonium (as well as some other elements that may be present in spent fuel) form stable fluorides that are volatile and can be separated from fission products so most of the radioactivity remains in the solid waste.

One of the many variations involves bubbling fluorine through fuel molten salt compositions to selectively remove uranium and protactinium. The main problems arise from the chemical reactivity of fluorine that requires special handling (NEA 2012).

The analytical techniques that are used to monitor all these fuel processing methods do not differ from the techniques described in the previous sections of this chapter. The main problems are during the initial stages of the process where radioactive fission products and a myriad of actinides complicate sampling and sample handling. There is usually no need to precisely determine the nonradioactive fission products and the radionuclides are best monitored by radiometric methods (gamma spectrometry mainly). In cases where the isotopic composition must be determined, the standard mass spectrometric methods are used. As no changes in the isotopic composition of uranium or plutonium occur during reprocessing, this type of measurement is only sparsely required.

2.3.7 Analysis of Waste in the NFC

Waste materials of the NFC must be managed to minimize their impact on the environment and human health (WNA 2012). The absolute amount of waste material is small relative to the amount of power produced from nuclear reactors, but as it contains highly radioactive nuclides treatment of waste and disposal must be carefully managed. It is customary to classify the waste materials according to their radioactivity as low-level waste (LLW), medium-level waste (MLW), and high-level waste (HLW). Another factor is the half-lives of the isotopes in the waste that determine the time that the waste is likely to remain hazardous. Finally, the physical barriers needed for the safe disposal also depend on the type of radioactivity: for alpha- and beta-emitting nuclides, prevention of leakage into the environment is sufficient, while for gamma and neutron emission, there is also a need to attenuate the radiation from the waste container. There are some general principles applied to treatment of radioactive waste:

1. Concentrate-and-contain strategy that involves reduction of volume of the waste material and placing it in sealed containers. This approach is also used for other toxic, nonradioactive, materials.
2. Dilute-and-disperse is the opposite approach where the sample is diluted in a large volume (to an activity level below legal disposal limits) and spread over a wide area (or dumped in a large water body). This approach too is used for other toxic, nonradioactive, materials.
3. Delay-and-decay tactics imply that the waste is stored for a period of time that allows the short-lived radionuclides to decay until the reactivity level is reduced. The HLW from SNF decays to 1‰ of its initial activity within 40–50 years, and it is calculated that after ~1000 years the level will be close to the natural radioactivity background. Figure 2.12 depicts the decay with time of the overall radioactivity, and the contribution of individual radionuclides, of the HLW from 1 ton of spent fuel.

The HLW from SNF (the fuel itself or the waste generated from reprocessing operations) contains most of the radioactivity (95%) but only a small fraction (3%) of the

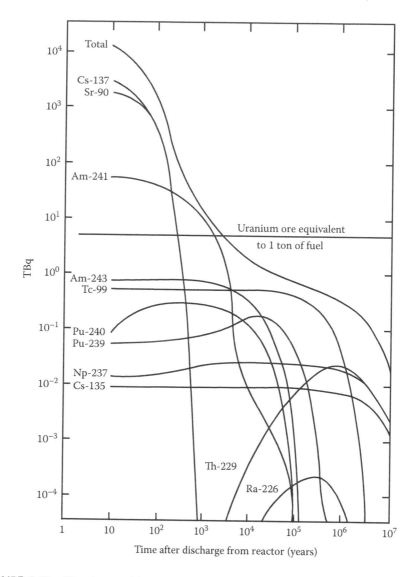

FIGURE 2.12 The decay with time of the overall radioactivity, and the contribution of individual radionuclides, of the high-level waste from one tone of spent fuel. (From the World Nuclear Association, http://www.world-nuclear.org/uploadedImages/org/info/Nuclear_Fuel_Cycle/activityhlw.gif, last accessed on July 27, 2014. With permission.)

volume of the waste, while the MLW possibly will contain 7% of the volume and 4% of the reactivity (WNA 2012). The MLW may be solidified in concrete or bitumen (tar) for disposal. However, the HLW requires cooling (due to the heat generated) and shielding because of gamma and neutron emission, and requires special handling. The SNF can be reprocessed so that fission products and useless actinides are separated from each other and from uranium and other products of commercial value. The active waste may be vitrified (combined with glass) or immobilized by incorporation in a crystal lattice of a naturally stable rock (Synroc (synthetic rock)) and placed in sealed stainless steel containers for eventual deep underground disposal. Used SNF that is not reprocessed must be encapsulated prior to disposal (usually as whole fuel assemblies) and transported to the burial site. To put things in proportion, the volume of the HLW from a typical commercial nuclear power plant is about 3 m³ year⁻¹ (WNA 2012).

Before making a decision on the optimal treatment of radioactive waste, it must be characterized. The analytical methods used for assaying HLW are similar to those used for SNF, that is, nondestructive assay based on measurement of the spontaneous gamma and neutron emission, and in cases where dissolution is possible detailed analysis of the elemental and isotopic composition.

Highlights: Treatment of the radioactive waste from nuclear reactors is one of the points that receive wide public attention and the disposal and burial of HLW in particular is a contentious issue due to the concerns about leakage to the environment. The technical solutions that are currently used to treat the waste that were listed earlier (concentrate-and-contain, dilute-and-disperse, and delay-and-decay) are not suitable for HLW, where safer solutions like vitrification or Synroc are sought. The characterization of the LLW and MLW waste is not as complicated as that of spent fuel but still greatly more complex than analysis of fresh fuel. Some of the procedures and methods used in other parts of the NFC are suitable for LLW and MLW. The composition of HLW must be determined in order to estimate the decay rate of the radioactivity and to classify the required protective measures that depend on the radionuclides and their products (emitters of alpha, beta, gamma, and neutrons).

2.4 DEPLETED URANIUM AND ITS CHARACTERIZATION

DU is formed in the tails stream of the enrichment process and contains <0.72% of ^{235}U (the natural content), usually 0.2%–0.4%. The chemical, physical, and mechanical properties of DU are identical to those of natural uranium (NU) and the radioactivity is slightly lower due mainly to the reduction in the ^{234}U content (usually about half to two-thirds of NU). The world inventory of DU is estimated at about 1.5 million tons (WNA 2012). The main production method of DU is by deconversion of depleted UF_6 containers, as mentioned earlier, that are the tails of enrichment facilities. The standard process involves reacting UF_6 with steam to produce UO_2F_2 that can then be reacted with hydrogen to produce HF and U_3O_8. The U_3O_8 is packed in containers for storage and the HF can be reused in the UCF. A different process converts the UF_6 into UO_2 and UF_4 and the latter can be reduced with magnesium or calcium to form uranium metal. In some facilities, the depleted-UF_6 (DUF_6) remains in storage in the original cylinders for future use as feed material for enrichment. The economic aspects depend on the price of uranium, the cost of conversion (and deconversion), the cost of enrichment (SWU), and the enrichment technology deployed (e.g., if laser-based technologies like MLIS become commercially viable).

Unfortunately, we could not find detailed specifications for DU that are similar to those publicly available from the ASTM, for example, for nuclear grade UF_6, UO_2, and uranium metal. So the analytical methods suitable for characterization of DU are based mainly on the methods applicable to the aforementioned substances. The most notable exception is presented at the end of this section (Trueman et al. 2004).

The high-density (19.1 g cm^{-3}), hardness, and relatively low price make DU the material of choice for advanced armor piercing munitions (kinetic penetrators) with the additional feature that uranium is pyrophoric so it may also serve as an incendiary weapon. Alloys of uranium with small proportions of molybdenum or titanium (staballoys) are often used to improve the mechanical properties and ease of melting and casting. For example, uranium alloys with 0.75% or 3.5% of titanium by weight are used as kinetic penetrators. The high density and hardness also make uranium, especially DU, as a good material for armor plating of tanks, shielding gamma and x-ray radiation, and also as ballast for aircraft (a Boeing 747 may have 400–1500 kg of DU). For each kilogram of uranium enriched to 5% ^{235}U from 11.8 kg of natural uranium, 10.8 kg of depleted (0.3% ^{235}U) uranium are produced. If the UF_6 used for enrichment comes from natural uranium, the analytical methods used to characterize UF_6, UO_2, or uranium metal are suitable for DU, but if UF_6 originating from reprocessed irradiated nuclear fuel was used then additional assays for Tc, Np, and Pu are required.

As mentioned earlier, DU may be used for re-enrichment (if the price of uranium increases significantly relative to the cost of enrichment) or for fueling reactors (especially if MOX fuel is considered), but at present, the likelihood of these applications is quite low. Uranium recovered from spent fuel may also be used for different applications, including manufacturing of DU munitions. In this case, the uranium may contain nonnatural isotopes like ^{232}U, ^{233}U, and ^{236}U formed in the reactor. Due to the high specific activity of ^{232}U and the fact that its daughter, ^{208}Tl, has an energetic gamma ray (Peurrung 1998), the full isotopic composition of uranium must be determined in this type of material.

A certified reference material (CRM-115) for characterization of the isotopic composition of DU metal was prepared and can be used for precise determination of the minor isotopes (Mathew et al. 2013). Modified total evaporation (MTE) with a TRITON TIMS device was used for highly accurate determination of the ^{235}U/^{238}U, ^{234}U/^{238}U, and ^{236}U/^{238}U ratios, and establishing that the ^{233}U/^{238}U ratio was below $5*10^{-9}$.

As mentioned earlier, specifications for DU whether it is intended for use in munitions, radiation shielding, or aircraft ballast are difficult to find. In any case, the analytical procedures for its characterization, described earlier for other uranium compounds, are valid also for DU. Dissolution of the metal samples in concentrated nitric acid (HF may be added if residues remain) is required for meticulous analysis of impurities by ICP-OES, ICP-MS, or/and other suitable analytical method. Conversion to U_3O_8 in a muffle furnace with steam for impurity determination by DC-arc is also an option. Determination of the H, C, N, O, and S content with dedicated instrumentation may also be carried out. On the other hand, the impurity requirements for DU are not as strictly controlled as they are for other nuclear materials. If the depleted uranium is alloyed, as mentioned earlier to improve the mechanical properties, the concentration of the alloying element must be determined according to specifications.

On the other hand, there is a wealth of publications on the determination of DU as a contaminant in soil, vegetation, and especially bioassays, as described in Chapters 3 and 4, respectively. One example of such a study is the characterization of DU in contaminated soil, where soil samples from the Balkan conflict areas were analyzed using alpha spectrometry for determination of ^{236}U, ^{237}Np, and ^{239,240}Pu after fractionation by a five-step sequential extraction procedure (Radenkovic et al. 2007). DU projectiles found in Kosovo were characterized, and as expected were composed mainly of DU with small amounts of titanium, and ^{236}U was also detected indicating that the uranium originated from reprocessed fuel. The analytical methods used included gamma spectrometry with HPGe for U, Th, and Pa isotopes, scanning electron microscopy with x-ray microanalysis (EDS), alpha spectrometry, and ICPMS to determine that the flakes of the fired projectile contained U and Ti and minor amounts of other elements (Pollanen et al. 2003).

Probably the most comprehensive published assay of DU used in armor penetrators was reported on the basis of analysis of an unfired CHARM-3 penetrator (Trueman et al. 2004). A sample from the penetrator was dissolved in 9 M HCl, spiked with ^{232}U as a yield monitor, and the uranium was separated from impurities on an ion-exchange resin. The isotopic composition of uranium was determined by mass spectrometric techniques. Actinides (^{241,243}Am and ^{237}Np) were determined in the uranium-free solution by gamma spectrometry and $^{239+240}$Pu and ^{238}Pu were measured by alpha spectrometry and their presence was confirmed by ICPMS. Technetium-99 was determined by ICPMS when rhenium was used as a carrier and interferences from iron were eliminated by precipitating with ammonia while ruthenium and molybdenum were removed by separation on a chromatographic resin. The content of these radioactive nuclides is summarized in Table 2.7.

TABLE 2.7
Radiochemical Analysis (Specific Activity) of Unfired Depleted Uranium U CHARM3 Penetrator

Radionuclide	Specific Activity and Uncertainty (2σ) (Bq g⁻¹)
U-238	12,343 ± 0.916
U-235	156.6 ± 0.461
U-234	2,382 ± 1.161
U-236	7.113 ± 0.113
Np-237	5.002 ± 0.267
Pu (239+240)	2.195 ± 0.098
Pu-238	1.059 ± 0.055
Am-243	6.678 ± 0.117
Am-241	2.166 ± 0.085
Tc-99	78.97 ± 0.795

Source: From Trueman, E.R. et al., *Sci. Total Environ.*, 327, 337, 2004. With permission.

TABLE 2.8

Nonactive Impurities in an Unfired DU CHARM-3 Penetrator[a]

Element	Concentration (mg kg⁻¹)	Uncertainty (%)	Limit of Detection[b] (µg kg⁻¹)
Al	753.4	2.02	3.73
Ca	76.1	2.27	0.43
Co	215.4	8.9	16.41
Cu	28.9	28.08	1.67
Ni	31.3	13.0	2.75
Pb	19.1	50.13	0.024
Si	1125.0	5.24	6.65
Ti	5886.9	1.17	0.52
Zn	216.7	5.05	2.26
Zr	39.0	39.85	0.03

Source: From Trueman, E.R. et al., *Sci. Total Environ.*, 327, 337, 2004. With permission.

[a] The concentration of the following elements was below the limit of detection: Ag, As, B, Ba, Be, Bi, Cd, Cr, Cs, Fe, Ga, In, K, Li, Mg, Mn, Mo, Na, P, Rb, S, Se, Sr, Tl, V.
[b] Calculated as three times the standard deviation of the blank.

Thirty-five nonradioactive elements were also determined by ICPMS and ICP-OES. Of these, titanium was the most abundant as it is used for alloying (0.75% as mentioned earlier), and some amounts of Al, Ca, Co, Cu, Ni, Pb, Si, Zn, and Zr, as seen in Table 2.8 (Trueman et al. 2004). Some of these traces were also found in DU fragments recovered from Kosovo (Pollanen et al. 2003).

Highlights: The tolerance for impurities and isotopic composition of DU is quite large compared with the very strict specifications for nuclear fuel or for the feed material of enrichment plants (in fact, we did not find such specifications for DU). After dissolution of the DU, the analytical techniques used are basically the same for all forms of uranium. In cases where DU is fabricated from reprocessed fuel, some additional tests are required to verify the radioactivity. One feature of DU is that it is alloyed with different elements in order to improve its physical properties (like hardness or density) without any concerns about the nuclear properties of the alloy.

2.5 SUMMARY

The industrial applications of uranium mainly arise from its major role in the NFC (front end, in the nuclear reactor, and back end) and in depleted uranium (DU kinetic penetrators, shielding and ballast for aircraft). The techniques used for prospecting uranium deposits vary according to the approach of the prospectors. Geologists seek promising minerals and geological unconformities, geochemists search for elevated uranium levels in water sources, botanists may look for certain vegetation patterns, physicists try to find radioactive signals through aerial or ground surveyor emanation of radon, and discovery of uranium deposits by remote imaging is also being advanced. The assay of the uranium content in the minerals is done by classical (gravimetry, titrimetry, and spectrophotometric methods) as well as by modern analytical methods (ICPMS, ICP-OES, nuclear techniques). The accompanying components in the mineral affect the ease of extracting the uranium, and the most suitable process for this, and therefore also need to be characterized by suitable analytical methods.

The rigorous specifications for nuclear grade materials that are used as nuclear fuel (mainly UO_2 and U metal or alloys) or as feed material for enrichment facilities (primarily UF_6) are described in great detail and require strict control. The focus of the analytical procedures is on impurities that affect the nuclear properties (mainly through absorption of neutrons), chemical properties (like corrosion resistance or those that may concentrate in the enrichment product), and physical and mechanical properties (like pellet strength, heat transfer). The isotopic composition of uranium plays an important role as the value of uranium strongly depends on the [235]U content.

At the back end of the NFC characterization of SNF or nuclear waste materials, focus on the separation and disposal of the radioactive materials—activation products and fission products—formed during irradiation of the fuel. The question of long-term storage of irradiated fuel elements or reprocessing them to separate the useful components (mainly uranium and plutonium) from the hazardous constituents has not yet been settled.

REFERENCES

Alonso, J.I.G., Thoby-Schulzendorff, D., Giovanonne, B. et al. (1994). Characterization of spent nuclear fuel dissolver solutions and dissolution residues by inductively coupled plasma mass spectrometry, *J. Anal. At. Spectrom.* 9, 1209–1215.

Aziz, A., Jan, S., Waqar, F. et al. (2010). Selective ion exchange separation of uranium from concomitant impurities in uranium materials and subsequent determination of the impurities by ICP-OES, *J. Radioanal. Nucl. Chem.* 284, 117–121.

B811. (2013). Standard specification for wrought zirconium alloy seamless tubes for nuclear reactor fuel cladding. West Conshohocken, PA: ASTM.

Barretto, P.M. (1981). Recent developments in uranium exploration, *IAEA Bull.* 23, 15–20.

Bevard, B.B., Wagner, J.C., Parks, C.V. et al. (2009). *Review of Information for Spent Nuclear Fuel Burnup Confirmation.* Oak Ridge, TN: ORNL.

Bhattacharjee, K., Banerjee, B., Bawitlung, L. et al. (2013). A study on parameters optimization for degradation of endosulfan by bacterial consotia isolated from contaminated soil, *Proc. Natl. Acad. Sci. India Sect. B: Biol. Sci.* 84, 657–667.

Burger, S., Riciputi, L.R., and Bostick, D.A. (2007). Determination of impurities in uranium matrices by time-of-flight ICP-MS using matrix matched method, *J. Radioanal. Nucl. Chem.* 274, 491–505.

C1022. (2010). Standard test methods for chemical and atomic absorption analysis of uranium ore concentrates. Conshohocken, PA: ASTM.

C1052. (2007). Standard practice for bulk sampling of liquid uranium hexafluoride. West Conshohocken, PA: ASTM.

C1075. (1997). Standard practices for sampling uranium-ore concentrates. West Conshohocken, PA: ASTM.

C1287. (2010). Standard test method for determination of impurities in nuclear grade uranium compounds by inductively coupled plasma mass spectrometry. West Conshohocken, PA: ASTM.

C1295. (2013). Standard test method for gamma energy emission from fission and decay products in uranium hexafluoride and uranyl nitrate solution. West Conshohocken, PA: ASTM.

C1334. (2010). Standard specification for uranium oxides with a ^{235}U content of less than 5% for dissolution prior to conversion to nuclear-grade uranium dioxide. West Conshohocken, PA: ASTM.

C-1347. (2008). Standard practice for preparation and dissolution of uranium materials for analysis. West Conshohocken, PA: ASTM.

C1413. (2011). Standard test method for isotopic analysis of hydrolyzed uranium hexafluoride and uranyl nitrate solutions by Thermal Ionization Mass Spectrometry. West Conshohocken, PA: ASTM.

C1431. (2010). Standard guide for corrosion testing of aluminum based spent nuclear fuel in support of repository disposal. West Conshohocken, PA: ASTM.

C1441. (2004). Standard test method for the analysis of refrigerant 114, plus other carbon-containing and fluorine-containing compounds in uranium hexafluoride via Fourier-Transform Infrared (FTIR) spectroscopy. West Conshocken, PA: ASTM.

C1462. (2013). Standard specification for uranium metal enriched to more than 15% and less than 20% ^{235}U. West Conshohocken, PA: ASTM.

C1477. (2008). Standard test method for isotopic abundance analysis of uranium hexafluoride by Multi-Collector, Inductively Coupled Plasma-Mass Spectrometry. West Conshohocken, PA: ASTM.

C1517. (2009). Standard test method for determination of metallic impurities in uranium metal or compounds by DC-Arc emission spectroscopy. West Conshohocken, PA: ASTM.

C1539. (2002). Standard test method for determination of technetium-99 in uranium hexafluoride by liquid scintillation counting. West Conshohocken, PA: ASTM.

C1561. (2003). Standard guide for determination of plutonium and neptunium in uranium hexafluoride by alpha spectrometry. West Conshohocken, PA: ASTM.

C1682. (2009). Standard guide for characterization of spent nuclear fuel in support of geologic repository disposal. West Conshohockcn, PA: ASTM.

C1689. (2008). Standard practice for subsampling of uranium hexafluoride. West Conshohocken, PA: ASTM.

C1703. (2013). Standard practice for sampling of gaseous uranium hexafluoride. West Conshohocken, PA: ASTM.

C696. (2011). Standard methods for chemical, mass spectrometric and spectrochemical analysis of nuclear grade uranium dioxide powders and pellets. West Conshohocken, PA: ASTM.

C761. (2011). Test methods for chemical, mass spectrometric, spectrochemical, nuclear, and radiochemical analysis of uranium hexafluoride. West Conshohocken, PA: ASTM.

C-787. (2011). Standard specifications for uranium hexafluoride for enrichment. West Conshohocken, PA: ASTM.

C888. (2008). Specifications for nuclear grade gadolinium oxide (Gd_2O_3) powder. West Conshohocken, PA: ASTM.

C922. (2008). Standard specifications for sintered gadolinium oxide-uranium dioxide pellets. West Conshohocken, PA: ASTM.

C967. (2013). Standard specifications for uranium ore concentrate. West Conshohocken, PA: ASTM.

C968. (2012). Test methods for analysis of sintered gadolinium oxide-uranium oxide pellets. West Conshohocken, PA: ASTM.

C996. (2010). Standard specifications for uranium hexafluoride enriched to less than 5% U-235. Conshohocken, PA: ASTM.

Campbell, L.W., Smith, L.E., and Misner, A.C. (2011). High energy delayed gamma spectroscopy for spent nuclear fuel assay, *IEEE Trans. Nucl. Sci.* 58, 231–240.

Chajduk, E., Bartosiewicz, I., Pyszynska, M. et al. (2013). Determination of uranium and selected elements in Polish dictyonema shales and sand stones by ICP-MS, *J. Radioanal. Nucl. Chem.* 295, 1913–1919.

Charlton, J.D. and Kotrappa, P. (2006). Uranium prospecting for accurate time efficient surveys of radon emissions in air and water with comparison to earlier radon and helium surveys. *Proceedings of the 2006 International Radon Symposium*, September 17–20, http://www.aarst.org/proceedings/2006/2006_01_Uranium_Prospecting_for_Accurate_Time-Efficient_Surveys_of_Radon_Emissions_in%20Air_Water_with_Comparison.pdf (accessed July 27, 2014).

Chaudhry, M.S., Qureshi, M.I., and Qureshi, I.H. (1978). Determination of uranium in ores using instrumental neutron activation analysis, *J. Radioanal. Chem.* 42, 427–434.

Chichester, D.L. and Sterbentz, J.W. (2011). A second look at neutron resonance transmission analysis as a spent fuel NDA. Idaho Falls, ID: Idaho National Laboratory, INMM 2011 INL/CON-11-20783.

DeGraw, J. (2012). June 23, 2012 Uranium contamination event and action level exceedence. Blind River, Ontario, Canada: Cameco Corporation. http://www.suretenucleaire.gc.ca/fra/pdfs/Regulatory_Action/2012/20120827-cameco-report.pdf (accessed July 27, 2014).

Dickson, B.L., Gulson, B.L., Snelling, A.A. et al. (1985). Evaluation of lead isotopic methods for uranium exploration, Koongarra Area, Northern Territory, Australia, *J. Geochem. Explor.* 24, 81–102.

Duval, J.S., Cook, B., and Adams, J.A.S. (1971). Circle of investigation of an airborne gamma-ray spectrometer, *J. Geophys. Res.* 76, 8466–8470.

EGADSNF. (2011). Spent nuclear fuel assay data for isotopic validation—State-of-the-art Report. Paris, France: OECD, NEA/NSC/WPNCS/DOC(2011)5.

Esteban, A., Cristallini, O., and Perrotta, J.A. (2008). UF6 sampling method using alumina. *29th ESARDA Annual Meeting, France, 2007 and 49th INMM Annual Meeting*. Nashville, TN, 2008, pp. 1–9, http://www.abacc.org.br/wp-content/uploads/2012/01/ESARDA_2007.pdf (accessed July 27, 2014).

Fajardo, Y., Ferrer, L., Gomez, E. et al. (2008). Development of an automatic method for americium and plutonium separation and preconcentration using a multisyringe flow injection analysis multipump flow system, *Anal. Chem.* 80, 195–202.

Feiveson, H., Mian, Z., Ramana, M.V. et al. (2010). *Spent Fuel from Nuclear Power Reactors: An Overview of a New Study by the International Panel on Fissile Materials*. Princeton, NJ: International Panel on Fissile Materials.

Grasty, R.L. (1979). Gamma ray spectrometric methods in uranium exploration—Theory and operational procedures. In Hood, P.J. (ed.), *Geophysics and Geochemistry in the search for Metallic Ores* (pp. 147–161). Ottawa, Ontario, Canada: Geological Survey of Canada, Economic Geology Report 31.

Gunther-Leopold, I., Kivel, N., Waldis, J.K. et al. (2003). Characterization of spent nuclear fuel by an online combination of chromatographic and mass spectrometric techniques. *Proceedings of the Seventh International Conference on Nuclear Criticality Safety*. Tokai, Japan: ICNC, October 2003, JAERI-Conf 2003-019.

Gunther-Leopold, I., Kivel, N., Waldis, J.K. et al. (2008). Characterization of nuclear fuels by ICP mass-spectrometric techniques, *Anal. Bioanal. Chem.* 390, 503–510.

Halevy, I., Ghelman, M., Yehuda-Zada, Y. et al. (2014). Radiation contamination estimation from micro-copters or helicopter airborne survey: Simulation and real measurements (pp. 54–57). *44th Journees des actinides Conference*. Ein-Gedi, Israel, May 2014.

Hightower, J.R., Dole, L.R., Lee, D.W. et al. (2000). Strategy for characterizing transuranics and technetium contamination in depleted UF6 cylinders. Oak Ridge, TN: ORNL/TM-2000/242.

Hitchen, A. and Zechanowitsch, G. (1980). Titrimetric determination of uranium in low grade ores by ferrous ion-phosphoric acid reduction method, *Talanta* 27, 383–389.

Hofman, G.L., Meyer, M.K., and Ray, A. (1998). Design of high density gamma-phase uranium alloys for LEU dispersion fuel applications. *Reduced Enrichment for Test Reactor Conference* (pp. 1–12). Sao Paolo, Brazil: Argonne National Laboratory.

Horwitz, E.P., Chiaizia, R., Dietz, M.L. et al. (1993). Separation and preconcentration of actinides from acidic media by extraction chromatography, *Anal. Chim. Acta* 281, 361–372.

Horwitz, E.P., Dietz, M.L., Chiarizia, R. et al. (1992). Separation and preconcentration of uranium from acidic media by extraction chromatography, *Anal. Chim. Acta* 266, 25–37.

IAEA. (2003). Guidelines for radioelement mapping using gamma ray spectrometry data. Vienna, Austria: IAEA-TECDOC-1363.

IAEA, OECD-NEA and IAEA. (2012). Uranium 2011: Resources, supply and demand. Vienna, Austria: IAEA.

IAEA341. (1992). Analytical techniques in uranium exploration and ore processing. Vienna, Austria: IAEA.

IAEA-NF-T-15. (2013). Advances in airborne and ground geophysical methods for uranium exploration, IAEA Nuclear Energy Series NF-T-15. Vienna, Austria: IAEA.

IAEA-TECDOC1363. (2003). Guidelines for radioelement mapping using gamma ray spectrometry data. Vienna, Austria: IAEA.

IAEA-TECDOC-1613. (2009). Nuclear fuel cycle information system—A directory of nuclear fuel cycle facilities. Vienna, Austria: IAEA.

IAEA-TECDOC-750. (1994). Interim guidance for the safe transport of reprocessed uranium. Vienna, Austria: IAEA.

Jordan, D.V., Orton, C.R., Mace, E.K. et al. (2012). Automated UF_6 cylinder enrichment assay: Status of the hybrid enrichment verification array (HEVA) project. POTAS stage II. Richland, WA: Pacific Northwest National Laboratory PNNL-21263.

Jovanovic, S.J. and Pan, P. (2013). Characterization of uranium in contaminated soil from port Hope, Ontario, Canada, *J. Nucl. Energy Sci. Power General Technol.* S1, 1–9.

Judge, E.L., Barefield, J.E., Berg, J.M. et al. (2013). Laser-induced breakdown spectroscopy measurements of uranium and thorium powders and uranium ore, *Spectrochim. Acta B* 83–84, 28–36.

Keegan, E., Richter, S., Kelly, I. et al. (2008). The provenance of Australian uranium ore concentrates by elemental and isotopic analysis, *Appl. Geochem.* 23, 765–777.

Khan, M.H., Warwick, P., and Evans, N. (2006). Spectrophotometric determination of uranium with arsenazo-III in perchloric acid, *Chemosphere* 63, 1165–1169.

Kilger, R. (2010). Criticality safety in the waste management of spent fuel from NPPs. Munich, Germany: Eurosafe Forum Organization, Eurosafe Homepage. http://www.eurosafe-forum.org/userfiles/2_03_Eurosafe2010%206_Kilger%20(GRS).pdf (accessed July 27, 2014).

Kim, Y.S., Han, B.Y., Shin, H.S. et al. (2012). Determination of uranium concentration in an ore sample using laser-induced breakdown spectroscopy, *Spectrochim. Acta B* 74–75, 190–193.

Kovbasenko, Y., Bilodid, Y., and Yeremenko, M. (2003). Comparative analysis of isotope composition of VVER-1000 spent fuel depending on their manufactory and operation conditions. *Seventh International Conference on Nuclear Criticality Safety* (p. 13). Tokai-Mura, Japan: ICNC2003.

Kraiem, M., Essex, R.M., Mathew, K.J. et al. (2013). Re-certification of the CRM 125-A UP2 fuel pellet standard for uranium isotope composition, *Int. J. Mass Spectrom.* 352, 37–43.

Kumar, R., Acharya, C., and Joshi, S.R. (2011). Isolation and analyses of uranium tolerant *Serratia marcescens* strains and their utilization for aerobic U(VI) bioadsorption, *J. Microbiol.* 49, 568–574.

Kumar, R., Nongkhlaw, M., Acharya, C. et al. (2013). Uranium (U)-tolerant bacterial diversity from U ore deposit of Domiasiat in north-east India and its prospective utilisation in bioremediation, *Microb. Environ.* 28, 33–41.

Lebrun, A. and Bignan, G. (2001). Nondestructive assay of nuclear low-enriched uranium spent fuels for burnup credit application, *Nucl. Technol.* 135, 216–229.

Lovley, D.R. and Phillips, E.J. (1992). Reduction of uranium by *Desulfovibrio desulfuricans*, *Appl. Environ. Microbiol.* 58, 850–856.

Madbouly, M., Nassef, M.H., Diab, A.M. et al. (2009). A comparative analysis of uranium ore using laser fluorimetric and gamma spectrometry techniques, *J. Nucl. Radiat. Phys.* 4, 75–81.

Mal'tsev, E.D. (1960). Determination of the optimum yield of enriched ore in radiometric enrichment of uranium ores, *Atomnaya Energiya* (translated) 8, 121–126.

Mandal, A., Biswas, A., Mittal, S. et al. (2013). Geophysical anomalies associated with uranium mineralization from Beldin Mine, South Purulia Shear Zone, India, *J. Geolog. Soc. India* 82, 601–606.

Mathew, K.J., Singleton, G.L., Essex, R.M. et al. (2013). Characterization of uranium isotopic abundances in depleted uranium metal assay standard 115, *J. Radioanal. Nucl. Chem.* 296, 435–440.

Minty, B. (1997). Fundamentals of airborne gamma-ray spectrometry, *AGSO J. Aust. Geolog. Geophys.* 17, 39–50.

Murty, B.N., Jagannath, Y.V.S., Yadav, R.B. et al. (1997). Spectrophotometric determination of uranium in process streams of a uranium extraction plant, *Talanta* 44, 283–295.

Moreno, J.M.B., Alonso, J.I.G., Arbore, P. et al. (1996). Characterization of spent nuclear fuels by ion chromatography-inductively coupled plasma mass spectrometry, *J. Anal. At. Spectrom.* 11, 929–935.

NEA. (2012). Spent nuclear fuel reprocessing flowsheet. Paris, France: OECD Nuclear Energy Agency, NEA/NSC/WPFC/DOC(2012)15.

Nyade, P.K., Wilton, D.H.C., Longerich, H.P. et al. (2013). Use of surficial geochemical methods to locate areas of buried uranium mineralization in the Jacque's Lake area of the Central Mineral Belt, Labrador, Canada, *Can. J. Earth Sci.* 50, 1134–1146.

Ollila, K. (2003). Air-oxidation tests with Gd-doped UO_2: Preliminary dissolution experiments with pre-oxidized Gd-doped UO_{2+x}. Olkiluoto, Finland: Posiva 2003-08.

Onishi, H. and Sekine, K. (1972). Spectrophotometric determination of Zr, U, Th and rare earths with arsenazo III after extrcation with theonyltrifluoroacetone and tri-n-octylamine, *Talanta* 19, 473–478.

Orton, C.R., Fraga, C.G., Christensen, R.N. et al. (2011). Proof of concept simulations of the multi-isotope process monitor: An on-line, non-destructive, near-real-time safeguards monitor for nuclear fuel reprocessing facilities, *Nucl. Instrum. Methods Phys. Res. A* 629, 209–219.

Orton, C.R., Fraga, C.G., Christensen, R.N. et al. (2012). Proof of concept experiments of the multi-isotope process monitor: An online, nondestructive, near-real-time monitor for spent nuclear fuel reprocessing facilities, *Nucl. Instrum. Methods Phys. Res. A* 672, 38–45.

Perkin-Elmer. (n.d.). Guide to the safe handling of radioactive materials in research. http://shop.perkinelmer.com/content/manuals/gde_safehandlingradioactivematerials.pdf (accessed July 27, 2014).

Peurrung, A.J. (1998). Predicting U-232 in uranium. Richland, WA: PNNL 12075.

Pitkin, J.A. and Duval, J.S. (1980). Design parameters for aerial gamma-ray surveys, *Geophysics* 45, 1427–1439.

Pollanen, R., Ikaheimonen, T.K., Klemola, S. et al. (2003). Characterisation of projectiles composed of depleted uranium, *J. Environ. Radioact.* 64, 133–142.

Priyadarshi, N. (2010). Geobotanical methods for prospecting uranium deposits. http://nitishpriyadarshi.blogspot.in/2010/09/geobotanical-methods-for-prospecting.html (accessed July 27, 2014).

Quemet, A., Brennetot, R., Chevalier, E. et al. (2012). Analysis of 25 impurities in uranium matrix by ICP-MS with iron measurement optimized by using reaction collision cell, cold plasma or medium resolution, *Talanta* 99, 207–212.

Rackley, R.I., Shockey, P.N., and Dahill, P.N. (1968). Concepts and methods of uranium exploration. Black Hills Area: South Dakota, Montana, Wyoming; *20th Field Conference
Guidebook* (pp. 115–124), http://archives.datapages.com/data/wga/data/069/069001/115_
wga0690115.htm (accessed July 27, 2014).

Radenkovic, M.B., Kandic, A.B., Vukanac, I.S. et al. (2007). Chemical and radiochemical
characterization of depleted uranium in contaminated soils, *Russ. J. Phys. Chem. A*
81, 1448–1451.

Renshaw, J.C., Halliday, V., Robson, G.D. et al. (2003). Development and application of an
assay for uranyl complexation by fungal metabolites, including siderophores, *Appl.
Environ. Microbiol.* 69, 3600–3606.

Richmond, M.S. (1965). Analysis of uranium concentrates at the National Bureau of Standards
260-8. Washington, DC: National Bureau of Standards.

Richter, S., Alonso, A., De Bolle, W. et al. (2005). Update on REIMEP-15: Isotopic ratios of
uranium in UF_6. Report EUR 21562 EN. Geel, Belgium: IRMM, JRC.

Richter, S., Kuhn, H., Truyens, J. et al. (2013). Uranium hexafluoride (UF6) gas source mass
spectrometry for certification of reference materials and nuclear safeguards measurements at IRMM, *J. Anal. At. Spectrom.* 28, 536–548.

Sahu, P., Mishra, D.P., Panigrahi, D.C. et al. (2013). Radon emanation from low-grade uranium ore, *J. Environ. Radioactivity* 126, 104–114.

Sarangi, S.A.K. (2000). Uranium and its measurement in ore by radiometric method, *J. Mines
Metals Fuels Annu. Rev.* 48, 264–266.

Sarka, A., Alamelu, D., and Aggarwal, S.K. (2009). Laser induced breakdown spectroscopy
for determination of uranium in thorium–uranium mixed oxide fuel materials, *Talanta*
78, 800–804.

Satyanarayana, K. and Durani, S. (2010). Separation and inductively coupled plasma optical
emission spectrometric (ICP-OES) of trace impurities in nuclear grade uranium oxide,
J. Radioanal. Nucl. Chem. 285, 659–665.

Sindelar, R. (2010). Materials performance and aging considerations for power reactor (PR) and
research reactor (RR) spent nuclear fuel in storage. *IAEA International Conference for
Management of Spent Fuel from Nuclear Power Reactors* (p. ppt). Vienna, Austria: IAEA.
http://www-ns.iaea.org/meetings/rw-summaries/vienna-2010-mngement-spent-fuel.asp
(accessed July 27, 2014).

Singhal, R.K., Sharma, P.K., Bassan, M.K.T. et al. (2011). Comparative determination of uranium in rock phosphates and columbite by ICP-OES, alpha and gamma spectrometry,
J. Radioanal. Nucl. Chem. 288, 149–156.

Smith, L.E., Anderson, K.K., Ressler, J.J. et al. (2010). Time-spectral analysis method for
spent fuel assay using lead slowing-down spectroscopy, *IEEE Trans. Nucl. Sci.* 57,
2230–2238.

Smith, L.E. and Lebrun, A.R. (2011). Design, modeling and viability analysis of an online uranium enrichment monitor, *IEEE Nuclear Science Symposium Conference Record N20-5*
(pp. 1030–1037). Vienna, Austria.

Souza, A.L., Cotrim, M.E.B., and Pires, M.A.F. (2013). An overview of the spectrometric
techniques and sample preparation for determination of impurities in uranium nuclear
fuel grade, *Microchem. J.* 106, 194–201.

Trueman, E.R., Black, S., and Read, D. (2004). Characterisation of depleted uranium (DU)
from an unfired CHARM-3 penetrator, *Sci. Total. Environ.* 327, 337–340.

Wang, T., Khangaokar, T., Long, W. et al. (2014). Development of a kelp-type structure module in a coastal ocean model to assess the hydrodynamic impact of seawater uranium
extraction technology, *J. Marine Sci. Eng.* 2, 81–92.

Whitten, W.B., Reilly, P.T.A., Verbeck, G. et al. (2004). Mass spectrometry of UF6 in a micro
ion trap. *Seventh International Conference Facility Operations—Safeguards Interface*
(pp. 1–6). Charleston, SC: Oak Ridge National Laboratory, TN.

WNA. (2012, December). Waste management: Overview. Retrieved from: http://www.world-nuclear.org/uploadedImages/org/info/Nuclear_Fuel_Cycle (accessed July 27, 2014).

WNA. (2013a, September). Processing of used nuclear fuel. Retrieved from: http://world-nuclear.org/info/Nuclear-Fuel-Cycle/Fuel-Recycling/Processing-of-Used-Nuclear-Fuel/#.Ul-dT8xBQ5g (accessed July 27, 2014).

WNA. (2013b, October). World Nuclear Association. Nuclear fuel fabrication. Retrieved from: http://world-nuclear.org/info/Nuclear-Fuel-Cycle/Conversion-Enrichment-and-Fabrication/Fuel-Fabrication/#.UlzfCVAsAfg (accessed July 27, 2014).

Wogman, N.A., Brodzinski, R.L., and Van Middlesworth, L. (1977). Radium accumulation in animal thyroid glands a possible method for uranium and thorium prospecting, *J. Radioanal. Chem.* 41, 115–125.

Wolf, S.F., Bowers, D.L., and Cunnane, J.C. (2005). Analysis of high burnup spent nuclear fuel by ICP-MS, *J. Radioanal. Nucl. Chem.* 263, 581–586.

Zoellner, T. (2010). *Uranium: War, Energy and the Rock That Shaped the World.* London, U.K.: Penguin Books.

3 Determination of Uranium in Environmental Samples

Uranium is a radioactive element that occurs naturally in low concentrations (a few parts per million) in soil, rock, and surface and groundwater.

3.1 INTRODUCTION AND GENERAL OVERVIEW

Uranium is a naturally occurring element that is abundant in soil, plants, freshwater, and seawater and can also be found in various biota samples. Its concentration in soil and plants varies widely between geographical locations according to the composition of the minerals in the environment and in natural water sources. In addition to natural sources, incidental releases of uranium that originate from anthropogenic activities, mainly those involving the uranium nuclear fuel cycle, affect its abundance in the environment. The International Atomic Energy Agency (IAEA) has published a number of technical reports in which many facets of the behavior of uranium in the environment, particularly transfer factors between environmental media, were described and discussed (TECDOC-1616 2009; IAEA 2010). Anthropogenic activities that may cause discharge of uranium compounds include uranium mining (as well as secondary uranium deposits such as phosphate and gold mines), uranium conversion facilities and isotope enrichment plants, nuclear fuel fabrication, nuclear power plants, reprocessing facilities, and nuclear waste treatment processes and their repositories. A schematic presentation, from the inaugural lecture delivered by Dr. Frank Winde in South Africa, is shown in Figure 3.1 (Winde 2013). The release of uranium from natural and anthropogenic sources into the environment is followed by its transport through air, water, and soil that could lead to human exposure through the food chain, drinking water, and inhalation from air.

Needless to say that uranium originating from the nuclear fuel cycle operations is frequently not of natural isotopic composition. However, there are also variations in the isotopic composition of naturally occurring uranium compounds, as mentioned in Chapter 1, with minute differences in the $^{235}U/^{238}U$ ratio (Hiess et al. 2012), but significant variations were observed in the $^{234}U/^{238}U$ ratio, mainly in freshwater sources from different origins (see Frame 1.2).

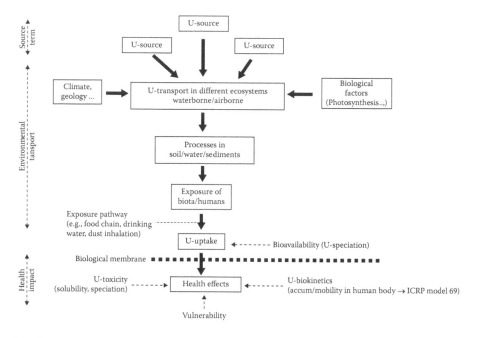

FIGURE 3.1 A schematic presentation of the transport of uranium from the natural and anthropogenic sources, through the environment leading to human exposure. (From Winde, F., *Uranium pollution of water: A global perspective on the situation in South Africa, Inaugural Lecture Presented at the North-West University*, Vaal Triangle Campus, South Africa, February 22, 2013. With permission.)

3.1.1 TRANSPORT OF URANIUM IN THE ENVIRONMENT

Once uranium is released into the environment through natural or anthropogenic processes, it can be transported by water, air, or soil until it is either immobilized by chemical reactions and physical interactions or will remain in a mobile form for an extended period. The following section is broadly based on the works of Winde on the environmental and health effects of uranium-bearing gold mine tailings in South Africa and can serve as a demonstration of the complexity of trying to follow the environmental fate of uranium (Winde 2013).

While many heavy metals are soluble only under acidic conditions and precipitate in alkaline or neutral conditions, the geochemical mobility of uranium is complex because there are soluble compounds, including soluble complexes, over a wide pH range. Under reducing conditions, like those pertaining to rich organic matter, highly soluble hexavalent uranium ions are converted to tetravalent ions that may lead to deposition of insoluble oxides. This process may be reversed if the sediment is exposed to atmospheric oxygen. Dissolved uranium may also be removed by co-precipitation, for example, in iron-rich water iron hydroxide may be deposited with uranium if the pH increases. The deposition–dissolution processes may be repeated if the conditions change periodically (seasonably or even diurnally). For example, it was found co-precipitation decreases at night, leading to elevated dissolved

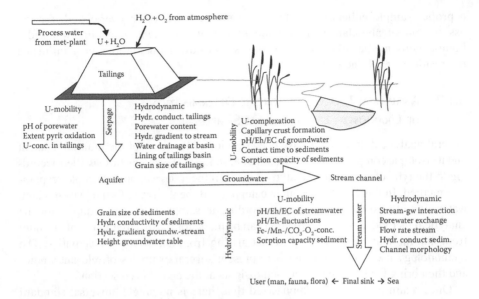

FIGURE 3.2   Factors and processes governing the physical transport and chemical mobility of dissolved uranium from gold mine tailings along the aqueous pathway. (From Winde, F., *Uranium pollution of water resources in mined-out and active goldfields of South Africa—A case study in the Wonderfonteinspruit catchment on extent and sources of U-pollution and associated health risks, International Mine Water Conference*, Pretoria, South Africa, 2009, pp. 772–782. With permission.)

uranium levels in water while in the daytime the dissolved uranium level decreases. The deposition of calcite ($CaCO_3$), and co-precipitation of uranium, depends on the CO_2–$CaCO_3$ equilibrium that in turn is affected by fluctuations in the CO_2 atmospheric concentration and by the water temperature. Thus, interpretation of uranium concentration measurements should be carried out with caution due to these fluctuations (Winde 2002). Figure 3.2 shows a schematic diagram of the aqueous transport processes of uranium from gold mine tailings in South Africa (Winde 2009).

Soil contaminated with uranium-bearing particles (mainly uraniferous tailings) is found in the waste of gold mines in South Africa. In some cases, the tailings were pumped into natural caves and sinkholes or accumulated in tailings dams (in South Africa mining, as explained by Winde, this refers to hydraulically deposited waste dumps in the shape of pyramids that are stable once excess slurry water drains out). The tailings deposits may be eroded and the particles transported by wind, especially during the dry winter season, and affect urban residents or they add contamination to water sources (Winde 2013). Unlike polluted water that may travel over quite great distances, there is usually a gradient of uranium-bearing particle deposition with distance.

The processes described earlier can serve as an example of the complexity of following the environmental fate of uranium. There are many factors that affect the transport of uranium in the environment so that even an accurate analytical measurement of a given sample at best represents that sample. Thus, devising a proper sample collection strategy is a prerequisite, often overlooked, for the assessment of abundance of uranium in the environment. Even in supposedly homogenous media, like water reservoirs, periodical fluctuations may influence the results as explained earlier.

3.1.2 ANALYTICAL PROCEDURES FOR THE DETERMINATION OF URANIUM IN ENVIRONMENTAL SAMPLES

Several analytical procedures have been developed for assay of the uranium content and its isotopic composition in environmental samples. These can be classified according to the type of sample, the analytical measurement device, or the sample preparation method. In this chapter, the main category will be the type of sample (soil, plant, water, and others) with subdivision according to sample treatment required for the analytical method. Another distinction can be made for the differentiation of uranium from natural sources as opposed to that arising from anthropogenic activities. The methodology used in this chapter is to present several examples of relevant studies and then bring forth the highlights and insights at the end of each section.

Once again, it should be emphasized that there is no single universal standard method, or a global organization that dictates procedures, for the determination of uranium in environmental samples. For example, the US-EPA has a method for acid digestion of soil, sediment, and sludge samples for the analysis of several elements (though not uranium) by atomic absorption (EPA method 3050B), or a microwave-assisted acid digestion of siliceous and organically based matrices that is also intended for ICPMS analysis (EPA method 3052) and lists no less than 15 methods for the determination of uranium in drinking water. The ASTM has a standard test method for the analysis of total and isotopic uranium and total thorium in soil samples by inductively coupled plasma mass spectrometry (C1345 2008). As mentioned earlier, the approach adopted here is to present some examples of studies for each topic and then attempt to derive some general insights.

First, we would like to mention a number of general reviews and overviews that are not method specific but provide excellent background material for a better understanding of the analytical determination of uranium in environmental samples.

A review of the environmental chemistry of uranium is a good starting point (Zavodska et al. 2008). Some of the key points that are relevant to the determination of uranium in environmental samples based on this review are summarized later. As noted in Chapter 1, the fact that about 5% of all known minerals contain uranium means that a great diversity in its behavior is expected. This is influenced mainly by the valence state of uranium: in a reducing environment, the U(IV) form is dominant, but in an oxidizing environment, the more soluble and mobile U(VI) state is prevalent. Another feature that affects the environmental fate of uranium, especially in the U(VI) state, is its tendency to form complexes with carbonates, sulfates, phosphates, and chloride ions and the effect that the pH has on the stability of these complexes. High-molecular-weight organic substances (humic acids) strongly bind uranium, especially under acidic conditions, so their presence may inhibit the migration of uranyl ions, that is, immobilize the uranium compounds. Iron mineralogy affects the redox conditions, so its presence and oxidation state also play an indirect role in the mobilization of uranium. In addition, temperature, pressure, and solution composition (pH, redox potential, ionic strength, and complex-forming ligands) determine the aqueous chemistry of uranium (Zavodska et al. 2008). A table that summarizes the analytical methods and typical detection limits that are used for the determination of uranium in a variety of environmental samples was also presented in that review and is reproduced here as Table 3.1 (Zavodska et al. 2008). Note the large variety of the environmental samples and of the analytical methods used for the determination of uranium in those types of samples.

A comprehensive overview of preconcentration techniques for uranium (VI) and thorium (IV) prior to analysis was published (Prasada Rao et al. 2006). The multitude of off-line techniques that were reviewed includes liquid–liquid extraction, liquid membranes, ion exchange, extraction chromatography, flotation, absorptive electrochemical accumulation, solid-phase extraction (SPE), and ion imprinting polymers. In addition, online preconcentration methods for uranium, thorium, and mixtures of the two are also briefly surveyed. This overview includes over 100 references and is a good source for finding a suitable preconcentration technique with regard to the enrichment factor, retention and sorption capacity, method validation, and types of real samples. The review article focused on samples in which the uranium was already in solution so that digestion procedures for solid samples were not discussed (Prasada Rao et al. 2006).

Another relevant general review summarizes the knowledge on the behavior of ^{238}U series radionuclides in soils and plants and is intended to provide a comprehensive source of information for environmental impact studies (Mitchell et al. 2013). The summary of the data on plant to soil concentration ratios that depends on the specific soil and type of plant and the distribution of uranium within the parts of the plant is especially important. The dependence of the sorption of dissolved uranium compounds on the type of soil (like the clay content) and the parameters mentioned earlier (pH, complex forming agents, anions, presence of iron, organic matter, etc.), based mainly on studies of the K_d (distribution factor) of spiked soil samples, is discussed. It is noted that in general the uranium concentration in plants is several orders of magnitude lower than in soil, but some plants may efficiently absorb uranium and translocation within the plant is quite common (Mitchell et al. 2013). These features, and especially the soil-to-plant transfer factors, will be discussed in Section 3.4 that deals with the uranium content in plants and soil and the relation between them.

The methods that can be used for the determination of uranium isotopes in environmental samples were briefly surveyed (Borylo 2013). The methods were divided into radiometric methods that include different techniques of neutron activation analysis, liquid scintillation and alpha spectrometry and nonradiometric methods that include ICPMS and its variations and TIMS, and methods for assaying the amount of uranium (without isotopic composition) like ICP-OES, atomic absorption, and x-ray-based methods. The types of environmental samples listed were plants, mosses, water, soil, phosphates, sediments, and surface water. Sample preparation procedures were mentioned only in passing, stating that digestion (or mineralization) with concentrated acids (HNO$_3$, HCl, and HF) should be followed by anion exchange on suitable resins. For alpha spectrometry, uranium was purified and deposited on steel disks (Borylo 2013).

TABLE 3.1
Summary of the Analytical Methods Used for the Determination of Uranium in a Variety of Environmental Samples

Sample Matrix	Analytical Method	Detection Limit	Accuracy
Air	ICPMS (total U)	0.1 μg dm^{-3}	—
	α-Spectroscopy	5.55 * 10^{-4} Bq	—
	INAA	0.03 μg on filter	—
Rainwater	α-Spectroscopy (isotope quant.)	0.02 dpm dm^{-3} for ^{238}U in solution	68%
Drinking water	Fluorimetry (total U)	<20 μg dm^{-3}(direct) 0.1 μg dm^{-3}(pure)	104% (cleaned)
	Gross α-counting	0.037 Bq dm^{-3}	92.6%
	LIF	0.08 μg dm^{-3}	100% <1 μg dm^{-3}
Natural waters	Spectrophotometry	0.1 μg dm^{-3}	100% <1 μg dm^{-3}
Water	Fluorimetry (total U)	5 μg dm^{-3}	117.5% at 6.3 μg dm^{-3} 97.7%–108%
	α-Spectroscopy (isotope quant.)	0.02 dpm dm^{-3} for ^{238}U in solution	>80%
	NAA (total U)	3 μg dm^{-3}	103% (average)
	Pulse laser phosphom.	0.05 ppb	
Groundwater	FI-ICPMS (isotope quant.)	3 ng dm^{-3} for ^{238}U	±1.8%
	Spectrometry (total U)	1.2 μg dm^{-3}	—
Water and waste	ICPMS (total U)	0.1 μg dm^{-3}	105%–110%
Seawater	XRF (total U)	0.56–0.64 μg dm^{-3}	—
	Cathodic stripping voltammetry (total U)	0.02–0.2 nmol dm^{-3}	—
Soil, sediment, and biota	α-Spectroscopy (isotope quant.)	0.3 μg/sample	67%
Minerals	LIF	—	—
Building materials and lichens	α-Spectroscopy (isotope quant.)	0.3 μg/sample	54%–73%
Vegetation	ICPMS (total U)	0.1 μg dm^{-3}	—
	LIF (total dissolved U)	0.05 mg kg^{-1} plant ash	—
Process water	LIF (total dissolved U)	0.01 μg dm^{-3}	—
	Ion chromatography, spectrophotometric determination of U(VI)	0.04 mg dm^{-3}	—
Rocks, minerals, biologic materials	Spectrophotometric	0.062 mg dm^{-3} with back extraction	99%–103%
Coal ash	ICP-AES	29 μg dm^{-3}	98%
Sediment and pore water	ICPMS	40 pg dm^{-3}	99%
Field survey	Scintillation detector	200–500 dpm/100 cm^2	—

Source: Zavodska, L. et al., Environmental chemistry of uranium, HU ISSN 1418-7108: HEJ Manuscript no.: ENV-081221-A, pp. 1–19, 2008. With permission.

Last but not least, an excellent comprehensive document that covers practically all facets of environmental behavior of uranium was published by the Canadian Council of Ministers of the Environment (Environment 2007). The chemical and physical properties of uranium were reviewed and its distribution in the environment and bio-accumulation in various flora and biota were discussed. For example, guidelines for the permissible uranium concentration in soil were set according to the intended land use. For agricultural use and commercial land use, the maximum uranium concentration was 33 mg kg^{-1}, for residential and parkland uses it was 23 mg kg^{-1}, and a value of 300 mg kg^{-1} was set for industrial land use. This document also contains many tables that summarize the toxicological effects of uranium on humans and the uranium content in several food products, vegetation, soil, water, etc. In addition, a summary of the analytical methods that are used for the determination of uranium in a variety of environmental samples, very similar to Table 3.1, is given (Environment 2007).

3.1.3 ANALYTICAL METHODS FOR THE DETERMINATION OF URANIUM

The analytical methods used for the determination of uranium in environmental samples are basically the standard methods reviewed in brief in Chapter 1. The main differences are in the sample preparation procedures required for the analysis of the variety of environmental samples that include soil of different types, sediments, diverse types of vegetation, water from different sources with a wide range of acidity, salinity, suspended matter, etc. In addition, the environmental samples may include airborne particulate matter, vapors, and gases, as well as special samples involved in the food chain that may affect humans. Finally, the interplay of uranium (and other contaminants) between the environmental compartments—for example, the transfer factors of uranium from soil-to-plant or from vegetation to food products (e.g., free-range grazing cattle) are also part of the media that need to be characterized.

A general overview of atomic spectrometric techniques, including atomic absorption spectroscopy (FAAS and ETAAS), atomic emission spectrometry (ICP-OES), and mass spectrometry (ICP-MS), for the determination of uranium was published (Santos et al. 2010). The advantages and limitations of each technique were discussed and compared and the complexity and costs were also considered. In addition, use of preconcentration and separation to improve performance was also described.

As seen in the specific examples given later, numerous analytical methods are used for the determination of uranium in environmental samples. The most popular among them are ICPMS and alpha spectrometry, but neutron activation analysis, gamma spectrometry, and XRF are often deployed and even simple spectrophotometric (like colorimetric aresnazo-III) techniques are sometimes still used. For the precise determination of total uranium and its isotopic composition, isotope dilution (ID) methods can be used. One example is a comparison of ID-TIMS and ID-SIMS for isotope ratios in soil standards where two separation and preconcentration chromatographic techniques were also compared (Adriaens et al. 1992).

3.2   URANIUM CONTENT IN SOIL

The uranium content in soil can be determined directly by some analytical methods that are mainly based on nuclear techniques (variations of neutron activation analysis, gamma spectrometry, x-ray fluorescence, or laser-ablation ICPMS), but the common, popular, and more accurate methods require digestion and dissolution of the entire soil sample or at least rely on leaching the uranium out of the sample matrix. In principle, the methods used for assaying uranium in minerals (see Chapter 2) are also suitable for soil characterization, but uranium is usually present in the latter only as a low-level impurity, usually below 100 µg U g^{-1}. We shall first overview the procedures deployed for the treatment of soil samples prior to analysis and refer to the analytical devices used for the measurement of the uranium content and isotopic composition in these studies.

Unlike the situation with soil samples, it should be noted that the uranium content in sediments depends on many factors such as the organic matter content, the grain composition, and size distribution, as well as their chemical affinity to uranium. In some cases, the uranium compounds coat the grain surface and may be washed away, leading to an erroneous estimate of the uranium content.

3.2.1   DIGESTION AND SAMPLE PREPARATION

Digestion of soil samples for analytical purposes can be carried out in closed vessels (usually assisted by a microwave oven) or open crucibles (either as acid digestion or with a flux—mixture of salts—to *open out* the minerals). The details vary considerably and depend on the type of soil, the availability of proper equipment, and the analytical technique that will subsequently be used for the measurement, as demonstrated by the examples discussed later. After discussing some specific examples, we will try to derive the general guidelines and insights for the treatment of soil samples for the determination of the uranium content.

Three digestion procedures to prepare soil samples for alpha spectrometry were described and compared (Amoli et al. 2007). The soil samples were first air dried and sieved by a 2 mm mesh to remove stones and plant root fragments and then dried at 110°C for 24 h. For hot-plate open vessel acid digestion, the dried samples were then refluxed in concentrated nitric acid for 4 h at 150°C and the solution was filtered after cooling. An alternative procedure involved placing the dried samples in a closed vessel with concentrated nitric acid that was then irradiated in a microwave oven for 30 min at 120 psi and filtered. A third alternative placed the dried samples in a mixture of HF and HNO_3 for 24 h prior to the microwave irradiation procedure. This allows the gases formed in the initial digestion stage to escape, thus preventing the buildup of high pressures in the closed vessel in the microwave oven. In all cases, the filtrate was evaporated to dryness and the residue dissolved in 2.5 M HNO_3 and then passed through a TRU-Spec® (Eichrom) extraction chromatography column. The uranium was eluted, evaporated to dryness, dissolved in 9 M HCl and purified on an anion-exchange (Bio-Rad AG1x8) column, and prepared for the alpha spectrometric analysis by deposition on stainless steel disks. It was noted that in general soil samples may contain organic carbon and other impurities that affect the recovery of uranium and the analytical performance. The main conclusion of the study was that the microwave digestion procedure with HF-HNO$_3$ is the preferred method for this type of sample preparation (Amoli et al. 2007).

A study of the uranium content in soil samples in India used acid digestion in open vessels for UV fluorimetry determination (Srivastava et al. 2013). The samples were first dried at 40°C and ground to pass through a 300 μm sieve. Organic matter was destroyed by incineration at 600°C for over 6 h and then the ashed samples were digested in 1:1 HNO$_3$:HF mixture for 6 h. After evaporation to dryness, HClO$_4$ was added to complete the removal of organic matter and then concentrated HNO$_3$ was added. Once again the sample was evaporated to dryness and the residue dissolved in 4 M HNO$_3$ and extracted with ethyl acetate. The solution was evaporated to dryness and digested with concentrated HNO$_3$ and evaporated to dryness yet again. Finally, the residue was dissolved in dilute HNO$_3$ and the uranium content was determined by UV fluorimetry.

A different procedure for digestion of soil, rock, and sediment samples involved adding a ^{232}U spike and closed vessel microwave digestion in a 5:5:3 mixture of concentrated acids HNO$_3$:HCl:HF. The temperature and pressure were gradually increased to 210°C and 320 psi for 1 h (Popov 2012). The uranium was co-precipitated with iron hydroxide and the precipitate was filtered and dissolved in HCl. The Fe^{+3} was reduced to Fe^{+2} by ascorbic acid and then liquid–liquid extraction was used to bring the uranium into the organic phase (triisooctylamine/xylene), followed by stripping with 0.2 M HCl. The aqueous phase was evaporated to dryness and incinerated at 550°C for 3 h to destroy traces of organic compounds. The cooled residue was dissolved in 12 M HCl, evaporated to dryness, and dissolved once again in 7.5 M HCl. The iron was removed from the aqueous phase by repeated extractions with diisopropyl ether and then evaporated to dryness. The residue was dissolved with nitric acid and then uranium sulfate salts were precipitated by adding H$_2$SO$_4$ and NaOH. The salt was then dried and incinerated for 2 h at 550°C to remove excess sulfuric acid as SO$_3$. Finally, the uranium was dissolved in 0.05 H$_2$SO$_4$ and electrodeposited on a stainless steel disk for alpha spectrometry (Popov 2012). This method used liquid–liquid extraction rather than chromatographic resins to purify, separate, and preconcentrate the uranium from the environmental sample matrix.

A different approach to determine uranium and plutonium in soil samples was based on fusion with borates (Figure 3.3). Samples were first dried at 110°C, ignited at 600°C and then isotopic spikes (^{236}U and ^{242}Pu) were added. A eutectic mixture of 80% lithium metaborate and 20% lithium tetraborate was mixed with the pre-ignited soil sample and placed in a Pt-Au (95:5) dish in a furnace at 1200°C for 30 min. The melt was swirled periodically and then the melt and dish were (carefully) transferred to a beaker with distilled water. The platinum dish and cover were removed and the mixture was acidified with concentrated nitric acid. Polyethylene glycol (PEG) was added as a flocculating agent and the mixture was stirred for 4 h at 40°C. The silica and excess boric acid that remain in the mixture and form a colloidal solution must be removed and this was done by filtration under suction. The uranium and plutonium were then separated by use of a chromatographic extraction column (UTEVA®, Eichrom)—a multistep procedure that will not be described here. The uranium was

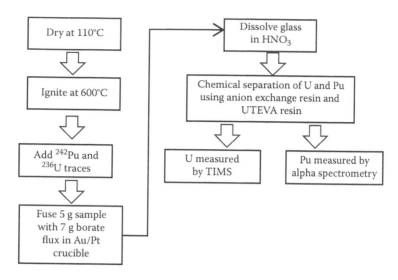

FIGURE 3.3 Schematic flow chart of the borate fusion method for the determination of uranium and plutonium in soils samples. (Adapted from Croudace, I. et al., *Anal. Chem. Acta*, 371, 217, 1998. With permission.)

measured by TIMS and the plutonium by alpha spectrometry (Croudace et al. 1998). The authors critically discuss this borate fusion method in comparison to an acid digestion method (leaching with *aqua regia*) and another fusion procedure that uses fluoride-pyrosulfate for the analysis of uranium in two NIST CRMs (Rocky Flats soil No. 1 and Ocean sediment). The comparison is shown in Table 3.2 and evidently the recovery and reproducibility of the acid leaching method is inferior to the two fusion methods (Croudace et al. 1998).

Another variation on the acid digestion procedure used a mixture of hydrofluoric acid, nitric acid, and perchloric acid in a sealed Teflon bomb heated at 150°C for 4 h

---

TABLE 3.2

Comparison of the Three Methods: *Aqua Regia* Leaching, Fluoride–Pyrosulfate Fusion, and Borate Fusion for the Determination of Uranium [Bq kg⁻¹] in NIST CRM Soil Samples

Sample	Double *Aqua Regia* Leach	Fluoride–Pyrosulfate Fusion	Borate Fusion
NIST 4353 (Rocky Flat soil)	20.60 ± 0.09	39.78 ± 0.14	37.90 ± 0.16
NIST 4353 (Rocky Flat soil)	20.70 ± 0.09	40.01 ± 0.19	38.32 ± 0.22
NIST 4357 (marine sediment)	6.29 ± 0.01	15.15 ± 0.01	14.46 ± 0.03
NIST 4357 (marine sediment)	6.78 ± 0.01	15.24 ± 0.02	14.26 ± 0.03

Source: Adapted from Croudace, I. et al., *Anal. Chem. Acta*, 371, 217, 1998. With permission.

Note: Reported errors are at the 95% uncertainty limit and are based on counting statistics for alpha spectrometry or the calculated errors from 5 blocks of 10 measurements on the TIMS.

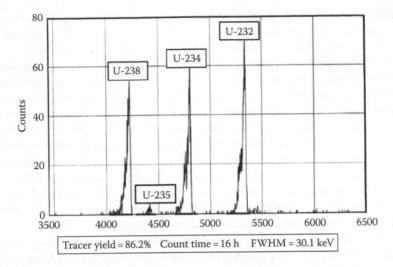

FIGURE 3.4 Alpha spectrum of uranium (with a ^{232}U spike) after separation from other actinides in a soil sample. (From Maxwell, S.L. et al., *Anal. Chim. Acta*, 701, 112, 2011. With permission.)

for the determination of uranium, americium, plutonium, and thorium in soil samples by alpha spectrometry (Serdeiro and Marabini 2004). Isotopic spikes (^{232}U, ^{242}Pu, ^{243}Am) were added as tracers and after digestion the samples were evaporated to dryness and then dissolved in nitric acid. Ferrous sulfamate and ascorbic acid were added and then separation was carried out on UTEVA® (Eichrom) chromatographic column followed by subsequent separation and purification with a TRU® (Eichrom) column. The actinides were electroplated of stainless steel disks in ammonium sulfate media at pH = 2.5.

A rapid method for the determination of actinides in samples collected from bricks and concrete based on addition of a ^{232}U spike, fusion, chromatographic separation, and alpha spectrometry was developed (Maxwell et al. 2011). The alpha spectra shown in Figure 3.4 for the uranium isotopes clearly indicate that separation and purification were very efficient.

A similar microwave digestion preparation procedure was used to determine the uranium content and its isotopic composition in soil samples collected in the vicinity of the Tokai-Mura plant in Japan after the criticality accident in 1999. The samples were oven-dried at 80°C to a constant weight and then powdered. Samples were placed in sealed Teflon pressure decomposition vessels with HNO_3, HF, and $HClO_4$ and digested in a microwave oven, then evaporated to dryness and dissolved in 2% HNO_3 prior to ICPMS analysis (Yoshida et al. 2001). This analytical method did not require separation on chromatographic columns.

A simplified approach to treatment of soil sample was used in a study of samples collected from the relocation zone in Belarus (4–60 km from Chernobyl). Sample collection was carried out with a coring device up to a depth of 20 cm (Boulyga and Becker 2002; Boulyga et al. 2002). The soil samples were dried to a constant weight at 105°C ± 5°C, homogenized and ashed at 600°C ± 50°C for 1 h. Vegetation was incinerated separately at 550°C ± 50°C for 2 h and combined with the soil samples. The ashed samples were leached with 8 M HCl and then passed through an anion-exchange column with AV-17 resin. Uranium was eluted with 8 M HNO_3 and further purified by extraction with diethylether. Accurate isotopic analysis, particularly for the $^{236}U/^{238}U$ ratio, was carried out by three types of ICPMS devices (a sector-field instrument, a quadrupole-based device with an optional hexapole collision cell, and a multicollector device with an ultrasonic nebulizer). The detection limits of the $^{236}U/^{238}U$ ratio for the three devices were reported as $5 * 10^{-6}$, $6 * 10^{-7}$ and $<3 * 10^{-7}$, respectively (Boulyga and Becker 2002). The same sample preparation methodology was used to determine burn-up of irradiated reactor fuel based on the uranium isotopic composition found in soil samples in Belarus (Mironov et al. 2005). The extraction of uranium from the incinerated soil samples with 9 M HCl supposedly preferably leaches the uranium of Chernobyl fuel origin (anthropogenic source) relative to the hexavalent natural (cosmogenic source) uranyl ions. The isotopic composition was determined with a sector-field ICPMS with a low-flow microconcentric nebulizer (MCN) that was equipped with membrane desolvator (Aridus®, Cetac). The resultant burn-up estimate was quite consistent with a value of ~9.4 MWd/kg U for soil samples collected 7–24 km from the power plant (Mironov et al. 2005).

Fractionation with a series of different extraction solutions is used sometimes to differentiate between the uranium species in soil samples. There are several variations of the fundamental procedure (Schultz et al. 1998). One of the schemes involves a five-stage process (Lotfy et al. 2012). In this study, the method used for the determination of the uranium content in each fraction was a low background alpha/beta counting system (LB-α/β-CS). In the first stage, acid soluble uranium was extracted from the soil sample with 1 M HNO_3 by sonication followed by centrifugation and the liquid phase was retained for analysis. The residue was then extracted with 1 M $MgCl_2$ and centrifuged for the determination of exchangeable cations. Carbonate-bound uranium was then extracted from the residue with 1 M acetic acid–sodium acetate buffer solution at pH = 5. Then uranium bound to iron oxide and manganese oxide was extracted with 0.04 M $NH_2OH \cdot HCl$ in 25% acetic acid. In the fifth stage, uranium bound to organic matter was extracted with 0.02 M HNO_3 and 30% H_2O_2 at 85°C followed by an additional aliquot of 30% H_2O_2. Finally, ammonium acetate in 20% HNO_3 was added and the sample was diluted with deionized water for the determination of the residual uranium. The results for the uranium extractable fraction in two types of Egyptian soils (sand and clay) with and without addition of 200 mg kg^{-1} as a spiked solution were analyzed. In the original sample, the water-soluble fraction was very low (0.15% and 0.25%), the exchangeable fraction was 7.41% and 5.23%, the carbonate associated uranium was slightly higher at 9.76% and 7.22%, the Fe-Mn-oxide fraction and organic-bound uranium together constituted 37% and 30%, and the residual fraction was the largest at 45.63% and 57.42%. The addition of the uranium spike (called treated soil in the article) slightly changed the water-soluble fraction, decreased the exchangeable fraction, and had some small effects on the distribution of uranium in the other fractions (Lotfy et al. 2012).

A slightly different procedure for the treatment of soil samples and determination of the uranium fractionation was carried out for the study of soil samples in Slovenia (Strok and Smodis 2010). The samples were first dried in an oven at 80°C to

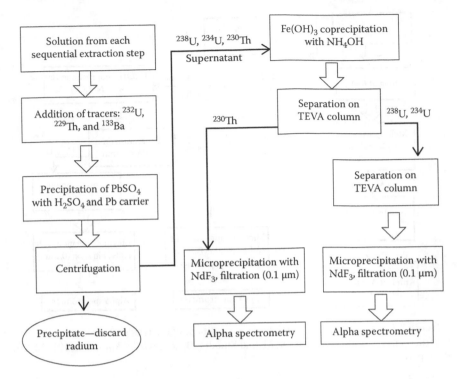

FIGURE 3.5 Schematic presentation of the radiochemical separation procedure used for the determination of radionuclides by alpha spectrometry. (From Strok, M. and Smodis, B., *J. Environ. Radioact.*, 101, 22, 2010. With permission.)

a constant weight and then sieved through a 2 mm mesh. Sequential extraction steps included exchangeable fraction (0.4 M $MgCl_2$ at pH = 5), organic matter (5%–6% NaOCl at pH = 7.5), carbonates (1 M sodium acetate in 25% acetic acid at pH = 4), oxides of Fe/Mn (0.04 M $NH_2OH \cdot HCl$ at pH = 2), and finally digestion of the residue (NaOH fusion followed by dissolution with $HNO_3/HCl/HF/HClO_4$). All steps were carried out at room temperature, except for the organic matter step at 96°C and the NaOH fusion that was performed at 900°C. Finally, the alpha spectrometry was measured after deployment of the radiochemical separation procedure shown schematically in Figure 3.5 (Strok and Smodis 2010). The procedure for the determination of radionuclides in the residue remaining at the end of the fractionation procedure described earlier is shown in Figure 3.6 (Strok and Smodis 2010). The same procedure was used in a later study to measure the radionuclide content in soil and plants and to determine the soil-to-plant transfer factors (Strok and Smodis 2013).

The alpha activity of uranium in soil samples from Slovakia was determined after a relatively simple sample preparation procedure (Donoval and Matel 2001). The soil samples were dried, ground, and ashed at 550°C, leached with HNO_3 + HCl and HNO_3 + HF. This was followed by solvent extraction, chromatographic separation of uranium from thorium, deposition with NdF_3 and alpha spectroscopy for the determination of the $^{238}U/^{234}U$ ratio.

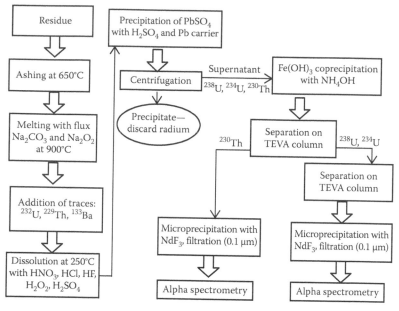

FIGURE 3.6 Schematic presentation of the treatment of the residue remaining at the end of the fractionation study. (From Strok, M. and Smodis, B., *J. Environ. Radioact.*, 101, 22, 2010. With permission.)

In an investigation of uranium-containing sediments, the samples were air-dried and digestion was carried out with a 5:2 mixture of concentrated HNO_3:H_2SO_4 in a microwave oven at 496 K and P = 2.6 MPa for one type of sediments and slightly lower temperature and pressure (449 K and 1.36 MPa) for another type. Uranium was then determined by ICPMS, and uranium fractionation by sequential extraction was measured by UV-Vis spectrophotometry (Zavodska et al. 2009).

Another variation of the microwave digestion procedure was used for soil and vegetation samples collected from the Mediterranean area in Spain (Vera-Tome et al. 2003). The common procedure for pretreatment of soil samples—drying to a constant weight at 80°C and sieving with 2 mm mesh—was followed by microwave digestion with HF:HNO_3 (3:6 mL) per 0.5 g of soil. This procedure deviates from those mentioned earlier when 2 mL of $HClO_4$ are added and the samples were then evaporated to fuming in order to eliminate excess fluoride and then two further attacks by 8 mL of *aqua regia* were carried out before radiochemical separation and alpha spectrometry for uranium and thorium assays (Vera-Tome et al. 2003). The soil-to-plant transfer factors that were found in this study will be discussed in the relevant section later.

One of the simplest and easiest soil sample preparation methods was described in a study of the uranium and thorium content in soil and plants grown near an abandoned lead–zinc–copper mine in Turkey (Sasmaz and Yaman 2008). Soil samples surrounding roots of plants were collected at 30–40 cm depths, dried at 100°C for 4 h, and ground using hand mortars. A 1:1:1 mixture of HCl-HNO_3-H_2O was added to the sample (6 mL for 1 g) and heated for 1 h so that sample constituents, except silica, were dissolved. The uranium concentration in the soil samples was 1.1–70.3 mg kg^{-1} with lower values found in the plants, as discussed in the following section (Sasmaz and Yaman 2008). Another simplified method for the determination of uranium in phosphate rock and phosphor-gypsum used liquid-extraction and alpha spectrometry (Andreou et al. 2012). The rock samples were spiked with ^{232}U as a tracer and treated with 8 M HNO_3 (10 mL for 0.1 g) and then extracted into 30% tributylphosphate (TBP) in dodecane. The TBP was washed with 0.1 M Na_2CO_3 to remove mono- and di-butyl-phosphate that may be present in the TBP. The uranyl ions were back-extracted from the organic phase with water, and the sample was evaporated to dryness and redissolved in ammonium sulfate for electrodeposition and alpha spectrometry counting for 15 h.

An example of a method developed for sequential flow injection (SIA) for separation of uranium and thorium from NORM (naturally occurring radioactive material) samples was based on chromatographic separation of the radioactive elements prior to analysis by alpha spectrometry (Mola et al. 2014). A gypsum standard sample was used for method validation and sludge samples were analyzed by the same method. Sample treatment included drying in an oven at 110°C followed by crushing in a ball mill and sifting in 250 μm sieve. The samples were incinerated at 550°C for 12 h and after cooling isotopic tracers (^{232}U and ^{229}Th) were added and the samples were dried under an IR lamp. Digestion was carried out with *aqua regia* (9 mL for 0.3 g sample) in a microwave oven for 30 min at ~215°C. The cooled solution was passed through a syringe with a filter to remove insoluble matter (silicates and sandy clay) and the eluted solvent was evaporated to dryness and redissolved in HNO_3 and $Al(NO_3)_3$ prior to separation of uranium from thorium on a chromatographic UTEVA® resin column. After separation each of the fractions containing uranium or thorium was evaporated to dryness, redissolved in sulfuric acid, and electroplated on a stainless steel disk for alpha spectrometry (Mola et al. 2014).

The ASTM method for the determination of uranium and its isotopic composition as well as thorium in soils employs a high-temperature incineration followed by treatment with HF as a digestion method for the removal of organic matter and silicates (C1345 2008). According to this method, ICPMS is intended to replace the more labor-intensive TIMS and alpha spectrometry methods so that sample preparation is simplified. In addition, other actinides may also be determined by ICPMS without further separation. The method also elaborates the calibration procedure and mass bias correction. The soil sample is dried, crushed (or shaken and tumbled) for homogenization, and then placed in a furnace at 650°C ± 50°C for at least 4 h. Bismuth (the internal standard) and an isotopic spike of uranium are added. The sample is then treated with concentrated HF and after the reaction subsides this is followed by addition of concentrated nitric acid. The sample is then placed on a hot plate at 180°C ± 20°C and stirred while being evaporated to dryness. After cooling H_2O_2 is added to the residue and the samples are once again evaporated to dryness and another aliquot of H_2O_2 is added. For the determination of uranium, 6 M HNO_3 is added and the sample is stirred on a hot plate at 120°C ± 10°C until complete dissolution is achieved. After filtering, ICPMS analysis for total uranium and $^{235}U/^{238}U$ is carried out. For complete isotopic composition of the minor isotopes, separation and preconcentration of uranium can be performed by extraction chromatography on a TRU® (Eichrom) column.

Highlights: First, a representative sample should be selected. Due to the potential variability of the uranium content from spot to spot and with depth, this could be

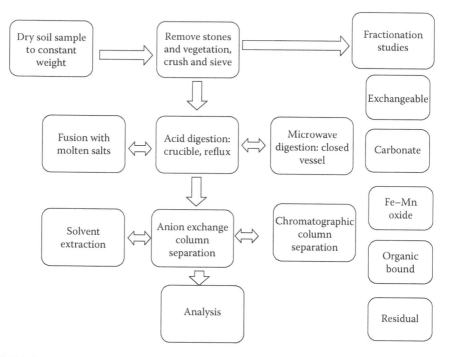

FIGURE 3.7   Schematic presentation of the stages for the determination of uranium in soil samples by a generic procedure.

quite a challenge. Furthermore, as the digestion methods are usually suitable for small-size samples (up to a few grams in most procedures), judicious sample selection is required. The digestion methods that precede the analytical measurement of uranium in soil samples can be viewed as variations on a central theme that includes a number of common steps shown in Figure 3.7:

a. Drying the soil samples to a constant weight in an oven at 80°C–110°C for several hours. In some procedures, the oven temperature is as low as 40°C.
b. Removal of stones and vegetation, grinding and sieving, usually through 2 mm (or smaller) mesh, sometimes followed by powdering to a smaller grain size.
c. In cases where fractionation (or speciation) is desired, the common procedure is used to separately determine the exchangeable fraction, the carbonate-associated uranium, the Fe-Mn-oxide fraction, organic-bound uranium, and the residual fraction. There are some variations on these stages.
d. Acid digestion with a mixture of mineral acids (HNO_3, HCl, HF, H_2SO_4, and $HClO_4$) and H_2O_2 is optional in some procedures. There are variations with different quantities of acids per gram soil and a diversity of combinations. This can be carried out in an open crucible although nowadays digestion in Teflon (or PFA) closed vessels in a microwave oven is more commonly used.
e. In some cases, fusion is deployed to totally dissolve tough residues (mainly silicates) and the fused sample is then dissolved in dilute nitric or hydrochloric acid.

f. Depending on the analytical method, radiochemical separation or solvent extraction can be carried out. For alpha spectrometry, this is a prerequisite and for ICPMS it is optional. In many cases, especially when alpha spectrometry is used, isotopic tracers are added as spikes for quantification and recovery estimates. In order to obtain more accurate isotopic composition measurements by mass spectrometry, preconcentration of uranium can be deployed.

3.2.2 SOIL SAMPLE PREPARATION FOR METHODS THAT DO NOT REQUIRE EXTENSIVE TREATMENT

Determination of uranium in soil samples can be carried out by nondestructive analysis (NDA) methods that do not require separation of uranium (needed for alpha spectrometry or TIMS) or even digestion of the soil for analysis by ICPMS, ICPAES, or some other spectroscopic methods. These NDA methods can be divided into passive techniques that utilize the natural radioactive emission (gamma and x-ray) of the uranium and progeny radionuclides or active methods where neutrons or electromagnetic radiation are used to excite the uranium and the resultant emissions (gamma, x-rays, or neutrons) are monitored. In many cases, sample preparation is simpler for these nondestructive methods but the requirement of a neutron source (from a nuclear reactor in many cases) or a radioactive source (x-ray or gamma) and relatively complex calibration and data interpretation procedures make the use of these techniques competitive only in some applications. In addition, the detection limits are usually inferior to the mass spectrometric techniques and the isotopic composition is not readily obtainable.

A comprehensive review of the sample pretreatment strategies for total reflection x-ray fluorescence (TRXRF) analysis was published in the form of a tutorial review surveying several preparation methods for a variety of samples (De La Calle et al. 2013); however, specific determination of uranium is only episodic in that review.

The ASTM has published a method, based on x-ray fluorescence (XRF), for the determination of uranium and thorium in soil or rock samples (C1255 2011). The samples are dried and ground and the content of uranium is determined by a wavelength-dispersive x-ray fluorescence (WDXRF) device. The method is suitable for a concentration range of 20–1000 µg U (or Th) g^{-1} soil, and the limit of detection of 20 µg g^{-1} is pronounced to be sufficient for clean-up verification.

Gamma spectrometry was used to determine the content of natural uranium and DU in surface soil samples collected in Kosovo and Greece (Anagnostakis et al. 2001). No sample treatment was carried out. The gamma spectrum of hermetically sealed (to prevent escape of radon) plastic containers was measured at least 3 weeks after being sealed. The ^{235}U content was estimated from the photo-peaks at ~186 and ~92.5 keV with a low-energy germanium (LEG) detector. The ^{235}U and ^{226}Ra peaks at 185.75 and 186.25 keV were resolved under these conditions. The ^{238}U activity was indirectly determined from the 63.29 keV peak emitted from its ^{234}Th progeny. The method could distinguish between natural and depleted uranium on the basis of the ratio between ^{235}U and ^{238}U content in the samples.

Methods based on neutron activation analysis (INAA and DNAA) were used for the determination of uranium in a variety of geological samples from Egypt (El-Taher 2010). The samples were crushed and sieved so that the particle diameter was between 63 and 125 μm and were dried to constant weight at 105°C. The powdered samples were placed in polyethylene capsules and irradiated for 6 h with a neutron flux of $7 * 10^{11}$ n cm^{-2} s^{-1} and measured with HPGe detector after cooling for 2 days. As expected, the uranium level in phosphate rock samples was considerably higher than in soil samples collected from two other regions in Egypt.

Qualitative determination of uranium in surface soils by a combination of x-ray diffraction (XRD) and wavelength-dispersive XRF (WDXRF) was described as an *easy-and-quick* method for appraising the uranium content in topsoil samples near a uranium deposit in Portugal (Figueiredo et al. 2011). XRD was used to identify the main mineral phases and XRF to roughly estimate the uranium concentration in as-collected samples. The only preparatory steps were drying the samples in an oven at 40°C before sieving and using the fraction of <150 μm for the x-ray analysis. The region of interest part of the WDXRF spectra of two soils samples (lower trace without uranium and upper trace with about 350 μg U L^{-1}) are shown in Figure 3.8 (Figueiredo et al. 2011).

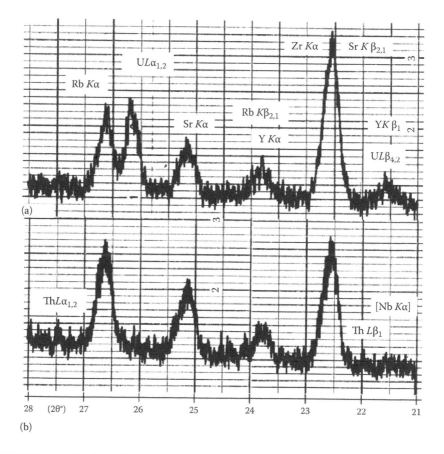

FIGURE 3.8 Region-of-interest WDXRF spectra of two soil samples. (a) Upper trace with a significant uranium content and (b) lower trace with no measureable uranium. (From Figueiredo, M.O. et al., *J. Geochem. Explor.*, 109, 134, 2011. With permission.)

A combination of sample collection methodologies (soil, water contained in pores and free water), sample treatment procedures, and analytical techniques (ICP-AES, gamma spectrometry, extended x-ray absorption fine structure (EXAFS), XRD, alpha spectrometry, and KPA) was deployed in a study of an unusual natural accumulation of uranium in an organic-rich soil in Switzerland (Regenspurg et al. 2010). The simple procedure relevant to this section involved drying the soil samples in an oven at 105°C for 24 h to determine the water content. The dried sample was homogenized, roots and stones were removed, and the sample was sieved through a 2 mm mesh prior to the determination of the uranium content by gamma spectrometry. U-238 was determined via the Th-234 and Pa-234 gamma lines at 63 and 1001 keV, respectively, and U-235 from the 135, 165, and 205 keV lines, as well as from the 186 keV gamma line (Ra-226 also contributes to this line). Uranium, together with several other elements, was also determined by ICP-OES after digestion and by alpha spectrometry after radiochemical separation (Regenspurg et al. 2010). In addition, sequential extraction (fractionation) was also carried out and the results suggested that a large fraction of the uranium was associated with the organic matter in these soil samples. A schematic summary of these steps is shown in Figure 3.9.

A comparative study of the thorium and uranium content in soil samples by gamma spectrometry and neutron activation analysis (NAA) showed a good correlation between the two methods (Anilkumar et al. 2012). The soil samples were dried, powdered, and packed in 250 mL plastic containers, counted for 50,000 with a p-type coaxial HPGe detector that was calibrated. Based on the assumption that ^{238}U was in equilibrium with its daughters, the intensity of the gamma lines at 352 and 609 keV from ^{214}Pb and ^{214}Bi, respectively, were used to assay the uranium in the sample. The drawbacks of NAA were noted: sample preparation is time consuming and several stable nuclides are transformed into radioactive isotopes after neutron capture, while analysis by passive gamma spectrometry is simpler.

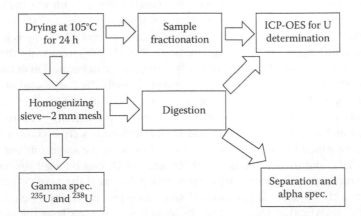

FIGURE 3.9 Preparation of soil samples for analysis by gamma spectrometry, ICP-OES, and alpha spectrometry. (Based on data from Regenspurg, S. et al., *Geochim. Cosmochim. Acta*, 74, 2082, 2010.)

The uranium content in soil samples collected in the Lesser Himalaya area was determined by gamma spectrometry (Ramola et al. 2011). The samples were collected from a depth of 60 cm and after removal of vegetation and stones were dried at 110°C for about a day. The samples were crushed to a fine powder and sieved and the <150 μm fraction was selected for analysis. Samples were placed in Marinelli beakers and stored for 1 month to establish secular equilibrium prior to measurement for 24 h of the gamma spectrum with liquid nitrogen–cooled HPGe detector. Here too, the ^{214}Bi and ^{214}Pb photopeaks were used to estimate the ^{238}U content in the samples.

A comparison of neutron activation analysis techniques for the determination of uranium concentrations in geological and environmental samples was published (Landsberger and Kapsimalis 2013). Short-lived and medium-lived NAA methods using thermal and epithermal neutrons were compared with Compton suppressed gamma-ray counting. The samples selected for this comparative study were reference materials (coal fly ash, coal, phosphate rock, and two Canadian uranium ores). The ores with high uranium content were not irradiated and only used for the passive gamma counting. Interferences that arise from chlorine, manganese, and sodium adversely affect the analysis of low levels of uranium. The limits of detection, for example, for the coal fly ash were between 0.25 μg U g^{-1} for Compton suppressed detection after irradiation with epithermal neutrons and 0.70 μg U g^{-1} for irradiation with thermal neutrons without Compton suppression (Landsberger and Kapsimalis 2013).

Highlights: The passive and active radio-analytical methods that are used to determine the uranium content in soil and sediment are simpler than the digestion methods required for mass spectrometric techniques and alpha spectrometry, but are usually less sensitive and larger uncertainties are associated with them. The basic sample preparation procedure involves drying the sample, removing foreign matter (stones and vegetation), and sieving it through a mesh, sometimes after pulverizing the sample. The sensitivity of gamma spectrometry depends on the background level and efficiency of the counting system, the counting time, and the detectors. In addition, as ^{238}U does not have very energetic gamma rays, the assay is based on the radiation emanating from its daughters (^{214}Pb and ^{214}Bi), requiring the establishment of secular equilibrium. Gamma spectrometry can be used also for assaying ^{235}U with its 186 keV line, but the contribution of ^{226}Ra at a similar energy must be accounted for. In general, the neutron activation analysis methods have better limits of detection than gamma spectrometry but usually require sample preparation and a neutron source. In addition, interferences from stable elements that are transformed into radioactive isotopes complicate the spectrum and sample handling. XRF and its variations (WDXRF and XRD) can be used for *easy and fast* screening with little or no sample preparation, but their limits of detection are satisfactory only for some applications (like clean-up verification). One advantage of the radio-analytical methods is that the sample size is usually in the order of a few hundred grams or even a kilogram, making it easier to prepare a composite representative sample.

3.3   URANIUM CONTENT IN PLANTS AND VEGETATION

Plants can serve as bioaccumulators of uranium, a topic that has recently received special attention after the accidents at the Chernobyl and Fukushima nuclear power plants. Frame 3.1 verbally quotes parts of a notice published in one of the leading scientific journals in the field of environmental monitoring (Caldwell et al. 2012).

Several of the methods mentioned earlier for the determination of the uranium content in soil are also suitable for assaying the uranium content in plants. However, there are two major differences: the uranium concentration in plants is usually significantly lower than in soil and the fraction of organic matter and the moisture content in plants are much higher than in soil. These two factors dictate somewhat different sample preparation methodologies and more sensitive analytical devices are preferred as shown in the following examples. Some of the review articles mentioned earlier also discuss the uranium content, uptake (transfer factor), and distribution in plants (Zavodska et al. 2008; Mitchell et al. 2013). One of the most comprehensive sources listing the uranium content in many different types of plants is the Canadian report mentioned earlier (Environment 2007). An additional important source can be found in the report published by the IAEA (IAEA 2010).

A microwave-based digestion technique was employed for samples of Douglas fir bark from France (Jauberty et al. 2013). The samples were first pretreated with 0.1 M HNO_3 to remove natural contaminants and then 1 g samples were placed in sealed PTFE-TFM tubes with 5 mL of concentrated HNO_3 and 2 mL HCl. The microwave power was gradually increased from 200 to 850 W and the pressure reached 6 MPa and temperature 260°C. After cooling, the samples were diluted to 25 mL and the uranium content in 1 mL aliquots was determined by the photometric (colorimetric) arsenazo-III method. The agreement between this method and results obtained by gamma and alpha spectrometry was good, and the limit of detection for the plant samples for the photometric method was in the mg U g^{-1} range. The possible interference from iron was examined and found to have little influence on the uranium assay.

FRAME 3.1   ENVIRONMENTAL IMPACT (CALDWELL ET AL. 2012)

Caldwell provided the justification for publishing the article (Caldwell et al. 2012) and stated that after the accidents at Fukushima and Chernobyl there is "an urgent need to characterize the bio-availability and transport potential of radionuclides such as uranium that are associated with the nuclear fuel cycle. This information cannot be solely inferred from abiotic sampling when considering the interacting chemical, biological and physical processes as they influence contaminant behavior within an ecosystem." The accumulation of radionuclides in plants reflects the level of contamination and is indicative of the bioavailability of these contaminants in the food chain. Furthermore, certain plants can serve as selective biomonitors of radionuclides.

Source: Caldwell, E.F. et al., *J. Environ. Monit.*, 14, 968, 2012.

The distribution of uranium in the different compartments of Scots pine trees growing in the soil of a uranium mining heap in Belgium was studied (Thiry et al. 2005). Samples were collected from the surrounding soil and from 35-year-old trees to determine the phytomass and uranium levels. The sampling compartments of the trees were the disks cut from the trunk (base at 0 cm, 1.3 m, at 3 m aboveground, and every 3 m up to the top), the crown (old needles, twigs, and branches), and living roots and needles that recently fell to the ground. The radial distribution of uranium in the disks cut from the trunk was also determined. All samples were washed, air-dried, and calcinated at 550°C. The ashes were dissolved in concentrated HCl and diluted tenfold with 2% HNO_3 prior to ICPMS analysis in which thorium was used as an internal standard. The results are summarized in Table 3.3 (Thiry et al. 2005).

In another study, the uranium content in four types of plants and plant components growing in a boreal forest in Finland a microwave digestion procedure was

TABLE 3.3

Uranium Content and Biomass of Tree Components and Humus in the Pine Stand Growing on U-Mining Dump H382 and Estimation of the Total U Amount and Its Relative Distribution between Tree Components and at Stand Level

Sample	Mean U Content (μg kg^{-1})	Biomass (t ha^{-1})	U Mineral Mass (g ha^{-1})	U Distribution Aboveground (%)	In Stand (%)
Needles 1 year	133	3.60	0.48		
Needles >1 year	337	4.92	1.66		
Total foliage			2.14	58	
Twigs 1 year	46	0.75			
Branches	26	15.90			
Total branches			0.45	12	
Total crown			*2.59*	*70*	
Dead branches	63	9.18	0.58	16	
Outer bark	30	5.98	0.18	25	
Inner bark	18	3.66	0.07	2	
Sapwood	2	115.06	0.23		
Heartwood	1	12.36	0.02		
Total stemwood			0.25	7	
Total trunk			*1.07*	*30*	
Total tree			**3.66**	**100**	0.5
Litter + O layers	3935	50.00	197		29.5
Roots	11020[a]	42.80[b]	472		70
Total			**673**		**100**

Source: From Thiry, Y. et al., *J. Environ. Radioact.*, 81, 201, 2005. With permission.

[a] Average weighted U concentration in the root system, that is, considering the respective importance of different root classes.

[b] Average total biomass estimated from the literature for pine stands (Vogt 1996).

deployed (Roivainen et al. 2011). Soil samples were collected and four plant species (May lily, Fern, Rowan, and Norway spruce) growing near the sampling sites were taken. The plant samples were washed with purified water and oven-dried at 60°C for 24 h and then milled and dried again for 24 h at 105°C. The dried samples were digested in nitric acid in a microwave oven and analyzed by ICPMS. The results showed that the uranium content in the coarse roots was significantly higher than in the leaves in the four species. On a dry weight basis, the roots-to-leaves concentration ratio was between 5 (for the lily) and 25 (for the Rowan) and even higher ratios were observed in the fine roots.

In another study, the treatment procedure for plant samples included washing the roots with tap water to remove soil, followed by rinsing with distilled water and drying in an oven at 100°C for 30 min and then at 60°C for 24 h. Dried plants were ashed first at 250°C and then at 500°C for 2 h. Ashed samples were digested in nitric acid and then by a 1:1:1 mixture of HCl-HNO_3-H_2O (6 mL for 1.0 g ashes) at 95°C—similar to the procedure used for soil digestion in the same study that was mentioned earlier (Sasmaz and Yaman 2008).

A very interesting study on the uptake of uranium by plants grown in a rigorously controlled environment (hydroponics) used ICP-AES for quantification of the uranium content in the whole plants and plant sections (Ebbs et al. 1998). Sample treatment was simple: rinsing with deionized water followed by drying and digestion with nitric acid at 180°C for 2 h. This was followed by treatment with 1:1 HNO_3:$HClO_4$ at 220°C until the sample was completely digested and analyzed.

The uranium content in plants and mushrooms grown on a uranium-contaminated site in Germany was determined by ICPMS (Baumann et al. 2014). All samples were sent to the laboratory immediately after collection and processed there. First, the samples were rinsed with deionized water and then cut with a knife and separated into liquid sap and solid residue by centrifugation. Birch leaves were hacked by a food processor and centrifuged, but the amount of sap was not always sufficient for analysis. The residues were digested in a closed Teflon vessel in a microwave oven for half an hour. The uranium content in the sap and digested samples was determined by ICPMS. Some samples were also analyzed by time-resolved laser-induced fluorescence (TRLFS) in a He-cryostat that is suitable for speciation studies.

A study of uranium uptake by native and cultivated plants in the St. Petersburg area used instrumental neutron activation analysis (INAA) technique (Shtangeeva 2010). After sampling, the plants were washed in tap water to remove soil and dust and then air-dried at room temperature to constant weight. The samples were placed in ultrapure quartz ampoules and irradiated for 24 h in a $1 * 10^{14}$ n cm^{-2} s^{-1} thermal neutron flux. Details of the gamma counting were not given. The results showed that plants grown in radionuclide-enriched soils contained elevated uranium levels in their roots while in the upper plant parts the concentration was lower. The soil type affected the uptake of uranium as deed the plant brand.

The uranium content in plants can serve for monitoring contaminants in the environment (Caldwell et al. 2012). Samples of plants, soil, sediments, water, and common biota were collected from five distinct sites in the vicinity of a uranium processing facility with the objective of studying transport pathways and selecting the plants that are efficient bioaccumulators of uranium. Plant root samples were dipped in water to remove soil and sediments without detaching the fine hairs on the roots. Vegetation samples were dried to a constant weight in an oven at 60°C and ground to particles of 1 mm or less. Soil and sediment samples were treated in a similar fashion. An isotopic tracer of ^{232}U was added. All samples were dissolved by acid digestion for liquid scintillation alpha spectrometry. The uranium was then chemically separated through ion exchange and was precipitated with NdF_3 on stainless steel disks for alpha spectrometry. The results were reported on a dry weight basis. The facility in question handles natural uranium only so the isotopic composition of uranium was not determined. The analysis of total uranium and other elements was carried out with ICPMS equipped with a cross-flow nebulizer and numerous internal standards were used. The uranium concentration was determined in several types of plants in the contaminated study area and in a control site. Only some representative values are shown in Table 3.4.

Evidently, the highest uranium content was found in aquatic moss samples in the contaminated area, reaching concentrations of ~1% on a dry weight basis. As in other studies, the concentration of uranium in root samples was significantly higher than in aboveground portions of the terrestrial plants, ranging from a factor of 10

TABLE 3.4
The Uranium Concentration in Several Types of Plants in the Contaminated Study Area and in a Control Site

Common Name	Plant Tissue	Collection Site	Total Uranium (mg kg^{-1})
Aquatic moss	AP	Storm/process mix	9240
Aquatic moss	AP	Process water	2900
Aquatic moss	AP	Storm water	12,500
Aquatic moss	AP	Outfall	2480
Sunflower (tickseed)	AGP	Stream outfall	1.68
Sunflower (tickseed)	RT	Stream outfall	16.0
Reed (giant)	AGP	Outfall swamp	12.90
Reed (giant)	AGP	Stream outfall	3.75
Reed (giant)	RT	Outfall swamp	5050
Reed (giant)	RT	Stream outfall	209
Nutgrass (yellow)	AGP	Below outfall	5.37
Nutgrass (yellow)	RT	Below outfall	286
Nettle (stinging)	AP	Stream outfall	25.70
Nettle (stinging)	AGP	Control	0.25
Nettle (stinging)	RT	Control	2.8
Lenspod white-top	AGP	Below outfall	6.24
Lenspod white-top	AGP	Stream outfall	4.70
Lenspod white-top	RT	Control	5.48
Goldenrod (showy)	AGP	Stream outfall	5.12
Goldenrod (showy)	RT	Stream outfall	148

Source: Adapted from Caldwell, E.F. et al., *J. Environ. Monit.*, 14, 968, 2012. With permission.
Notes: AGP, aboveground portion; AP, all portions; RT, root.

for the sunflower to several hundred for the giant reed, nutgrass, and goldenrod, for example. The concentration of uranium in the root of the control sample of the Lenspod white-top was similar to the concentration in the aboveground portion of the same plant grown in the contaminated area (Table 3.4).

A rapid separation method of actinides and radiostrontium in vegetation samples that can be used in an emergency was developed (Maxwell et al. 2010). Vegetation samples were placed in zirconium crucibles, ^{232}U was added as a yield tracer for uranium, and the covered crucibles were placed in a preheated furnace. After 10 min at 600°C, the temperature was raised to 700°C for 2 h for 5 g samples. After cooling, 10 mL of a 1:1 mixture of $HNO_3:H_2O_2$ was added for digestion of the ashed samples and then evaporated to dryness and placed again in the furnace. After cooling, 15 g sodium hydroxide was added and the crucibles were once again placed in the furnace for 10 min at 600°C. The residue in the cooled crucibles was dissolved in water and 125 mg of iron (III) nitrate and 4 mg lanthanum (as a nitrate) were added and a hydroxide precipitate was formed. After several more stages, which will not be elaborated here, the solution containing the actinides was ready for chromatographic separation on three consecutive columns (TEVA, TRU, and DGA-resin) and finally, the separated actinides were determined by alpha spectrometry (Maxwell et al. 2010).

Highlights: The analytical methods used for determining the uranium content in plants and vegetation and the sample preparation techniques follow the same principles detailed earlier for soil and sediment samples. A generic flowchart of preparation of vegetation samples for analysis is shown in Figure 3.10. These steps include

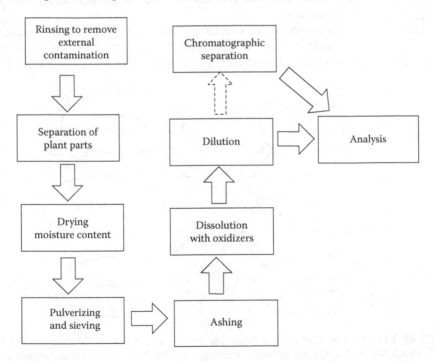

FIGURE 3.10   A generic flow chart of preparation of vegetation samples for analysis.

rinsing to remove external contamination, drying followed by pulverizing and sieving, incineration to ash, dissolution with oxidizing acids, dilution, and analysis.
Chromatographic separation is optional. However, due to the relatively high content
of organic matter in the plants, the ashing (wet ashing or incineration) step is important. In addition, due to the high moisture content in some parts of the plant, the
uranium content is always reported relative to the dry weight of the sample. In some
cases, the dry matter content is also reported so that the uranium content in the original sample may also be calculated. Most studies found that the highest concentration
of uranium was in the roots, but this could depend on the type of plant.

3.4   SOIL-TO-PLANT TRANSFER FACTORS

The soil-to-plant transfer factor, F_v, is defined as the dry weight ratio between the
uranium content in the plant and the soil (10 cm for pasture crops and 20 cm for
other crops) (TECDOC-1616 2009). The soil-to-plant transfer factors of uranium are
of interest from two main aspects: the possibility of uranium entering the food chain
through edible plants (or indirectly through meat products from farm animals that
consume such plants) and the other aspect is the potential for phyto-remediation, that
is, using plants to absorb uranium (or other heavy metals) and preconcentrate it from
soil or water sources and then process the plants. A schematic diagram of the factors
that govern the transfer of radionuclides from soil to the roots of plants is shown in
Figure 3.11. These include the type of plant, soil properties, cultivation technology,
properties of fallout waste and time after fallout, and the physical and chemical
properties of the radionuclides.

The amount of dry matter in plants varies with the type of plant and the segment
of the plant that is studied. Seeds usually contain over 85% of dry weight while
vegetative mass varies from a few percent, for example, as little as 5% in zucchini to

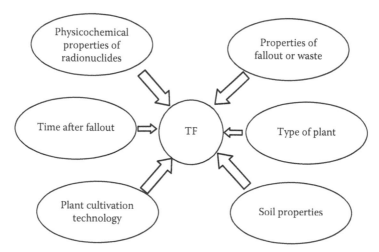

FIGURE 3.11   Factors influencing radionuclide root uptake. (Based on Sanzharova et al. in
TECDOC-1616, IAEA, Quantification of radionuclide transfer in terrestrial and freshwater
environments for radiological assessments, Technical, Vienna, Austria, 2009.)

TABLE 3.5
Relative Concentration of Uranium in Flora Varying with the Geological Origin of the Soil (n = 440)

Geological Origin of the Soil	Relative Index
Granite weathering soils	100
Weathering soils of Rotliegende	75
Muschelkalk weathering soils	74
Loess	72
Phyllite weathering soils	67
Slate weathering soils	64
Boulder clay	64
Weathering soils of Buntstandstein	55
Keuper weathering soils	52
Gneiss weathering soils	50
Pleistocene sands	46
Least significant difference	9

Source: From Anke, M. et al., *Chemie der Erde*, 69 (Suppl. 2), 75, 2009. With permission.

as much as 39% found in standard crested grass (TECDOC-1616 2009, pp. 14–15). Naturally, this parameter will also affect the uranium (and other radionuclides) content in the plant material. The properties of the soil (pH, organic matter content, ionic strength, and sand and clay content) also influence the behavior of uranium in the soil and its transfer factor to the plant.

A comprehensive study of uranium transfer from soil to plants, animals, and man in the food chain was published (Anke et al. 2009). The uranium content in vegetation in Germany was found to be much lower than in soil and the amount varied considerably between different types of soil and this was reflected in the concentration in plants and mineral waters. Table 3.5 shows the relative uranium content in flora in different types of soil (geological variation) in different geographical locations in Germany.

As far as plants are concerned, leafy plant species accumulated more uranium than tubes, stalks, and grains. It was also stated that the uranium content decreased significantly with the age of the plants. The uranium content in several types of plants grown in uranium mining areas and control sites is shown in Table 3.6 in units of µg U g^{-1} dry matter (DM).

The difference in the uranium concentration in some plant species between the control site and the mining area is evident. For example, in clover, rape seed, and chives, the uranium content was three to seven times higher in the mining area, while in onions, tomatoes, apples, and peeled potatoes, the uranium concentration was quite similar in both areas. These results reflect the differences in uranium content in soil as well as the different uptake, or soil-to-plant transfer, factors (Anke et al. 2009).

TABLE 3.6

Uranium Content of Several Wild and Cultivated Plants, Vegetables, and Fruits from a Control Site and a Uranium Mining Area in μg U g^{-1} Dry Matter (n = 374)

Plant Species	Number of Samples (Control:Mining)	Uranium Control Site	Uranium Mining Site	Effect of Site (%)
Meadow red clover	19:11	7.4 ± 2.0	57 ± 38	770
White clover	25:11	15 ± 10	68 ± 23	453
Rape seed	10:6	0.71 ± 0.10	2.2 ± 0.8	310
Chives	23:7	10 ± 4	29 ± 12	290
French beans	12:6	3.1 ± 1.2	8.1 ± 2.3	261
Lettuce	20:6	34 ± 18	73 ± 14	215
Wheat grain	16:14	0.58 ± 0.13	1.2 ± 0.8	207
Parsley	13:6	28 ± 8	54 ± 21	193
Carrots	18:8	4.4 ± 1.5	8.2 ± 3.7	186
Potatoes, peeled	26:14	2.3 ± 1.4	2.9 ± 1.9	126
Cucumber	15:8	7.0 ± 2.6	8.5 ± 4.5	121
Apple	25:9	2.7 ± 1.7	2.8 ± 1.5	104
Tomato	9:9	5.2 ± 1.2	5.0 ± 2.4	96
Onion	22:6	5.2 ± 3.1	4.3 ± 3.1	83

Source: From Anke, M. et al., *Chemie der Erde*, 69 (Suppl. 2), 75, 2009. With permission.

In a study of the transfer of uranium, aluminum, and manganese in the water–soil–plant system, potatoes were grown under controlled conditions near an abandoned uranium mining area in Portugal (Neves et al. 2012). Three types of soil were selected: two had a sandy-loam texture and were acidic with low clay content and organic matter but had a higher uranium content than the surrounding areas, and the third type had higher salinity and organic content but normal uranium content. In this study, irrigation water quality varied between tap water (TP), not contaminated water (NCW), and contaminated water (CW) with uranium concentrations of 1.3 ± 0.2, 17.0 ± 3.0, and 1035 ± 5 μg L^{-1}, respectively. The method used for assaying the uranium in the plants involved thorough washing with tap water followed by rinsing with distilled water. The samples were dried at 40°C (wet and dry weights were recorded) and finely ground. The samples were then ashed and digested by HNO_3–H_2O_2 prior to ICPMS analysis. Soil sample were air-dried, sieved (<2 mm), acid digested, and available soil fractions were analyzed by ICPMS. The uranium concentration was determined in peeled and unpeeled potato tubers and in the aerial part (leafy part) of the potato plant. The resultant concentrations found in the potato samples are summarized in Table 3.7.

It is not surprising that plants that were grown on contaminated soil (soil B) and irrigated with contaminated water (CW) showed the highest uranium content; however, the ratio of 100 compared with the tubers grown in *normal* soil and irrigated

TABLE 3.7

Uranium Content (μg kg⁻¹) in Potato Tubers (Unpeeled and Peeled) and Aerial Parts Grown in Three Types of Soil (A, B, and C) and Irrigated with Noncontaminated Water (NCW), Contaminated Water (CW), and Tap Water (TP)

	Soil A		Soil B		Soil C
	CW	NCW	CW	NCW	TP
Tuber with peel	121 ± 49	69 ± 25	589 ± 147	302 ± 88	6.8 ± 1.7
Tuber peeled	7.7 ± 2.0	6.5 ± 1.2	19.9 ± 2.1	17.8 ± 2.7	1.9 ± 0.2
Aerial part	2309 ± 647	1090 ± 184	4464 ± 1467	2860 ± 1326	856 ± 197

Source: Adapted from Neves, M.O. et al., *Sci. Total Environ.*, 416, 156, 2012. With permission.

with tap water is quite high. The big difference between the peeled and unpeeled tubers shows that most of the contamination is in the peel (external). The large uranium content found in the aerial parts compared to the tubers is also interesting (Neves et al. 2012).

Highlight: The transfer factors of uranium from the soil to the plant show a strong dependence on the type of soil, uranium content in the soil, the type of plant and irrigation (for a comprehensive list, see Environment 2007). In addition, as mentioned earlier, the distribution of uranium within the plant parts has a tendency to concentrate in the roots, and in the peels of tubers but not inside the peeled potatoes.

3.5 URANIUM CONTENT IN NATURAL FRESHWATER SOURCES AND OCEANS

In Chapter 4, several analytical methods and sample preparation techniques for the determination of uranium in drinking water are discussed, so in order to avoid too much duplication in this section focus will be on surface water and oceans. Guidelines for elemental analysis in water samples by ICPMS are basically similar to the description given here (Fernandez-Turiel et al. 2000). Samples should be collected in plastic (e.g., high-density polypropylene) bottles that have been carefully washed with dilute nitric acid and then rinsed three times by the sample. The bottles should be filled to the top and acidified with ultrapure nitric acid to form a 1% HNO_3 solution. However, acidification must be avoided when speciation studies of uranium are also carried out as the oxidizing nitric acid may alter the valence state of uranium. Filtering should be carried out prior to analysis to remove suspended and particulate matter. The samples should be stored in a cool (4°C) dark place. In some cases, the differentiation between uranium that is dissolved in water and uranium that is bound to particles and in a colloidal phase is required. In this case, separation of the particulate matter and colloids can be carried out by selective filtering from the water-soluble uranium.

3.5.1 PRECONCENTRATION OF URANIUM FROM AQUEOUS SAMPLES

Modern analytical techniques usually have sufficient sensitivity to determine the concentration of uranium in aqueous environmental samples and in most cases mass spectrometric techniques can also provide isotopic composition data. However, in some samples, especially where the precise content of minor uranium isotopes is required then preconcentration, separation, and purification can improve the accuracy of the measurement. Several methods have been developed for this purpose based on solid phase extraction (SPE), electro-analytical selective absorption techniques, liquid-extraction, ion-exchange and chromatographic columns, co-precipitation, and selective sorption. Other methods, like single-drop microextraction, are being developed and may serve for microanalysis (Jain and Verma 2011). Some of these techniques are discussed in the context of the specific sample preparation procedures throughout the book, so in this section only a few select methods will be discussed.

An SPE method for online preconcentration of uranium in water and effluent samples is based on amberlite XAD-4 resin functionalized with β-nitroso-α-naphtol (Lemos and Gama 2010). A preconcentration factor of 10 was achieved and with a colorimetric arsenazo-III spectrophotometry a detection limit 1.8 μg L^{-1} was reported. The online system included a mini-column through which the solution was passed for 180 s. In the elution step, the uranium was desorbed from the column and mixed with the colorimetric reagent and introduced into the spectrometer. The parameters were optimized and the effects of several interferences were examined. Recovery of spiked tap water, well water, and effluent samples was close to 100% (Lemos and Gama 2010).

An electrochemical method for sample pretreatment coupled online with ICPMS for the determination of uranium in sub-nanogram per liter concentrations was developed (Pretty et al. 1998a). Uranium was preconcentrated and separated from the sample matrix by accumulation on an anodized glassy carbon electrode at −0.15 V followed by stripping at +1.15 V. The limit of detection depended on the accumulation time and 0.12 ng L^{-1} for 10 min accumulation was reported. The accumulation of uranium was effective even in saline media (like seawater). A thin-layer electrochemical cell was combined with a flow-injection system and an ICPMS instrument for online determination of uranium in aqueous samples. The same group also developed a technique, based on adsorptive stripping voltammetry, for matrix elimination and preconcentration of uranium (Pretty et al. 1998b). Propyl gallate was used as a chelating agent for the accumulation of uranium on a thin-film mercury electrode and 24-fold signal enhancement was achieved with 10 min accumulation (Figure 3.12).

Calix[6]arene-modified electrodes (bound to a gold surface by cysteamine 4-sulfonic) were also used for preconcentration and detection of uranyl ions in aqueous solutions (Becker et al. 2008). The calixarene derivative has a strong affinity toward uranyl ions (super uranophile) and selectively extracts the uranium (see bottom frame of Figure 3.13).

Another SPE procedure for preconcentration of uranium in natural waters used octadecyl silica membrane disks modified by tri-n-octylphosphine oxide (TOPO) (Shamsipur et al. 1999). The high extractive properties of organic phosphorus compounds for uranyl ions are well known (e.g., TBP in the nuclear fuel cycle),

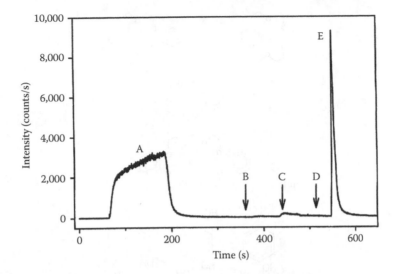

FIGURE 3.12 Mass spectrometric response for $^{238}U^+$ during an adsorptive stripping voltammetry-ICPMS experiment. (A) Injection of U-propyl-gallate complex with accumulation potential at −0.15 V; (B) potential step to −1.4 V; (C) potential step to −1.0 V; (D) potential step to 0.0 V; (E) stripping peak signal produced on injection of 1% HNO_3. (From Pretty, J.R. et al., *Int. J. Mass Spectrom.*, 178, 51, 1998b. With permission.)

and the method exploited this feature in a solid easy-to-use form. The influence of several interfering ions was tested, and the recovery of uranium from water with TOPO was found to be 85% for uranyl ions. A later publication by the same group used this SPE method, with addition of a small amount of piroxicam, for spectrophotometric determination of uranium (VI) in natural waters (Sadeghi et al. 2003).

A different approach to SPE devices for preconcentration of uranium was based on a microcolumn with alumina that was modified with sodium dodecyl sulfate (SDS) and 1-(2-pyridylazo)-2-naphtol (PAN) (Shemirani et al. 2003). The extraction parameters were optimized and a preconcentration factor of 150 was reportedly obtained with 99.8% recovery of uranium (VI). EDTA masking was used to eliminate or minimize interferences that may affect the arsenazo-III photometric method.

Crown ethers have also been used for selective extraction of uranyl ions from aqueous media (Mohapatra and Manchanda 1998). Although the focus of that study was on separation of uranium from fission products, the method can be applied for preconcentration of uranyl ions from acidic aqueous media. Four types of crown ethers were examined: benzo-15-crown-5, 18-crown-6 and dibenzo-18-crown-6, and the larger dibenzo-24-crown-8 (Figure 3.13a). The latter was not effective and the dibenzo-18-crown-6 had the best separation factor for uranium from most fission products (Mohapatra and Manchanda 1998).

A method for preconcentration and separation of ultra-trace amounts of uranium from aqueous samples (mineral water, rivers, wells, springs, and seawater) based on combining SPE with liquid–liquid microextraction was described (Dadfarnia et al. 2013). The water samples were filtered (0.45 μm) and acidified to pH ~2 with

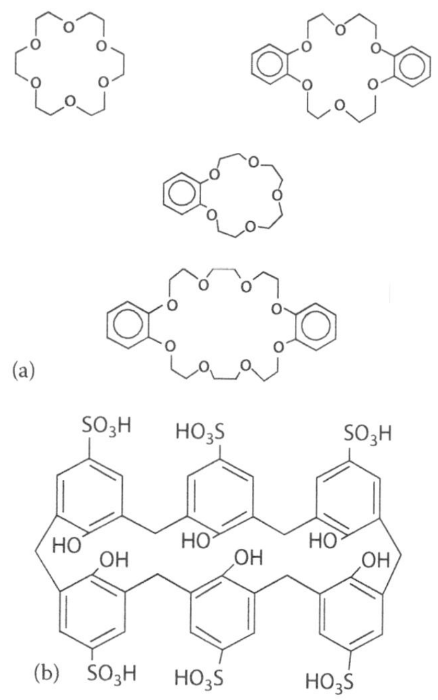

(a)

(b)

FIGURE 3.13   The chemical structure of two families of compounds that selectively attach uranyl ions. (a) Four types of crown ethers: benzo-15-crown-5, 18-crown-6 and dibenzo-18-crown-6 and the larger dibenzo-24-crown-8. (From Mohapatra, P.K. and Manchanda, V.K., *Talanta*, 47, 1271, 1998. With permission.); (b) 4-sulfonic calyx[6]arene (4-s-cal). (From Becker, A. et al., *J. Electroanal. Chem.*, 621, 214, 2008. With permission.)

nitric acid. The detailed procedure for preparation of the SPE column packed with γ-alumina and the sample treatment and loading can be found in the original paper (Dadfarnia et al. 2013). The eluent with the uranium was neutralized in a benzoate buffer solution so that a uranium benzoate complex was formed. Malachite green solution was added and ethanol and chloroform were injected. The uranium-benzoate-malachite complex was extracted into the chloroform drop that after centrifugation settled at the bottom of the test tube. It was then injected into quartz micro-flow-cell of a spectrophotometer for analysis at 621 nm. Recovery of uranium spikes from a variety of samples was close to 100%.

In recent years, novel techniques called diffusive gradient in thin films technique (DGT) were developed for obtaining speciation data and preconcentration of uranium from aqueous samples (Li et al. 2006). The method employs a bottom layer of a binding phase and an upper layer of a permeable hydrogel covered with a filter membrane. The metal ions diffuse through the membrane and upper layer and are captured by the bottom layer that contains a resin embedded in a gel (usually acrylamide), as shown schematically in Figure 3.14 (Gregusova and Docekal 2011).

Only labile species are thus trapped and accumulate on the bottom layer. One of these methods used a new resin gel based on Spheron-Oxin® chelating ion-exchanger anchored 8-hydroxyquinoline functional groups (Gregusova and Docekal 2011).

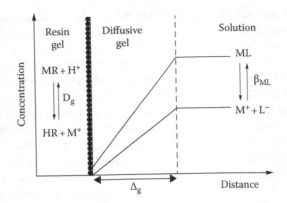

FIGURE 3.14 Scheme of the effect of competitive formation of the metal uptake in DGT technique. Metal ions (M^+) interact with the ligand (L^-) forming a diffusible complex (ML). The distribution of the metal between the resin and embedded functional groups affects the metal uptake. (From Gregusova, M. and Docekal, B., *Anal. Chim. Acta*, 684, 142, 2011. With permission.)

The performance of the device was compared with the conventional Chelex-100-based resin gel for uranium in water that contains carbonates. The efficiency of the DGT device was determined by placing it in solution containing uranium and equilibrating it for up to 48 h. The DGT probe was then removed and the elution efficiency of uranium bound to the gel was determined by repeatedly soaking the disks in 1 M HNO_3. The DGT technique is especially suitable for assessing the uranium content in high-salinity natural waters (Gregusova and Docekal 2011). The performance of three absorbents (Chelex-100, MnO_2, and Metsorb) used as the binding layer in DGT devices was compared for extended periods in saline water samples (Turner et al. 2012). The main feature that was tested was the linearity of accumulated uranium as a function of time, and under high-salinity conditions (like seawater) the MnO_2 showed the best performance.

Highlights: Various procedures have been deployed to selectively preconcentrate uranyl ions from aqueous media. These processes also separate the uranium from potential interferences and alleviate matrix effects. The selection of an explicit technique depends on the required limits of detection and the analytical device used for measurement of uranium. The use of DGT probes seems to be gaining popularity, especially for sampling saline media. These methods may be helpful in environmental analysis despite the fact the limit of detection of modern analytical techniques is well below the uranium level found in almost all realistic samples.

3.5.2 URANIUM IN OCEAN WATER AND SEAWATER

The generally accepted value for the concentration of uranium in seawater or ocean water is ~3.3 μg L^{-1} (3.3 ppb). Slightly different values were also reported, but these could be due either to the uncertainty in the analytical technique or to variations with geographic locations, effects of anthropogenic activities, and depth profiles. In recent years, special interest has been given to the isotopic composition of uranium and the contribution of anthropogenic activities (like the presence of [236]U isotope) (Christl et al. 2012). At trace levels, in environmental samples, the $^{236}U/^{238}U$ ratio is best determined by accelerator mass spectrometry (AMS). In one study of ocean water samples, the uranium was first co-precipitated with iron hydroxide, followed by chromatographic resin separation. For AMS measurement, the eluted uranium was co-precipitated again with iron hydroxide, oxidized at 650°C, mixed with niobium powder, and pressed into the AMS target holder (Christl et al. 2012). The control of the blank level ratio for these measurements was challenging and was found to be $(2.1 \pm 0.3) * 10^{-11}$, corresponding to $(0.17 \pm 0.03) * 10^6$ ^{236}U atoms/kg ocean water. One of the main conclusions of this study was that even at a depth of 4000 m the ^{236}U content indicated anthropogenic sources. Another AMS study of the depth profile of ^{236}U in the Japan Sea showed (Sakaguchi et al. 2012b) that its concentration decreased from $(13.3 \pm 3) * 10^6$ atoms kg^{-1} in the surface water to $(1.6 \pm 0.3) * 10^6$ atoms kg^{-1} close to sea floor (a depth of 2800 m). Sample preparation involved filtering the seawater sample, adding a tracer (^{233}U for uranium) and co-precipitation with iron hydroxide. The sample was heated for 3 h with stirring and then allowed to stand for 12 h. The pH was adjusted with nitric acid to pH = 1 (uranium and plutonium dissolve) and ammonium phosphomolybdate was added to co-precipitate cesium (a step not required for uranium determination). The sample was stirred for 1 h and stood for 24 h, followed by centrifugation. Uranium and plutonium in the supernatant were co-precipitated with iron hydroxide and the supernatant was removed. The precipitate was redissolved in HCl and separated by anion-exchange chromatography and then uranium was determined by AMS (Sakaguchi et al. 2012b). A similar method was used by the same group to determine the release of uranium, plutonium, and cesium in environmental waters following the Fukushima Daiichi accident (Sakaguchi et al. 2012a).

A method for the precise determination of the $^{234}U/^{238}U$ ratio in the open ocean was described (Andersen et al. 2010). Uranium was extracted and separated from the seawater matrix by iron co-precipitation, followed by two chromatographic resins (TRU-spec and U-TEVA-spec). Analysis was carried out by a multicollector ICPMS for precise isotope composition determination. The results showed a deviation of $146.8 \pm 0.1‰$ (almost 15%) of the marine $^{234}U/^{238}U$ ratio from that expected on the basis of secular equilibrium.

Highlights: The main interest in studies of uranium content in oceans or seawater is to determine the effect of anthropogenic activities. Therefore, the presence and abundance of the minor isotopes, particularly ^{236}U, must be accurately determined, and this requires preconcentration and separation of the uranium and the use of accelerator mass spectrometry (AMS) for the analysis. Other mass spectrometric techniques (ICPMS or TIMS) can also be used but with inferior performance. The high salinity of ocean water introduces a matrix effect that could bias ICPMS measurements of the uranium content, so the separation and preconcentration methods described earlier may be needed for precise quantification of uranium (an internal standard can also be used for this purpose).

3.5.3 Uranium in Surface Waters

First, it should be emphasized again that collection of water samples must be carefully controlled and executed (Rathore 2013). The samples should be free of suspended matter and sediments and filtered before collection and preservation. In addition, unacidified water samples should not be stored and should be analyzed on the same day (or acidified and/or stored under suitable conditions). In many cases, the concentration of uranium in surface water is well above the detection limits of modern analytical techniques (most notably, ICPMS) so no special treatment or preconcentration is necessary, except for precise determination of the minor isotopes. A few examples of analysis of uranium in surface water are presented later.

The uranium content and the $^{234}U/^{238}U$ ratio in several surface water sources and wells in the southern and northern parts of Israel were determined by flow injection-ICPMS with an internal standard (^{209}Bi) to correct for the matrix effect (Halicz et al. 2000). The samples were collected in plastic bottles, acidified and filtered, and analyzed without any further sample treatment. The water samples flushed the 100 µL loop of the flow injection system and after filling the loop the sample was transferred to an ultrasonic nebulizer (USN) for the $^{234}U/^{238}U$ isotope ratio determination and to a cross-flow nebulizer (that is less sensitive to matrix effects) for the uranium content measurements. Method validation and calibration were performed with analytical standards, including ocean water standard (NASS-3) and isotope ratio results were compared with TIMS measurements of carbonate samples. The highest $^{234}U/^{238}U$ ratio was 196 ppm and the lowest 64 ppm (in equilibrium the expected ratio is 54.8 ppm), and in one of the sources of the Jordan River the ratio was 125 ppm. The uranium concentration ranged from 0.2 µg L^{-1} in the Jordan River sources to 12 µg L^{-1} in one of the drilled wells in the southern Negev.

The uranium content in surface waters in a lake in Turkey was determined by ICPMS (Zorer and Sahan 2011). The water samples were simply acidified (0.1 M HNO$_3$), diluted by a factor of 100 and measured without any chemical pretreatment. The concentration of uranium in 18 samples ranged from 82 to 113 µg L^{-1}, so the concentration in the diluted samples was about 1 µg L^{-1} in all cases. Isotopic composition measurements were not reported in this study.

An electroanalytical method based on adsorptive stripping voltammetry that is claimed to be *faster and cheaper* was developed for the determination of uranium in river water samples and a seawater standard (Grabarczyk and Koper 2011). A hanging mercury drop electrode was used and accumulation was executed by pulsing the potential. Recovery of uranium spikes from river water samples was close to 100% with an RSD of about 5% at 1, 5, and 10 nmol L^{-1} (238, 1190, and 2380 ng L^{-1}, respectively).

The uranium activity ratio in water and fish samples from pit lakes in Kazakhstan and Tajikistan was determined by ICPMS, alpha spectrometry, and radiochemical neutron activation analysis (RNAA) (Stromman et al. 2013). The samples were collected from lakes that were formed in open pits in abandoned uranium mines. Fish in those lakes are occasionally consumed by the local population and domestic animals drink water from those lakes that contain high uranium levels (1 and 3 mg L^{-1}). The $^{234}U/^{238}U$ activity ratio was found to increase from about 1 to 1.5 a few kilometers downstream from the lakes. The uranium content in fish will be discussed in Section 3.6. Physical and chemical properties (conductivity, temperature, pH, and ionic strength) of water samples were measured in situ and samples for analysis were collected in precleaned plastic bottles, acidified with HNO$_3$ to pH = 1 sent to the laboratory. The uranium content and the $^{234}U/^{238}U$ ratio were measured by ICPMS, alpha spectrometry, and RNAA.

Uranium levels in Cypriot groundwater samples were determined by ICPMS and alpha-spectroscopy (Charalambous et al. 2013). This is quite a typical example of the utilization of these two analytical methods and also demonstrates the difference in sample preparation requirements. The study encompassed 81 groundwater samples from existing wells in Cyprus. For ICPMS analysis, all the water samples were digested in a microwave oven and analyzed according to standard procedures. Some samples were selected for alpha-spectroscopy and preparation included preconcentration and separation by cation exchange (Chelex-100 resin) and electrodeposition on a stainless steel disk. In general, the uranium levels and $^{234}U/^{238}U$ activity ratios were higher in the wells that were drilled in sedimentary rock formations compared to wells in igneous rocks.

Highlights: The determination of the uranium content and isotopic composition in surface water samples is normally easier than in ocean water due to the generally lower salt content in the former. Matrix effects could cause a bias in the uranium content measurements that can be corrected with an internal standard or separation of uranium and care must be taken to avoid precipitation or absorption of uranium in the sample container. The $^{234}U/^{238}U$ ratio may yield interesting geological information.

3.6 URANIUM CONTENT IN AIR AND OTHER ENVIRONMENTAL SAMPLES

For the general public, exposure to uranium by inhalation is due mainly to breathable particulate matter that contains traces of uranium. Populations near nuclear facilities where uranium is mined or processed may also be exposed to aerosols that contain significant levels of uranium and the studies focus mainly on the size distribution of inhalable particles and their dissolution properties in lung fluid (see Chapter 4). UF_6 is the only significant gaseous uranium compound and its release into the atmosphere is followed almost immediately by hydrolysis into UO_2F_2 aerosols (Kips et al. 2007) so inhalation of gaseous uranium compounds is highly unlikely. Thus, methods for the determination of uranium in air focus on sampling particulate matter and aerosols and analysis of these samples.

One such method used XRF for analysis of uranium and thorium captured on membranes or spun glass air filters (Pilz et al. 1987). The membrane filters were found to be more efficient because the captured dust or aerosol particles settled on the surface of the filter while in the spun glass filters the dust penetrated the pores decreasing the sensitivity. A different approach to determine actinides (Pu, Th, Am, and U) in air filters and drinking water was presented (Thakur et al. 2011). Fixed air samplers (FAS) were used to collect particulate matter from an effluent air stream at the Waste Isolation Pilot Plant (WIPP) that is a deep underground geologic nuclear repository in New Mexico. The filters were 47 mm diameter membrane Versapor, and the airflow through them was ~170 L min^{-1} for 24 h or over a weekend. The filters were digested in a microwave oven in closed vessels with an acid mixture (HNO_3, HCl, and HF) to a maximum temperature of 190°C for 10 min. The digested samples were evaporated to dryness and the residue digested with $HClO_4$ to destroy the black particulates that originate from diesel exhaust and to remove organic matter and fluorine traces. In addition, high-volume air samplers (~1.13 m^3 min^{-1}) were positioned further away to trap suspended particles. These were 20 × 25 cm glass fiber filters, positioned 5 m aboveground level, and they were operated to accumulate particulates for 3–6 weeks. The filters were ashed at 500°C in a muffle furnace for 6 h. Then isotopic tracers and an iron carrier were added and the ash was treated with HNO_3:HF to complete decomposition of silica. Then fluoride ions were removed by treatment with $HClO_4$ and HNO_3 and after evaporation to dryness this process was repeated until all fluorides were completely removed. Actinide separation by co-precipitation and separation on chromatographic resins followed and after deposition on stainless steel disks alpha spectrometry was used to determine the separated actinides (Thakur et al. 2011). Another method that used ICPMS for analysis of actinides in airborne particulate matter deployed polypropylene filters (Lariviere et al. 2010). Method development included use of NIST standard reference materials (SRM-1648 and SRM-2783) that were deposited on the filter that was then ashed at 550°C for 4 h in a porcelain crucible. Isotopic tracers were added and the ash was dissolved in concentrated HNO_3 in a microwave oven at 200°C for 30 min. Online separation was carried out with chromatographic resins, and the analytical measurements were made by sector field-ICPMS. A complete system is shown in Figure 3.15, and the results for uranium that was separated and preconcentrated on U/TEVA chromatographic column are depicted in Figure 3.16 (Lariviere et al. 2010).

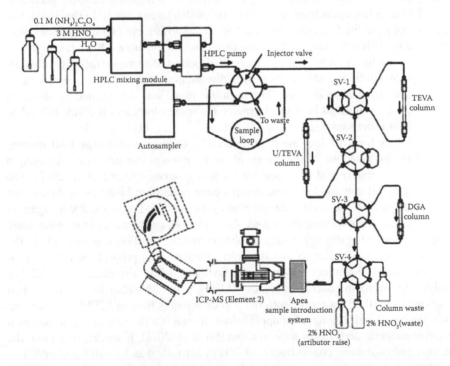

FIGURE 3.15 Schematic of the system used for online separation of actinides (U, Pu, and Am) collected on a polypropylene filter. Separation of uranium was carried out with U/TEVA chromatographic resin and the analytical measurements were made by sector field-ICPMS. (From Lariviere, D. et al., *Anal. Methods*, 2, 259, 2010. With permission.)

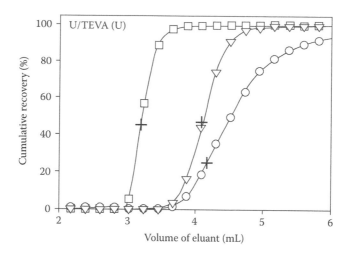

FIGURE 3.16 Recovery of uranium from U/TEVA chromatographic resin using H_2O (○), 0.01 M $(NH_4)_2C_2O_4$ (▽), and 0.1 M $(NH_4)_2C_2O_4$ (□). (From Lariviere, D. et al., *Anal. Methods*, 2, 259, 2010. With permission.)

A special case was concerned with measurement of uranium aerosols generated when DU munitions penetrate armored vehicles (Parkhurst 2003). In this study, DU penetrators were fired at a tank and another armored vehicle and the aerosols were collected by different means (filter cassettes, cascade impactors, five-stage cyclone, and a moving filter). Wipe samples were also collected from surfaces and deposition trays. Analysis included particle size distribution, morphology, uranium oxide phases, and dissolution in vitro. The uranium mass was determined by different methods: radioanalytical beta spectrometry for some filters and ICPMS, ICP-AES, and KPA for other samples.

A novel method that does not require filters or expensive analytical instrumentation for in situ detection of trace aerosols of uranium above uranium ores using a handheld photometer and solid reagent kit was described (Yang et al. 2013). The solid reagent kit consisted of three components: potassium bisulfate to control the acidity, arsenazo-III as a chromatographic agent, and EDTA as a masking agent to minimize interferences from other ions. Aerosols containing traces of uranium were trapped by bubbling through a nitrate solution and the solution was mixed with the solid kit described earlier so that the color intensity was proportional to the uranium content. The results compared quite favorably with those obtained by ICPMS analysis for uranium concentrations of ~0.1 mg U m^{-3}. Another innovative system developed a method for online introduction of aerosols into an ICPMS instrument. This has been used, among other applications, to monitor the uranium concentration in environmental aerosols in a clean room (Su et al. 2014). It was reported that the background equivalent concentration of ^{238}U in aerosols was $1.1 * 10^{-9}$ g U m^{-3}.

There are other environmental samples in which the uranium content, and especially its isotopic composition in environmental samples, is a useful indicator of different phenomena. For example, the uranium content and $^{238}U/^{235}U$ isotope ratio in shallow carbonate sediments can serve as paleoredox proxy (Romaniello et al. 2013).

The concentration of uranium in ancient carbonates differed significantly from more recent carbonate sediments and small differences were also found in the $^{238}U/^{235}U$ and $^{234}U/^{238}U$ isotope ratios. The carbonate samples (corals, calcifying algae, mollusks, and ooids), as well as core samples, were collected in the field, air-dried, and transferred to the laboratory. Samples were rinsed with distilled water and dried in an oven at 105°C and large samples were powdered. The samples were ashed at 750°C for 24 h that allowed most of the carbonate to escape as CO_2, thus preventing violent dissolution with acid digestion that used 3 M HNO_3. After dilution, analysis was carried out with ICPMS, and in some cases, a double $^{233}U-^{236}U$ spike was used for accurate isotopic composition measurements (Romaniello et al. 2013).

An interesting type of bioaccumulation is the study of bacteria that selectively absorb uranium as shown in a review article (Pollmann et al. 2006). One study found that a gram-negative bacterium, *Pseudomonas* MGF-48, rapidly accumulated uranium from electroplating effluent, reaching a level of 17.4% of uranium per dry weight of the bacteria. This may have a potential for bio-remediation of uranium polluted aqueous effluents (Malekzadeh et al. 2002). A detailed study of the mechanism by which the bacteria accumulate and sequester uranium used transmission electron microscopy, energy-dispersive x-ray analysis, FTIR, XRD, and atomic force microscopy (Kazy et al. 2009). The conclusion was that high uranium (and thorium) accumulation by the *Pseudomonas* sp. bacterium could be due to a combined effect of ion exchange–complexation and microprecipitation mechanism (see also Frame 2.1).

Bioaccumulation of uranium in marine birds was studied (Borylo et al. 2010). Dead bodies of three species of birds (sea birds, wintering birds, and migratory birds) were collected and dissected. A ^{232}U tracer was added to the biological samples that were mineralized, the uranium was separated on an anion exchange resin, and electrodeposited on stainless steel disks for alpha spectrometry. The distribution of the uranium content and the $^{234}U/^{238}U$ activity ratio among the different organs (liver, muscles, feathers, skeleton, skin, and rest of viscera) and in the whole organism were determines for several birds (Borylo et al. 2010).

The concentration of uranium in several fish species that were found in heavily contaminated water (above 1 mg U L^{-1}) and the distribution among their organs were studied (Stromman et al. 2013). The fish were caught by gillnet fishing in Kurday and supplied from fishermen using rods in Taboshar pit lake and reference samples were obtained from fish industries at a different (uncontaminated) lake. Not surprisingly, the concentrations of uranium in gill, liver, muscle, and bones in fish from the Pit Lake were higher than in the reference samples, especially the kidney and gills samples that contained about 9 mg kg^{-1} dry weight (Stromman et al. 2013).

Some studies that can be included in the environmental context are concerned with the transfer of heavy metals, including uranium, through the food chain to human diet. For example, the assessment of uranium and other metals in bovine meat and organs was carried out in South Africa (Ambushe et al. 2012). Muscle, liver, kidney, fat, and bone samples were collected at an abattoir near a mining area. The samples were cut into small pieces and freeze-dried for 24 h, then homogenized in a blender and powdered. The samples were digested with nitric acid in a microwave oven at a maximum pressure of 350 psi and temperature of 210°C. Then diluted with water and internal standards of Ga, In, Tl, and Th were added prior to

ICPMS analysis. The results showed that uranium was concentrated mainly in the kidneys and bones. The source of the uranium was attributed to the animals grazing in areas that are polluted with heavy metals. The conclusion was that heavy metals, including uranium, may bioaccumulate in bovine meat, but the level was low and posed no hazard to humans, except perhaps through consumption of some internal organs (Ambushe et al. 2012). A study in France examined the transfer of uranium, thorium, and decay products from grain, water, and soil to chicken meat and eggs (Jeambrun et al. 2012). Sample preparation of the eggs included removal of the shell and homogenizing the albumen and yolk, and the chickens were plucked and eviscerated and after drying at 80°C the bones were removed from the meat. The eggs, chicken, and grain samples were ashed at 480°C. Soil samples were sieved at 2 mm and lyophilized and then crushed into fine powder. Water samples were acidified but not filtered. Uranium and thorium were determined by ICPMS measurements after spiking with ^{233}U and ^{229}Th. Digestion of solid samples was carried out by reflux at 80°C for 3 days with concentrated HNO_3, HCl, and H_2O_2 with a few drops of HF. The samples were evaporated to dryness and redissolved in 3 M HNO_3. Uranium was purified and preconcentrated using TEVA chromatographic resin. Alpha spectrometric analysis of uranium for isotopic composition determination was performed after deposition on a stainless steel disk with 300,000 s counting time. The conclusion was that for these chicken meat samples water was the main source of uranium activity while soil would be more important as a source of thorium contamination.

Highlights: The broad assortment of samples presented in this section indicates the variability encountered by environmental samples. Uranium content in aerosols is usually associated with health physics considerations in facilities where uranium is mined, milled, and processed or in neighboring communities that may be affected by these activities. In general, exposure from airborne uranium containing particulates is very rare so the environmental interest is rather limited. Bioaccumulation by bacteria is still something of anecdotal interest, although it may be applied to bioremediation. Uranium that may enter the food chain is of concern, and several studies have focused on its content in vegetation and meat products. The analytical methods for all these samples are not unique and even the sample preparation techniques are similar to those described in detail in the previous sections.

3.7 SUMMARY

The main interest in the analysis of uranium in environmental samples is its effect as radioactive toxic heavy metal on the flora and fauna and assessment of the potential risk to human life directly or through the food chain. Natural uranium is present in practically all types of environmental samples—plants, soil, water bodies, and even air. In addition, anthropogenic activities related mainly to releases and discharges from the uranium fuel cycle may contaminate nearby areas, and that pollution may spread by wind and water action to considerable distances from the source. In order to assess the uranium content in the environment, representative samples need to be gathered (see Frame 3.2)—a task that is much more complicated than generally expected due to the variability of the sampled media.

FRAME 3.2 SAMPLE COLLECTION FOR ASSAYING URANIUM IN THE ENVIRONMENT

Determination of the uranium level in the environment begins with the collection of representative air, water, soil, and plant samples. Each type of sample and the recommended sampling strategy differ from the others in a significant way. Unlike sampling for nuclear forensic purposes, where the objective is to locate and characterize the extraordinary particles or *hot spot*, environmental assays seek to obtain representative samples. The following summarizes in brief general aspects of environmental sampling that pertain also to collection of uranium-bearing samples.

Air sampling: Online monitoring of atmospheric gases, at best, can represent the momentary concentration of the analytes. Sampling a large volume of air over a long period of time and trapping the target analytes with a suitable cartridge or sorbent tube may serve as an integrator and yield the time-averaged concentration. The length of time that airborne particulate matter may stay in air depends mainly on the particle size. Particles with a diameter over 10 μm would be deposited within hours while small particles with a diameter below 1 μm may remain airborne for weeks and may be washed down by rain or snow. Sampling particulate matter in air can be divided into passive measures and active devices. Among the passive measures there are simple water-filled receptacles or deposition trays coated with Vaseline or grease that retain the particulate matter that is deposited upon them. A more sophisticated version deploys a moving sticky tape that collects the particles that are deposited on the exposed tape section. All these are suitable for trapping large particles that sink to the ground and the average concentration of the analyte (uranium in this case) is calculated dividing the total measured uranium by the product of the collection time and the surface area of the sampling device. The most common type of active air samplers deploys large-volume suction pumps and filters (usually consisting of polyester, fiberglass, or common fabric) that collects airborne particles. Another type uses a cyclone collector (also called an inertial collector) to gather the airborne particles, and these can be further classified according to their size. Wet scrubbers and electrostatic precipitators are not usually used for environmental air sampling. The concentration of the analyte in the airborne particulate matter is calculated by dividing the concentration on the filter or in the cyclone collector by the volume of sampled air. As mentioned earlier, these devices serve as integrators for the sampling period while transient concentrations cannot be determined by these methods.

Water sampling: Samples from surface waters, drilled wells, or streams should be collected in clean plastic containers that were washed with dilute nitric acid and rinsed a number of times with the sample. The sample should be filtered to remove particulate matter (unless it is required for analysis) and acidified with nitric acid (0.1%–1%). The container should be completely filled, leaving little or no room for ambient air, sealed tightly, and stored in a cool dark place.

The temperature and pH should be measured in the body of water in situ. Despite all these precautions, the provisos listed in Section 3.1 regarding the effects of temperature, pH, and atmospheric CO_2 concentration on the solubility of uranium compounds should be considered. Most water sources also vary with time and location as well as with seasonal and diurnal changes that could be quite large. In addition, the presence of suspended particulate matter in the water may also affect the measured uranium concentration as uranium ions may be absorbed on the particles that are removed from the sample by filtering. Conversely, the analytical results may be biased if the filtering allows some suspended particles (that may contain uranium) to remain in the sample. Therefore, for a truly representative picture of the uranium concentration in a body of water, a number of samples should be collected from different locations and depths and over a period of time. This is expensive and is not usually practiced. In some environmental studies, the sediments should also be sampled as they may be in equilibrium with the water and affect the uranium concentration.

Soil sampling: Soil sample collection for assessment of the uranium content in the environment is even more complex than water sampling as there could be large variations of the mineral composition and uranium content of the soil even within a few meters apart and significant changes with depth especially if contamination occurred. The standard procedure calls for collection of several samples in order to create a composite sample, followed by sieving to remove stones and vegetation and then pulverizing and homogenization of the sample. The total uranium content in the soil sample does not change because of temperature, moisture, light or biological and chemical processes but the composition (like valence) of uranium compounds may change upon storage. This could affect the fractionation measurements and change the ratio between the different selective extraction procedures. Thus, gathering a representative soil sample and characterizing the uranium is not a simple task.

Plant sampling: Collection of plant samples is even more complicated as in addition to the large variety of plants and geographic variations there is a distribution of the uranium content within the different compartments of the plant itself. One approach is to compare the same type of plant in different locations—usually a contaminated area and a control area—and determine the uranium content in the same compartments of the plant (like the tuber example given earlier). Another important point is clearly stating the basis for calculating the uranium content (whole plant, plant section, dry matter, or ashed sample).

Once these samples are collected, an arsenal of the analytical procedures for the determination of uranium is available. Some of these were described in this chapter, but given the variation in the types of samples, in the sample treatment procedures and in the analytical methods, there is no universal routine for all. In order to report meaningful results for the uranium content in vegetation and food products, a standard basis for comparison must be established. This could be either on the basis of dry matter or, preferably, on the basis of the uranium content in ashed samples. In both cases, the fraction of dry matter or ash content in the sample should also be given so that the uranium concentration in the untreated sample can be assessed (see Table 4.5 for several such examples).

REFERENCES

Adriaens, A.G., Fasset, J.D., Kelly, W.R. et al. (1992). Determination of uranium and thorium concentrations in soils: A comparison of isotope dilution-secondary ion mass spectrometry and isotope dilution-thermal ionization mass spectrometry, *Anal. Chem.* 64, 2945–2960.

Ambushe, A.A., Hlongwane, M.M., McCrindle, R.I. et al. (2012). Assessment of levels of V, Cr, Mn, Sr, Cd, Pb and U in bovine meat, *S. Afr. J. Chem.* 65, 159–164.

Amoli, H.S., Barker, J., and Flowers, A. (2007). Closed vessels microwave digestion method for uranium analysis of soils using alpha spectroscopy, *J. Radioanal. Nucl. Chem.* 273, 281–284.

Anagnostakis, M.J., Hinis, E.P., Karangelos, D.J. et al. (2001). Determination of depleted uranium in environmental samples by gamma-spectroscopic techniques, *Arch. Oncol.* 9, 231–236.

Andersen, M.B., Stirling, C.H., Zimmermann, B. et al. (2010). Precise determination of the open ocean $^{234}U/^{238}U$ composition, *Geochem. Geophys. Geosyst.* 11, 1–8.

Andreou, G., Efstathiou, M., and Pashalidis, I. (2012). A simplified determination of uranium in phosphate rock and phosphogypsum by alpha spectroscopy after its separation by liquid-extraction, *J. Radioanal. Chem.* 291, 865–867.

Anilkumar, R., Anilkumar, S., Narayani, K. et al. (2012). A comparative study of 232-Th and 238-U activity estimation in soil samples by gamma spectrometry and neutron activation analysis technique, *Radiat. Prot. Environ.* 35, 14–16.

Anke, M., Seeber, O., Muller, R. et al. (2009). Uranium transfer in the food chain from soil to plants, animals and man, *Chemie der Erde* 69(Suppl. 2), 75–90.

Baumann, N., Arnold, T., and Haferburg, G. (2014). Uranium content in plants and mushrooms grown on a uranium-contaminated site near Ronneburg in Eastern Thuringia/ Germany, *Environ. Sci. Pollut. Res.* 21, 6921–6929.

Becker, A., Tobias, H., Porat, Z. et al. (2008). Detection of uranium(VI) in aqueous solution by calix[6]arene modified electrode, *J. Electroanal. Chem.* 621, 214–221.

Borylo, A. (2013). Determination of uranium isotopes in environmental samples, *J. Radioanal. Nucl. Chem.* 295, 621–631.

Borylo, A., Skwarzec, B., and Fabisiak, J. (2010). Bioaccumulation of ^{234}U and ^{238}U in marine birds, *J. Radioanal. Nucl. Chem.* 284, 165–172.

Boulyga, S.F. and Becker, J.S. (2002). Isotopic analysis of uranium and plutonium using ICP-MS and estimation of burn-up of spent uranium in contaminated environmental samples, *J. Anal. Atom. Spectrom.* 117, 1143–1147.

Boulyga, S.F., Matusevich, J.K., Mironov, V.P. et al. (2002). Determination of $^{236}U/^{238}U$ isotope ratio in contaminated environmental samples using different ICP-MS instruments, *J. Anal. Atom. Spectrom.* 17, 958–964.

C1255, ASTM. (2011). Standard test method for analysis of uranium and thorium in soils by energy dispersive x-ray fluorescence spectroscopy. ASTM, Conshocken, PA.

C1345, ASTM. (2008). Standard test method for analysis of total and isotopic uranium and total thorium in soils by inductively coupled plasma mass spectrometry. ASTM, West Conshohocken, PA.

Caldwell, E.F., Duff, M.C., Ferguson, C.E. et al. (2012). Bio-monitoring for uranium using stream-side terrestrial plants and macrophytes, *J. Environ. Monit.* 14, 968–976.

Canadian Council of Ministers of the Environment. (2007). Canadian soil quality guidelines for uranium: Environmental and human health. Scientific Supporting Document PN 1371, Canadian Council of Ministers of the Environment.

Charalambous, C., Aletrari, M., Piera, P. et al. (2013). Uranium levels in Cypriot groundwater samples determined by ICP-MS and alpha-spectroscopy, *J. Environ. Radioact.* 116, 187–192.

Christl, M., Lachner, J., Vockenhuber, C. et al. (2012). A depth profile of uranium-236 in the Atlantic Ocean, *Geochim. Cosmochim. Acta* 77, 98–107.

Croudace, I., Warwick, P., Taylor, R. et al. (1998). Rapid procedure for plutonium and uranium determination in soils using a borate fusion followed by ion-exchange and extraction chromatography, *Anal. Chem. Acta* 371, 217–225.

Dadfarnia, S., Shabani, A.M.H., Shakerian, F. et al. (2013). Combination of solid phase extraction and dispersive liquid–liquid microextraction for separation/preconcentration of ultra trace amounts of uranium prior to fiber optic-linear array spectrophotometry determination, *J. Hazard. Mater.* 263, 670–676. doi:10.1016/j.hazmat.2013.10.028.

De La Calle, I., Cabaliero, N., Romero, V. et al. (2013). Sample pretreatment strategies for total reflection x-ray fluorescence analysis: A tutorial review, *Spectrochim. Acta B* 90, 23–54.

Donoval, M. and Matel, L. (2001). Determination of uranium and thorium in soils from Podunajske Biskupice. *Ninth International Conference SIS'01*, Bratislava, Slovakia.

Ebbs, S.D., Brady, D.J., and Kochian, L.V. (1998). Role of uranium speciation in the uptake and translocation of uranium by plants, *J. Exp. Bot.* 49, 1183–1190.

El-Taher, A. (2010). INAA and DNAA for uranium determination in geological samples from Egypt, *Appl. Radiat. Isot.* 68, 1189–1192.

Fernandez-Turiel, J.L., Llorens, J.F., Lopez-Vera, F. et al. (2000). Strategy for water analysis using ICP-MS, *Fresenius J. Anal. Chem.* 368, 601–606.

Figueiredo, M.O., Silva, T.P., Batista, M.J. et al. (2011). Uranium in surface soils: An easy-and-quick assay combining x-ray diffraction and x-ray fluorescence qualitative data, *J. Geochem. Explor.* 109, 134–138.

Grabarczyk, M. and Koper, A. (2011). How to determine uranium faster and cheaper by adsorptive stripping voltammetry in water samples containing surface active compounds, *Electroanalysis* 23, 1442–1446.

Gregusova, M. and Docekal, B. (2011). New resin gel for uranium determination by diffusive gradient thin films technique, *Anal. Chim. Acta* 684, 142–146.

Halicz, L., Segal, I., Gavrieli, I. et al. (2000). Determination of the 234U/238U ratio in water samples by inductively coupled plasma mass spectrometry, *Anal. Chim. Acta* 422, 203–208.

Hiess, J., Condon, D.J., McLean, N. et al. (2012). ^{238}U/^{235}U systematics in terrestrial uranium bearing minerals, *Science* 335, 1610–1614.

IAEA. (2010). Handbook of parameter values for the prediction of radionuclide transfer in terrestrial and freshwater environments. Technical reports series 472. Vienna, Austria: IAEA.

Jain, A. and Verma, K.K. (2011). Recent advances in applications of single-drop microextraction: A review, *Anal. Chim. Acta* 706, 37–65.

Jauberty, L., Droget, N., Decossas, J.L. et al. (2013). Optimization of the arsenazo-III method for the determination of uranium in water and plant samples, *Talanta* 115, 751–754.

Jeambrun, M., Pourcelot, L., Mercta, C. et al. (2012). Study on transfers of uranium, thorium and decay products from grain, water and soil to chicken meat and egg contents, *J. Environ. Monit.* 14, 2170–2180.

Kazy, S.K., DwSouza, S.F., and Sar, P. (2009). Uranium and thorium sequestration by *Pseudomonas* sp.: Mechanism and chemical characterization, *J. Hazard. Mater.* 163, 65–72.

Kips, R., Leenaers, A., Tamborini, G. et al. (2007). Characterization of uranium particles pro-
    duced by hydrolysis of UF6 using SEM and SIMS, *Microsc. Microanal.* 13, 156–164.
Landsberger, S. and Kapsimalis, R. (2013). Comparison of neutron activation analysis tech-
    niques for the determination of uranium concentrations in geological and environmental
    materials, *J. Environ. Radioact.* 117, 41–44.
Lariviere, D., Benkhedda, K., Kiser, R. et al. (2010). Rapid and automated sequential deter-
    mination of ultra-trace long-lived actinides in air filters by inductively coupled plasma
    mass spectrometry, *Anal. Methods* 2, 259–267.
Lemos, V.A. and Gama, E.M. (2010). An online preconcentration system for the determination
    of uranium in water and effluent samples, *Environ. Monit. Assess.* 171, 163–169.
Li, W., Zhao, J., Li, C. et al. (2006). Speciation measurements of uranium in alkaline waters
    using diffusive gradients in thin films technique, *Anal. Chim. Acta* 575, 274–280.
Lotfy, S.M., Mostafa, A.Z., and Abdel-Sabour, M.F. (2012). Fractionation of uranium forms as
    affected by spiked soil treatment and soil type, *Agric. Sci. Res. J.* 2, 59–64.
Malekzadeh, F., Farazmand, A., Ghafourian, H. et al. (2002). Uranium accumulation by a bac-
    terium isolated from electroplating effluent, *World J. Micobiol. Biotechnol.* 18, 295–300.
Maxwell, S.L., Culligan, B.K., Kelsey-Wall, A. et al. (2011). Rapid radiochemical determina-
    tion of actinides in emergency concrete and brick sample, *Anal. Chim. Acta* 701, 112–118.
Maxwell, S.L., Culligan, B.K., and Noyes, G.W. (2010). Rapid separation of actinides and
    radiostrontium in vegetation samples, *J. Radioanal. Nucl. Chem.* 286, 273–282.
Mironov, V.P., Matusevich, J.L., Kudrjashov, V.P. et al. (2005). Determination of uranium con-
    centration and burn-up of irradiated reactor fuel in contaminated areas in Belarus using
    uranium isotopic ratios in soil samples, *Radiochim. Acta* 93, 781–784.
Mitchell, N., Perez-Sanchez, D., and Thorne, M.C. (2013). A review of the behaviour of U-238
    series radionuclides in soils and plants, *J. Radiol. Prot.* 33, R17–R48.
Mohapatra, P.K. and Manchanda, V.K. (1998). Ion-pair extraction of uranyl ion from aqueous
    medium using crown ethers, *Talanta* 47, 1271–1278.
Mola, M., Nieto, A., Penalver, A. et al. (2014). Uranium and thorium sequential separation
    from norm samples by using a SIA system, *J. Environ. Radioact.* 127, 82–87.
Neves, M.O., Figueiredo, V.R., and Abreu, M.M. (2012). Transfer of U, Al and Mn in the
    water–soil–plant (*Solanum tuberosum* L.) system near a former uranium mining area
    (Cunha Baixa, Portugal) and implications to human health, *Sci. Total Environ.* 416,
    156–163.
Parkhurst, M.A. (2003). Measuring aerosols generated inside armoured vehicles perforated by
    depleted uranium ammunition, *Radiat. Prot. Dosimetry* 105, 167–170.
Pilz, N., Hoffmann, P., Lieser, K.H. et al. (1987). Determination of thorium and uranium in air
    filters by XRF analysis using excitation of K-lines by radionuclides, *Fresenius Z. Anal.
    Chem.* 329, 581–583.
Pollmann, K., Raff, J., Merroun, M. et al. (2006). Metal binding by bacteria from uranium
    mining waste piles and its technological application, *Biotechnol. Adv.* 24, 58–68.
Popov, L. (2012). Method for determination of uranium isotopes in environmental samples by
    liquid–liquid extraction with triisooctylamine/xylene in hydrochloric media and alpha
    spectrometry, *Appl. Radiat. Isot.* 70, 2370–2376.
Prasada Rao, T., Metilda, P., and Gladis, J.M. (2006). Preconcentration techniques for ura-
    nium (VI) and thorium (IV) prior to analytical determination—An overview, *Talanta*
    68, 1047–1064.
Pretty, J.R., Duckworth, D.C., and Van Berkel, G.J. (1998a). Electrochemical sample pretreat-
    ment coupled on-line with ICPMS: Analysis of uranium using an anodically conditioned
    glassy carbon working electrode, *Anal. Chem.* 70, 1141–1148.
Pretty, J.R., Van Berkel, G.J., and Duckworth, D.C. (1998b). Adsorptive stripping voltam-
    metry as a sample pretreatment method for trace uranium determinations by inductively
    coupled plasma mass spectrometry, *Int. J. Mass Spectrom.* 178, 51–56.

Ramola, R.C., Choubey, V.M., Prasad, G. et al. (2011). Radionuclide analysis in the soil of Kumaun Himalaya, India using gamma ray spectrometry, *Curr. Sci.* 100, 906–914.

Rathore, D.P.S. (2013). Letter to the editor: Query related to publication titled "A comparative analysis of uranium in potable waters using laser fluorimetry and ICPMS techniques" by Shenoy et al. 294, 413–417 (2012). doi:10.1007/s10967-012-1705-2, *J. Radioanal. Nucl. Chem.* 298, 721–723.

Regenspurg, S., Margot-Roquier, C., Harfouche, M. et al. (2010). Speciation of naturally accumulated uranium in an organic-rich soil of an alpine region (Switzerland), *Geochim. Cosmochim. Acta* 74, 2082–2098.

Roivainen, P., Makkonen, S., Holopainen, T. et al. (2011). Soil-to-plant transfer of uranium and its distribution between plant parts in four boreal forest species, *Boreal Environ. Res.* 16, 158–166.

Romaniello, S.J., Herrmann, A.D., and Anbar, A.D. (2013). Uranium concentrations and $^{238}U/^{235}U$ isotope ratios in modern carbonates from the Bahamas: Assessing a novel paleoredox proxy, *Chem. Geol.* 362, 305–316. http://dx.doi.org/10.1016/j.chemgeo.2013.10.002.

Sadeghi, S., Mohammadzadeh, D., and Yamini, Y. (2003). Solid-phase extraction—Spectrophotometric determination of uranium(VI) in natural waters, *Anal. Bioanal. Chem.* 375, 698–702.

Sakaguchi, A., Kadokura, A., Steier, P. et al. (2012a). Isotopic determination of U, Pu and Cs in environmental waters following the Fukushima Daiichi Nuclear Power Plant accident, *Geochem. J.* 46, 355–360.

Sakaguchi, A., Kadokura, A., Steier, P. et al. (2012b). Uranium-236 as a new oceanic tracer: A first depth profile in the Japan Sea and comparison with caesium-137, *Earth Planet. Sci. Lett.* 333–334, 165–170.

Santos, J.S., Teixeira, L.S.G., Dos Santos, W.N.L. et al. (2010). Uranium determination using atomic spectrometric techniques: An overview, *Anal. Chim. Acta* 674, 143–156.

Sasmaz, A. and Yaman, M. (2008). Determination of uranium and thorium in soil and plant parts around abandoned lead–zinc–copper mining area, *Commun. Soil Sci. Plant Anal.* 39, 2568–2583.

Schultz, M.K., Burnett, W.C., and Inn, K.G.W. (1998). Evaluation of a sequential extraction method for determining actinide fractionation in soils and sediments, *J. Environ. Radioact.* 40, 155–174.

Serdeiro, N.H. and Marabini, S. (2004). A rapid method for determination of uranium, americium, plutonium and thorium in soil samples. *Eleventh International Congress on the International Radiation Protection Association*, Madrid, Spain, May 23–28, 2004.

Shamsipur, M., Ghiasvand, A.R., and Yamini, Y. (1999). Solid-phase extraction of ultratrace uranium(VI) in natural waters using octadecyl membrane disks modified by tri-n-octylphosphine oxide and its spectrophotometric determination with dibenzoylmethane, *Anal. Chem.* 71, 4892–4895.

Shemirani, F., Abkenar, S.D., and Jamali, M.R. (2003). Preconcentration of trace uranium from natural water with solid-phase-extraction, *Bull. Chem. Soc. Jpn.* 76, 545–548.

Shtangeeva, I. (2010). Uptake of uranium and thorium by native and cultivated plants, *J. Environ. Radioact.* 101, 458–463.

Srivastava, S.K., Balbudhe, A.Y., Vishwa Prasad, K. et al. (2014). Variation in the uranium isotopic ratios $^{234}U/^{238}U$, ^{238}U/total-U and ^{234}U/total-U in Indian soil samples: Application to environmental monitoring, *Radioprot.* 48, 231–242.

Strok, M. and Smodis, B. (2010). Fractionation of natural radionuclides in soils from the vicinity of a former uranium mine Zirovski vrh, Slovenia, *J. Environ. Radioact.* 101, 22–28.

Strok, M. and Smodis, B. (2013). Soil-to-plant transfer factors for natural radionuclides in grass in the vicinity of a former uranium mine, *Nucl. Eng. Design* 261, 279–284.

Stromman, G., Rosseland, B.O., Skipperud, L. et al. (2013). Uranium activity ratio in water and fish from pit lakes in Kurday, Kazakhstan and Taboshar, Tajikistan, *J. Environ. Radioact.* 123, 71–81.

Su, Y.Y., Li, Z.M., Li, M. et al. (2014). Development of aerosol sample introduction interface coupled with ICP-MS for direct introduction and quantitative on-line monitoring of environmental aerosol, *Aerosol Sci. Technol.* 48, 99–107.

TECDOC-1616, IAEA. (2009). Quantification of radionuclide transfer in terrestrial and freshwater environments for radiological assessments. Technical, IAEA, Vienna, Austria.

Thakur, P., Ballard, S., and Conca, J.L. (2011). Sequential isotopic determination of plutonium, thorium, americium and uranium in the air filter and drinking water samples around the WIPP site, *J. Radianal. Nucl. Chem.* 287, 311–321.

Thiry, Y., Schmidt, P., Van Hees, M. et al. (2005). Uranium distribution and cycling in Scots pine (*Pinus sylvestris* L.) growing on a regevetated U-mining heap, *J. Environ. Radioact.* 81, 201–219.

Turner, C.S.G.., Mills, G.A., Teasdale, P.R. et al. (2012). Evaluation of DGT techniques for measuring inorganic uranium species in natural waters: Interferences, deployment time and speciation, *Anal. Chim. Acta* 739, 37–46.

Vera-Tome, F., Blanco Rodriguez, M.P., and Lozano, J.C. (2003). Soil-to-plant transfer factors for natural radionuclides and stable elements in a Mediterranean area, *J. Environ. Radioact.* 65, 161–175.

Vogt, K.A., Vogt, D.J., Palmiotto, P.A. et al. (1996). Review of root dynamics in forest ecosystems grouped by climate, climatic forest type and species, *Plant Soil* 187, 159–219.

Winde, F. (2002). Uranium contamination of fluvial systems—Mechanisms and processes. Part III. Diurnal and event-related fluctuations of stream chemistry—Pitfalls from mining affected streams in South Africa, Germany and Australia, *Cuadernos de Investifacion Geografica* 28, 75–100.

Winde, F. (2009). Uranium pollution of water resources in mined-out and active goldfields of South Africa—A case study in the Wonderfonteinspruit catchment on extent and sources of U-pollution and associated health risks. *International Mine Water Conference*, Pretoria, South Africa, pp. 772–782.

Winde, F. (2013). Uranium pollution of water: A global perspective on the situation in South Africa. *Inaugural Lecture Presented at the North-West University*, Vaal Triangle Campus, South Africa, February 22, 2013.

Yang, Y., Xiao, S., Zhang, Y. et al. (2013). In situ detection of trace aerosol uranium using a handheld photometer and solid reagent kit, *Anal. Methods* 5, 4785–4789.

Yoshida, S., Muramatsu, Y., Tagami, K. (2001). Determination of uranium isotopes in soil core samples collected on JCO grounds after the criticality accident, *Environ. Sci. Technol.* 35, 4174–4179.

Zavodska, L., Kosorinova, E., Lesny, J. et al. (2009). Physical–chemical characterization of uranium containing sediments, *Nova Biotechnologica* 9–3, 303–311.

Zavodska, L., Kosorinova, E., Scerbakova, K. et al. (2008). Environmental chemistry of uranium. HU ISSN 1418-7108: HEJ Manuscript no.: ENV-081221-A, pp. 1–19, http://heja.szif.hu/ENV/ENV-081221-A/env081221a.pdf (accessed July 30, 2014).

Zorer, O.S. and Sahan, T. (2011). The concentration of 238U and the level of gross radioactivity in surface waters of the Van Lake (Turkey), *J. Radioanal. Nucl. Chem.* 288, 417–421.

4 Exposure, Toxicity, and Biomonitoring of Uranium Exposure

"Alle Ding' sind Gift, und nichts ohn' Gift; allein die Dosis macht, daß ein Ding kein Gift ist" (English translation: "All things are poison, and nothing is without poison; only the dose permits something not to be poisonous.").

Paracelsus (Philippus Bombastus von Hohenheim, 1493–1541)

4.1 INTRODUCTION

This chapter deals with several aspects of the effects of uranium on human health. It begins with a description of the pathways through which humans may be exposed to uranium compounds, continues with a discussion of the toxicity of uranium, and concludes with a survey of the techniques used for determining exposure and assessment of dose incurred. As this treatise is concerned mainly with the analytical chemistry of uranium, each of these topics includes some representative examples of the analytical techniques and sample preparation procedures used. As will be shown later, there are many types of analytical instruments, measurement methodologies, and techniques, and there is no approach that is universally accepted as the definitive method. As always, the analyst should consider the available instrumentation, the information required, and the preparation method best suited for the specific type of sample. The main points in each section are summarized in the form of highlights just like in the other chapters.

4.2 EXPOSURE PATHWAYS

4.2.1 SOURCES OF ENVIRONMENTAL AND OCCUPATIONAL EXPOSURES

The route and site of exposure influence the effects of toxic chemicals on the body. In general, in the order of descending effectiveness and hazard, the routes for toxic agents are directly into the bloodstream, inhalation, intraperitoneal, subcutaneous, intramuscular, intradermal, oral, and dermal (Klaassen 1995, p. 15). This order would significantly differ for xenobiotics (substances that are foreign to the human body) that also have indirect health consequences, like radiological properties and genetic effects.

Uranium is a naturally occurring element and is quite abundant in soil, seawater, fresh water sources, and plants as described in the previous chapters. Consequently,

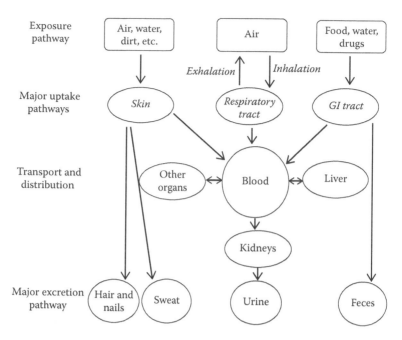

FIGURE 4.1 The main pathways for exposure to uranium compounds, the uptake and distribution routes, and the major excretion pathways.

humans are continually exposed to uranium compounds mainly through their diet and less significantly to uranium-containing aerosols through the air they breathe. In rare cases, almost solely in the nuclear industry or as a result of military action, exposure to uranium could also occur through injury and open wounds. Figure 4.1 schematically depicts these exposure pathways and the routes for transporting uranium from the organs that are exposed initially (gastrointestinal tract, lungs, and skin) to the main body organs. The body systems that may be impacted by exposure to depleted uranium (DU) can be seen in other publications (La Touche et al. 1987; Craft et al. 2004), and the biokinetic model of Leggett for embedded DU fragments is shown in Figure 4.2. For a comprehensive treatise on DU, the reader is referred to Miller (2007).

A fine distinction should be drawn between inhalation and ingestion of soluble and insoluble uranium compounds. Ingestion of insoluble uranium compounds has little health effects as practically all the material will be excreted in feces without entering the bloodstream and reaching vital organs, while if inhaled these particles may cause localized damage to lung tissues over an extended period. On the other hand, soluble uranium compounds entering the lungs or the digestive tract may be incorporated in the blood and absorbed by other organs or excreted from the body.

4.2.1.1 Daily Intake of Uranium through Ingestion

Ingestion of uranium contained in food and drinking water is the main route through which the general population is exposed to uranium compounds. The estimation of the average daily intake of uranium from food and drinking water based on several

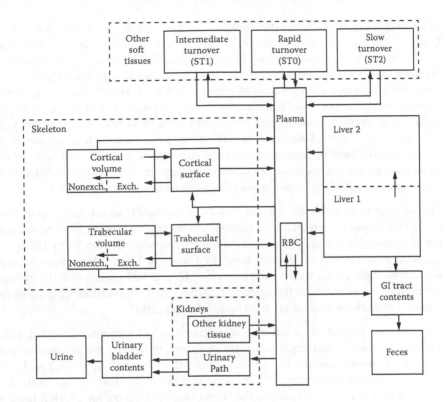

FIGURE 4.2 Biokinetic model for embedded fragments of depleted uranium. (From Leggett, R.W. and Pellmar, T.C., *J. Environ. Radioact.*, 64, 205, 2003b. With permission.)

studies, spans a wide range, depending on the location, source of drinking water, local regulatory limits and the adherence to them, and on the dietary habits of the study population. A large number of investigations in many countries have attempted to determine the amount of uranium consumed by the population in the study area and the results vary considerably. The results from several such studies are presented later, and the analytical methods used to assay the uranium content in drinking water and in food products are also described.

In many countries, there is a legal limit for the concentration of uranium in drinking water that is usually set between 15 and 30 µg L^{-1}. However, these limits apply to communal water supplies while people that obtain their drinking water from other sources like private wells or reservoirs that are not frequently monitored may actually regularly consume much larger amounts of uranium.

The United States: In several studies, the results varied according to the type of drinking water measured and the location of the study population. For example, one study reported a range of 0.07–1.1 µg of uranium ingested per day (µg day^{-1}) (EPA), another study found 2–3 µg day^{-1} (Singh et al. 1990), while yet slightly higher values reaching 4 µg day^{-1} were also reported (WHO 2004). According to one estimate, the average concentration of uranium in communal supplies of drinking water in the United States was given as 2.5 µg L^{-1} (EPA 2006), and the average daily intake of natural uranium was thus estimated as 3 μg day^{-1} (implying consumption of 1.2 L day^{-1}). On the other hand, a survey of uranium in drinking water in New York City found very low concentrations in the range of 0.03–0.08 μg L^{-1} (Fisenne and Welford 1986). As mentioned earlier, these values pertain to populations that receive their drinking water from communal supplies that are regularly monitored, but in isolated locations, where the main source of drinking water is from local wells, the actual intake could be much higher. For example, in a 1991–1992 survey of 186 wells in North Dakota, it was found that 48 of them contained uranium levels above 20 μg L^{-1} and in 26 of these the uranium concentration was over 100 μg L^{-1} (Roberts 2008). In an earlier more extensive survey, carried out between 1975 and 1977, 2864 wells were sampled and 82 contained uranium concentrations above 100 μg L^{-1}.

Canada: For some residents, the daily intake of uranium from drinking water alone could be as high as several hundred micrograms, as reported for the Nova Scotia region of Canada where uranium levels in drinking water were between 5 and 830 μg L^{-1} (Zamora et al. 1998). On the other hand, in a study of 130 drinking water sources in Ontario, the uranium level varied between 0.05 and 4.21 μg L^{-1} with the average uranium intake estimated as 0.8 μg day^{-1} (OMEE 1996). The Canadian guideline for uranium in drinking water is set at 20 μg L^{-1} (Canada 2000).

Germany: A study of the uranium concentration in 908 samples of bottled water and 163 samples of municipal tap water in Germany found a broad range of uranium levels from below the limit of detection (0.0005 μg L^{-1}) up to 16.0 μg L^{-1} in bottled water and up to 9.0 μg L^{-1} in communal tap water (Birke et al. 2010). The German limit was set according to the World Health Organization (WHO) limit of 15 μg L^{-1} for uranium in drinking water. A special case involves bottled water used for baby food so if the uranium level in bottled water is above 2 μg L^{-1} then the label on the bottle may not say that it is suitable for preparation of baby food. Detailed maps of the distribution of uranium in bottled and drinking water within Germany were presented and the level of uranium in German bottled water was compared with some neighboring countries in a box diagram (Figure 4.3). In another study, the uranium intake from food in three federal states was investigated and although regional differences were found the daily intake for most of the population was below 2 μg day^{-1} as shown in Figure 4.4 for men (Anke et al. 2009). In the same study, quite similar results were also reported for women in those three German federal states.

In this study, the samples were digested with hydrochloric acid, rhodium was used as an internal standard, and uranium measurement was by ICPMS. In addition to an extensive survey of the uranium content in drinking water (tap water and bottled mineral water), many edible plants and other food products were also tested. The average uranium concentration in tap water in five regions of Germany was in the range of 0.28–2.4 μg L^{-1} with a maximum of 8.6 μg L^{-1}. The range in the bottled mineral water samples varied from below the limit of detection (in this case, 0.015 μg L^{-1}) to 24.5 μg L^{-1} for one brand (Anke et al. 2009).

Israel: In a study on the uranium content in tap water and fresh water sources in several locations in Israel, the concentration in communal drinking water supplies

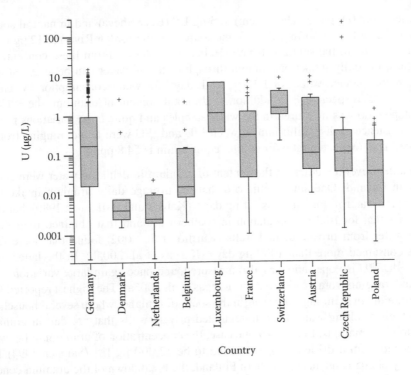

FIGURE 4.3 Comparison of uranium concentrations in bottled water in some European countries. (From Birke, M. et al., *J. Geochem. Explor.*, 107, 272, 2010. With permission.)

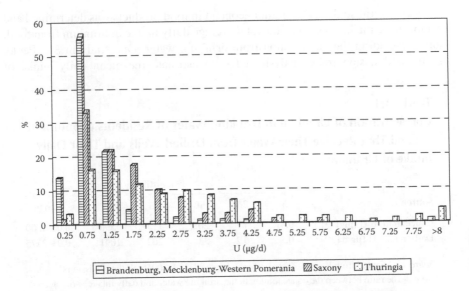

FIGURE 4.4 The distribution of uranium intake for men in three regions of Germany. (From Anke, M. et al., *Chemie der Erde*, 69(Suppl. 2), 75, 2009. With permission.)

varied from 0.8 µg L^{-1} (Jerusalem) to 5 µg L^{-1} (Beer-Sheva) and in natural sources from 0.2 µg L^{-1} (in springs that are the sources of the Jordan River) to 12 µg L^{-1} (in drilled wells in the south of Israel) (Halicz et al. 2000). From these concentration values, the daily intake of uranium through drinking water can be assessed using the recommended values of 1.2 and 1.6 L day^{-1} for water consumption by females and males, respectively. In addition to the total content of uranium, the ^{234}U/^{238}U isotope ratio was measured in the water samples and quite large variations proving the existence of disequilibrium between ^{234}U and ^{238}U were found, ranging from 64 to 183 ppm (expected ratio for secular equilibrium is 54.8 ppm).

Finland: Several studies on the content of uranium in drinking water were carried out in Finland. One study estimated that the average daily uranium intake from drinking water in Finland was 2.1 µg day^{-1} (Kahlos and Asikainen 1980), but other studies that focused on population in southern Finland that obtained their drinking water from private drilled wells (Kurttio et al. 2002) found that some families consumed more than 1000 µg day^{-1} (Karpas et al. 2005b). In the latter study, the ^{234}U/^{238}U isotope ratio was also measured, and once again large variations were found, ranging from 52 to 150 ppm (Karpas et al. 2005a). The highest reported level of uranium in drinking water was in the south of Finland, where several households get their drinking water from unregulated private wells that are dug in granitoid bedrock formations. In one extreme case, the concentration of uranium in tap water obtained from a drilled well was found to be 12,000 µg L^{-1} (Salonen 1994). In a survey of 205 residents of southern Finland, the breakdown of the uranium concentration in tap water was as shown in Table 4.1, where for the lowest 5% a mean value of 0.47 µg L^{-1} was found but the mean concentration of the top 5% exceeded 590 µg L^{-1} (Karpas et al. 2005b).

Brazil: In several studies, the uranium content in food products was determined and attempts were made to estimate the total average daily intake of uranium from food, usually excluding the contribution from drinking water. One study in Sao Paolo, Brazil, used fission track analysis (FTA) to estimate the mean daily intake of

---

TABLE 4.1

Concentration of Uranium in Drinking Water of Residents of Southern Finland That Receive Their Water from Drilled Wells and Their Daily Intake of Uranium

Source	Percentile					
	5	25	50	75	95	Range
Concentration U [µg L^{-1}]	0.47	4.7	20.2	117	590	0.03–1500
Daily intake U [µg day^{-1}]	0.37	6.2	33	202	1091	0.03–2775

Source: Adapted from Karpas, Z. et al., *Health Phys.* 88, 229, 2005b. With permission.

Note: The ratio between the concentration in the drinking water and daily intake varies according to the reported amount of water consumed.

TABLE 4.2
Estimated Daily Intake of Uranium (μg U day⁻¹) in Different Countries

Country	Total Daily Intake (μg U/day)	Source of Uranium Intake
Brazil (Sao Paolo)	0.97	Mean value of uranium
United States (New York)	1.30	content in various food types
United States (Chicago)	1.40	
United States (San Francisco)	1.30	
United Kingdom	1.00	
Japan (Yokohama)	0.10	
Vietnam	0.70	
Italy	3.90	Mean value of food diet and
Poland	1.80	drinking water
Ukraine	0.60	
Japan	0.70	
India	2.20	
Hong Kong	15.3	
France	1.00	
Russian Federation	3.50	
Worldwide	1.30	
Worldwide	1.30	

Source: Adapted from Garcia, F. et al., *Envrion. Intern.*, 32, 697, 2006. With permission.
Note: The daily intake found in several other studies is mentioned in the text.

uranium by residents and reported 0.97 μg day⁻¹ (Garcia et al. 2006). Their estimate was based on measurement of the uranium level in the main food products consumed by the local population but without taking into account the contribution from uranium in drinking water. By comparing their results to those obtained from several other studies (see Table 4.2), they concluded that the range of 0.7–1.4 μg day⁻¹ was typical for consumption of uranium in food (with exclusion of water) in many countries. Studies that included the contribution from drinking water ranged from 0.60 μg day⁻¹ in Ukraine to 15.3 μg day⁻¹ in Hong Kong, as shown in the lower part of Table 4.2.

Japan: A comprehensive study of the consumption of radionuclides in the diet (food products excluding drinking water) of Japanese population in 13 cities was carried out (Sugiyama and Isomurac 2007). The food samples were divided into 14 groups and the gamma emitting radionuclides were measured from the ashed food samples by a high-purity germanium (HPGe) detector for a counting time of $2 * 10^5 - 3 * 10^5$ s. The mean daily intake of ^{238}U for the population was estimated to be 12.6 ± 7.5 mBq day⁻¹ (~1.02 μg day⁻¹).

Pakistan: A study of the daily intake of uranium from food in a typical Pakistani diet used ICPMS for the analysis (Akhter et al. 2003). The range of daily intake was estimated to be 1.4–6.7 μg day⁻¹ with a geometric mean of 2.6 μg day⁻¹.

Morocco: A number of studies were carried out in Morocco to evaluate the annual committed effective doses from consumption of uranium and thorium in a typical Moroccan food basket (Misdaq and Bourzik 2004). FTA analysis was used to determine the content of U and Th in some popular food products (fruit, vegetables, and cereals), and the results were compared with those obtained by isotope dilution mass spectrometry (IDMS) finding between 0.26 and 0.56 $\mu g\ g^{-1}$ for most products tested. The final estimation was that the annual intake for members of the Moroccan population was 451 ± 27 Bq year^{-1} (note that 1.24 Bq day^{-1} is equivalent to ~100 μg day^{-1}) and 359 ± 20 Bq year^{-1} for ^{238}U and ^{232}Th, respectively. This study did not include the intake from drinking water. These values are about 100 times higher than the estimates from most other studies—and this is probably due to a calculation error.

Several other examples are presented in the following sections that deal specifically with the analytical aspects, but the point we wish to emphasize is that while the uranium concentration in food products appears to be quite similar in different locations the level of uranium in drinking water varies widely between geographical locations, even within the same region, and could also change with time at the same location. Therefore, attempts to assess the actual intake of uranium through water and food are at best a statistical exercise for the whole population with individual variations that could span several orders of magnitude.

4.2.1.2 Daily Intake of Uranium through Inhalation

The concentration of uranium in air is usually very low and this exposure pathway is negligible for the general population unless the people live in the vicinity of industrial plants or mines where uranium is processed or are employed in such facilities. The results of some studies in which the concentration of uranium in air was determined are discussed here.

A report by the World Health Organization (WHO/SDE/WSH/03.04/118 2004) gives the mean levels of uranium in ambient air as 0.02 ng m^{-3} in Tokyo, Japan (Hirose and Sugimura 1981), and 0.076 ng m^{-3} in New York City (Fisenne et al. 1987). Assuming a daily respiratory volume of 20 m^3 (see Table 4.3 for hourly inhalation rates) and a mean urban airborne concentration of 0.05 ng m^{-3}, then the daily intake of uranium from air would be about 1 ng. To put this in perspective, tobacco smoke (from two packets of cigarettes per day) contributes about 50 ng of inhaled uranium per day (Lucas and Markun 1970). It should be noted that the air inhalation rate depends on the activity level, age, weight, gender, and general physical condition, as shown in Table 4.3 (Wise 2012).

A special case of inhalation exposure to uranium would be in war zones in which DU munitions were used and where fine uranium oxide aerosols were produced through DU burning. Several studies have estimated the exposure through inhalation of uranium miners and the general consensus is that the main risk is from uranium progenies like radon and polonium. In a comprehensive report (Coons 2011), several studies are cited like the one that surveyed workers in uranium mining and milling operations in Grants, New Mexico (Boice et al. 2010). The conclusion of this study was "With the exception of male lung cancer, this study provides no clear or consistent evidence that the operation of uranium mills and mines adversely affected cancer incidence or mortality of county residents" (Boice et al. 2010). Perhaps the most intensively studied

TABLE 4.3
Inhalation of Fugitive Dusts

Inhalation Rates ($m^3\ h^{-1}$)

	Activity Level			
	Resting	Light	Moderate	Heavy
Adult male	0.7	0.8	2.5	4.8
Adult female	0.3	0.5	1.6	2.9
Average adult	0.5	0.6	2.1	3.9
Child age 6	0.4	0.8	2.1	2.4
Child age 10	0.4	1.0	3.2	4.2

Source: http://www.wise-uranium.org/rdcush.html.

case of exposure of uranium miners to radioactive substances (uranium and progeny) is the Wismut study that included tens of thousands of people that worked in former East Germany (then GDR (German Democratic Republic))—summarized in Frame 4.1.

It should be remembered that uranium-containing dust may be present in mining operations in which uranium is a minor component that is exploited as a by-product (e.g., in some phosphates excavations or gold mines), or even at trace levels without commercial value.

4.2.1.3 Exposure to Uranium through Skin and Injury

Workers in the nuclear industry or people involved in uranium mining or milling may be accidentally exposed to uranium through cuts or injuries. Fragments of DU munitions that are embedded in soft tissues may be slowly oxidized in the body and provide a continuous source that is absorbed by the bloodstream. Selected studies of these special populations are discussed later. However, the general population would not be normally exposed to uranium through this pathway.

Highlights: A clear distinction should be made between exposure pathways that lead to internal exposure mentioned earlier and external exposure to uranium that can have only minor health effects due to the low specific radioactivity of uranium. In summary, it could be stated that the general population may be exposed to uranium compounds mainly through ingestion of drinking water and food products and to some minor extent through inhalation. The exact extent of intake of uranium of any individual or even any community is difficult to estimate as large variations are found. Despite the fact that ingestion of food is the source, this type of exposure could be considered as environmental exposure. Exceptions to exposure of this type include people employed in mining operations (either directly in uranium mines or indirectly in mines where uranium is a major impurity) and the nuclear industry where inhalation or exposure through the skin wounds might also occur accidentally. In some cases, population residing in, or close to, areas that contain facilities in which uranium is processed or disposed (like mine tailings) could be exposed to uranium through their diet (contaminated drinking water and food products) or even

FRAME 4.1   WISMUT MINERS

The frantic race to obtain uranium for production of nuclear weapons intensified after dropping an atom bomb, based on enriched uranium, on Hiroshima (August 6, 1945) and a plutonium based bomb (plutonium produced in an uranium-fueled nuclear reactor) on Nagasaki 3 days later. The mineral-rich *Ore Mountains* in Central Europe were partly in Czechoslovakia (the famous Joachimsthal or Jachymov mines—origin of the minerals in which uranium, and later radium and polonium, were discovered) and partly in East Germany (DDR or GDR) that was occupied by the Soviet Union. A mining company that was given the codename Wismut (German for bismuth) was formed to develop uranium mining for the Soviet nuclear program. Until 1953, when the responsibility for operating the company was transferred to a joint Soviet-German Stock Company (SDAG—in the German acronym) most of the workers were compulsory or forced laborers (prisoners or conscripts) and working conditions could be described (quite euphemistically) as *poor*. For example, dry drilling of rocks with air flushing and without forced ventilation led to high dust concentrations as well as heavy workloads that included long hours without proper equipment and technical aid. After the mid-1950s, conditions were better and the workforce consisted mainly of civilians, and by the 1970s industrial hygiene greatly improved and ICRP recommendations were fulfilled (Enderle and Friedrich 1999; Zoellner 2010). The operation closed down in 1990, but the database of former Wismut employees that contained a total of about 400,000 people is the largest of its kind and served in many studies of the health effects on uranium miners [see, e.g., Grosche et al. 2006 and Bijwaard et al. 2011, and references within]. It is beyond the scope of this book to discuss all the detrimental health effects that these uranium miners suffered from, so here are the concluding remarks of one of the comprehensive studies (Enderle and Friedrich 1999): "Uranium mining at the Wismut enterprise in the GDR from 1946 to 1990 is an important historical example of considerable chronic exposure to radon progeny for several hundred thousands of people. Exposure was highest in the early post-war years. A total of 5000–6000 cases of bronchial carcinoma are already accepted as compensable occupational diseases to date." The author estimates that many cases of lung cancer arose during the early postwar years due to exposure to high dust levels and radionuclides.

In their study of lung cancer risk in a cohort of 59,001 Wismut workers that encompassed 1,801,630 person-years, the excess relative risk per working-level-month (WLM) was estimated as 0.21% (Grosche et al. 2006). The highest excess risk was observed 15–24 years after exposure in the youngest age group. Interestingly while the excess risk was proportional to the length of exposure (WLM), no significant association with lung cancer was found below exposure of 100 WLM.

Some remarks on these studies: There is little doubt that the lack of proper hygienic conditions in the early postwar days led to a large number of lung cancer and bronchial diseases [for a vivid description of those conditions, see (Zoellner 2010)]. The main causes were uranium containing dust and radon and their progenies, but there could also be some contribution to illness from arsenic and silicon in the breathable dust. External gamma radiation may have also played a role.

through inhalation. Populations that have been involved in military action may also be exposed to DU munitions either if injured by fragments of the material or through inhalation of uranium oxide aerosols formed when the DU is ignited in air.

4.2.1.4 Analytical Methods for the Determination of Uranium in Drinking Water

Some analytical methods were mentioned in brief in the previous sections of this chapter and in more detail in Chapters 1 through 3. There are several analytical methods that are commonly used for the measurement of the uranium concentration in drinking water, ranging from simple in-field methods to sophisticated laboratory methods. Some of the latter can also provide isotopic composition data. Analytical methods that are deployed for surface water and sea water in environmental studies (Chapter 3) are mostly also suitable for measurement of the uranium content in drinking water. Among these are colorimetric spectroscopy methods, electro-analytical methods, radiation monitoring, nuclear spectrometry, FTA, laser-induced fluorescence (LIF), solid fluorescence, thermal ionization mass spectrometry (TIMS), inductively coupled plasma optical emission spectrometry (ICP-OES), and inductively coupled plasma mass spectrometry (ICPMS). A comparison of some of the methods (photometry, laser fluorescence, liquid scintillation, gamma and alpha spectrometry) used for determining uranium in water was published (Tosheva et al. 2004).

The differences in sample size, sample preparation procedures, counting time, accuracy, isotope composition, and minimum detectable limits (MDL) should be noted. Some of these methods will be briefly surveyed here. The methods for determining uranium in urine are usually also suitable for measuring the concentration of uranium in water, and those that are described in detail in Section 4.4.1 will not be discussed here to avoid duplication.

Colorimetric Method (Photometry): A rapid, simple colorimetric method for the determination of uranium in groundwater and drinking water that can be used in the laboratory and in the field was described (Ratliff 2008). The first step is selectively trapping the uranium on a chromatographic resin (U/TEVA-2 in this example) followed by the formation of a colored complex with a pyridylazo indicator dye (Br-PADAP). At neutral pH, the complex absorbs light at 578 nm and is clearly visible. Quantification can be achieved with a spectrophotometer. The method can detect uranium concentrations above the US-EPA guideline for drinking water of 30 µg L^{-1}. In Section 4.4.1, another colorimetric method that is suitable for the determination of uranium in water and urine based on the formation of a complex with Arsenazo-III is presented.

Electroanalytical Methods: Many publications described the use of electro-analytical methods for measuring uranium in water and some examples are discussed here. A detailed review article with illustrative tables that summarize the electroanalytical methods for the determination of uranium can be found elsewhere (Shrivastava et al. 2014). The tables in that review article list the method, the principles of the measurement technique, the linear range, limit of detection, tolerance to interferences (where defined), and the field in which the method is applied.

Stripping voltammetry with a mercury electrode after controlled adsorptive preconcentration of uranium–cupferron complex was used for determining uranium concentration in ground water samples (Wang and Setiadji 1992). The advantages of the method are that the device can be used in the field and that it was claimed that it had a high throughput of 30–60 samples per hour. Two decades later, a modified version using a thin film of mercury on coated carbon nanotubes was described (Sahoo et al. 2013). A simple voltammetric method for determining uranium in water was described (Rajusth et al. 2013). The standard potential (E_0) for U(VI) – U (V) reduction is positive at +0.16 V so that the polarographic wave might overlap with the oxidation wave of mercury at this potential. Therefore, a medium of thioglycolic acid in acetate buffer was used to shift the reduction potential to a measurable negative voltage. Differential pulse polarography (DPP) was used to overcome other interferences. A linear calibration curve up to 20 μg mL^{-1} was shown with a limit of determination of 0.005 μg mL^{-1}. The results for uranium in waste water samples were validated by comparison with a spectrophotometric technique. In another study, a gold electrode was treated with modified 2-mecaptoethanol in order to obtain surface phosphate active sites that were specific for uranyl ions (Becker et al. 2009).

Inductively Coupled Plasma Optical Emission Spectrometry (ICP-OES or ICP-AES): This is one of the commonly deployed methods to determine the concentration of uranium in liquid samples, particularly when a multielement assay is required. A method involving solid phase extraction (SPE) on a mesoporous silica sorbent for 100-fold preconcentration of uranium and thorium from 100 mL aqueous samples prior to measurement with ICP-OES was described (Yousefi et al. 2009). Limit of detection was reported as 0.3 μg L^{-1}. A slightly different approach, using octadecyl-bonded silica in the presence 2,3-dihydro-9,10-dihydroxy-1,4-anthracenedion, for preconcentration prior to ICP-OES analysis was also presented (Daneshvar et al. 2009).

Pulsed Laser Phosphorimetry: This is the basis for an ASTM standard test method for measurement of total uranium in water following wet ashing when impurities or suspended materials are present (D5174 2013).

Alpha spectrometry can also be used to determine the total uranium content in water and the contribution of the alpha emitting ^{234}U and ^{238}U isotopes and is described as a *fast cost-effective method* (D6239 2009). The procedure involves adding a spike of ^{232}U to 200 mL of acidified drinking water, boiling it to remove radon and to reduce the volume to 50 mL. After adding DTPA (diethylene-triamine-pentaacetic acid) and adjusting the pH to 2.5–3.0, solvent extraction into the scintillator solvent is carried out and the sample is counted. The alpha spectrum clearly shows the peaks due to the three active isotopes ^{232}U, ^{234}U, and ^{238}U at energies of 5.3, 4.8, and 4.2 MeV, respectively.

Inductively Coupled Plasma Mass Spectrometry (*ICPMS*) is nowadays the method of choice in many studies due to its sensitivity, simplicity (sample preparation is rarely required), cost, throughput, and availability of isotopic composition information. Several examples of deployment of ICPMS for all kinds of assays will be presented throughout this chapter.

4.2.1.5 Analytical Methods for the Determination of Uranium in Food Products

Several analytical procedures involve ashing of the food product followed by measurement of the uranium content by nuclear methods: for example, gamma spectrometry (Sugiyama and Isomurac 2007) and FTA (Misdaq and Bourzik 2004; Garcia et al. 2006) have been reported. Alternatively, the ashed sample can be digested and the uranium content determined by the standard analytical techniques, mainly ICPMS, ICPAES, and alpha spectrometry (Barratta and Mackill 2001). The concentration of uranium in some popular beverages is shown in Table 4.4 (Anke et al. 2009). Note that the concentration in most of these beverages is lower than the level found in many tap water samples. It is common practice by soft drink manufacturers to use water that has been passed through ion-exchange resins that remove most of the dissolved minerals (including uranium) and compounds that may affect the flavor and odor of the product.

An important factor that should be considered when determining the uranium content in food or plant samples is the amount of dry matter that remains after the water content has been removed. Some of the examples shown in Table 4.5 demonstrate this point as the percentage of dry matter in the food sample varies from a few percent (like 5.6% in tomatoes) to 100% in salt and sugar for example (Anke et al. 2009). Thus, the uranium content should routinely be reported on the basis of dry matter or, if not, the basis for the analysis should be specifically noted.

The content of uranium in food products cannot usually be determined directly by nondestructive assay (NDA) methods due to the low concentration that is normally present in the sample so that sample preparation is usually necessary. One exception is the use of neutron activation analysis (Zikovsky 2006) to determine the uranium

---

TABLE 4.4
Uranium Content (μg U L⁻¹) in Some Beverages

Beverage	Number of Samples	Mean Uranium Concentration ($\mu g\ U\ L^{-1}$)
White wine	6	1.27 ± 1.05
Coke	6	0.92 ± 0.52
Lemonade	6	0.89 ± 0.51
Red wine	6	0.68 ± 0.35
Vermouth	6	0.39 ± 0.21
Fruit juice	6	0.36 ± 0.31
Beer	6	0.30 ± 0.28

Source: Anke, M. et al., *Chemie der Erde*, 69(Suppl. 2), 75, 2009. With permission.

TABLE 4.5
Percentage of Dry Matter and the Uranium Content (μg U kg^{-1} Dry Weight) in Selected Food Products

Food with Low Uranium Content			Food with High Uranium Content		
Food	Dry Matter (%)	Uranium (μg U kg^{-1})	Food	Dry Matter (%)	Uranium (μg U kg^{-1})
Apples	10	2.0 ± 0.3	Cucumbers	5.2	12 ± 5.0
Banana	18	1.1 ± 0.2	Red hot pepper	88	18 ± 7.0
Tomatoes	5.6	3.0 ± 0.9	Pepper sweet	88	19 ± 10
Oranges	13	2.6 ± 1.8	Lettuce	7.3	39 ± 27
Peas, green	21	3.3 ± 1.9	Carrots	7.0	8.0 ± 4.3
Potatoes, peeled	18	3.0 ± 0.6	Mixed mushrooms	6.0	105 ± 39
Sugar	100	1.0 ± 0.2	Table salt	100	10 ± 3.0
Milk chocolate	95	1.5 ± 0.5	Cinnamon	88	6.3 ± 2.6
Coffee	96	1.7 ± 0.4	Black tea	94	8.8 ± 2.5
Biscuits	97	5.2 ± 4.0	Dill	10	31 ± 13
Rolls	75	1.9 ± 1.0	Asparagus	4.6	53 ± 32
Toasted bread	68	2.8 ± 1.0	Sauerkraut	9.7	8.0 ± 4.9
Butter	85	0.7 ± 0.2	Hens eggs	25	16 ± 11
Chicken	31	2.6 ± 0.6	Bismarck herring	29	9.3 ± 4.9
Gouda cheese	58	2.9 ± 1.6	Soft cheese	45	7.9 ± 3.4

Source: Anke, M. et al., *Chemie der Erde*, 69(Suppl. 2), 75, 2009. With permission.

level in several types of food products that were purchased at local stores in Montreal, Canada. The samples were placed in plastic bags, irradiated for 1 min with a flux of $5 * 10^{11}$ n cm^{-2} s^{-1}, and after 1 min cooling were counted for 3000 s with a low-energy photon spectrometer. Another example is the use of epithermal neutron activation analysis (ENAA) for the determination of the uranium content in Nigerian food products (Kapsimalis et al. 2009). In this case, the fruit and vegetable sample preparation involved freeze-drying and pulverizing to a fine powder while dry food products were just homogenized. Approximately half gram quantities of each sample were placed in 1.5 mL polyethylene vials. Sample irradiation was by a flux of $2 * 10^{11}$ n cm^{-2} of epithermal neutrons for 10 min, followed by a cooling period of 15 min to allow decay of ^{28}Al and counting with a HPGe detector for 10 min. Sulfur powder was used as a flux monitor and NIST 1570 spinach leaves standard was used for quality control. Figure 4.5 shows a comparison of the gamma-ray spectrum of uranium in normal mode and with Compton suppression.

An example of the extensive sample preparation required for the determination of uranium in food products by alpha spectrometry was described (Barratta and Mackill 2001). The procedure included homogenizing the food products, followed by treatment at 100°C in order to dry the sample and then increasing the temperature slowly to 600°C to complete the ashing and incineration. The sample was then dissolved in nitric acid and the uranyl ions were extracted with ethyl acetate

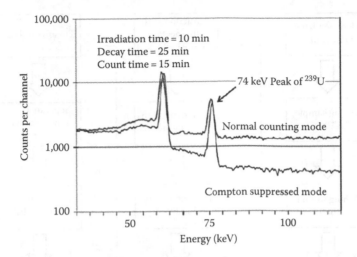

FIGURE 4.5 Gamma ray spectrum after excitation by epithermal neutron activation analysis (ENAA) for determination of the uranium content in Nigerian food products. Comparison of the gamma ray spectrum of uranium in normal mode and with Compton suppression. (From Kapsimalis, R. et al., *Appl. Radiat. Isot.*, 67, 2097, 2009. With permission.)

and aluminum nitrate. The uranium was stripped from the organic phase with water and the aluminum was removed by precipitation. The solution was evaporated to dryness, dissolved in hydrochloric acid, and then passed over an ion exchange column where uranium was separated from other actinides. Finally, the uranium is plated on a stainless steel disk and the alpha spectrum was measured. Figure 4.6 depicts a flowchart of these stages of sample preparation based on this work (Barratta and Mackill 2001) but can also serve as a schematic diagram of a generic approach for the preparation of food samples for analysis (in this case by alpha spectrometry but with some slight variation also by other analytical techniques).

Other common methods use ICPMS after digestion of the food product samples, as shown in Table 4.5 that is an excerpt from a much more extensive study of the uranium content in food products (Anke et al. 2009). One other example, discussed in more detail in Chapter 3, described a study of the uranium content in lettuce and tubers that involved first separation of the edible parts from the rest of the plant, gentle rinsing with water to remove soil particles and then acid digestion and analysis by ICPMS (Neves and Abreu 2009).

Highlights: The uranium content in drinking water from communal sources usually abides by the regulatory requirements and determination of the uranium concentration is straightforward by standard analytical techniques and sample preparation procedures. The basic procedure includes filtering to remove suspended particulate matter (if necessary) and acidification to prevent adsorption or precipitation of uranium. However, large variation in the uranium concentration and quality of drinking water from unregulated sources are known to exist so that additional sample preparation may be required, like solid phase extraction or separation. The variability in the physical and chemical properties of foodstuffs, like the fraction of dry matter, and

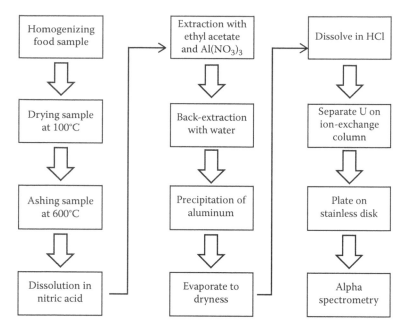

FIGURE 4.6 A flowchart of the stages of sample preparation based on the work of (Barratta 2001) that is also a generic approach for preparation of food samples for analysis (in this case by alpha spectrometry but with some slight variation also by other analytical techniques).

the range of uranium concentrations requires the application of different sample preparation procedures. The common features (with some exceptions) are drying and ashing of the food products, followed by digestion in order to prepare a liquid sample for the analytical device (e.g., as seen in Figure 4.6). In many cases, relatively simple analytical methods like colorimetric spectroscopy or electro-analytical techniques suffice, ICP-OES, and especially ICPMS are widely deployed for determining the uranium content in liquid samples. Alpha spectrometry may also be used, but sample preparation is more tedious. Some examples of nuclear methods for the determination of the uranium content in food without any sample preparations were also presented.

4.2.2 DISTRIBUTION OF URANIUM IN THE BODY (BIOKINETIC MODELS)

"The term toxicokinetics refers to the modeling and mathematical description of the time course of deposition, absorption, distribution, biotransformation and excretion of xenobiotics in the whole organism" (Medinsky 1996). Ideally, the physiological biokinetic models can predict the concentration of the foreign substance in different tissues and the correlation between them. Thus, analysis of a single sample of excreta or a tissue can serve to assess the concentration in other tissues or body organs. The older classical models treated the whole body as a single uniform compartment so that once the concentration of the xenobiotic material in blood was determined at several points of time after administration of the substance the kinetics of removal from the body could be assessed from a simple semilogarithmic plot (Medinsky 1996). In many cases, a nonlinear time relation was found so that a multicomponent model was required to describe the time dependence of elimination from the body. The different compartments had different time constants for equilibration with the concentration of the xenobiotic substance in blood.

In the case of uranium, once it enters the body it will be distributed among the body organs and tissues or excreted, depending on the exposure pathway (ingestion, inhalation, or through the skin), its chemical form (mainly its solubility and complexation) and physical characteristics (like particle size), and the body organ in which it settles. Most of the ingested uranium (whether insoluble or soluble) would be excreted through feces, but some fraction may enter the bloodstream. Inhaled insoluble particles could be expelled through expectoration or be retained in the lungs and a fraction would be dissolved by lung fluid and enter the bloodstream. Fragments of DU munitions may be embedded in soft tissues of the body and slowly dissolve and enter the bloodstream—a process that could take several years.

The biokinetic models deal mainly with the fate of the uranium fraction that enters the bloodstream, regardless of the pathway through which this occurred. Other models focus on the initial stage of exposure, that is, the mechanism through which the uranium (ingested, inhaled, or through skin) enters the bloodstream. The models schematically represent the body organs as compartments as shown in Figure 4.2 for embedded DU fragments. Figure 4.7 shows a similar model that focuses on the distribution of ingested uranium and includes an excretion pathway through hair. Each compartment may also be subdivided into smaller sections, as shown, for example, in the subcompartments of the skeleton in Figure 4.7 (Li et al. 2009).

In these models, the bloodstream is in contact (directly or indirectly) with each compartment and equilibrium exists between the uranium content of the compartment and the uranium concentration in the blood. Elevated levels of uranium in the bloodstream would lead to deposition of uranium in the compartment. The exchange rate of this process is shown in the model of Figure 4.7. If the uranium concentration in the blood is low, then uranium could be transported from the compartment (body organ) to the blood and eventually be removed from the body (excreted) either by the kidneys and bladder (urine), the intestine (feces), contained in keratin of the hair or nails, or even through exhaled breath or perspiration.

The World Health Organization (WHO) report surmises that the information on the kinetics of uranium metabolism arises mainly from laboratory animal tests and studies of humans that chronically consume uranium (in their diet) or people who have been accidently exposed. It was found that the absorption of ingested uranium depends on the solubility of the compound (Berlin 1986), the food content in the intestine (Sullivan et al. 1986; La Touche et al. 1987), and the presence of oxidizing agents (Sullivan et al. 1986). Several studies have calculated that the average uptake of ingested uranium in the human gastrointestinal tract is around 1%–2%, while the remainder is excreted in feces (Wrenn et al. 1985; Zamora et al. 1998; Karpas et al. 2005b). One exception is the German study that found uptake factors above 5% for female and male subjects (Anke et al. 2009). The absorption of starved female rats was 0.17% but increased to 3.3% when an oxidizing agent (trivalent iron) was also present (Sullivan et al. 1986). The absorption also increased from 0.06% to 2.8%

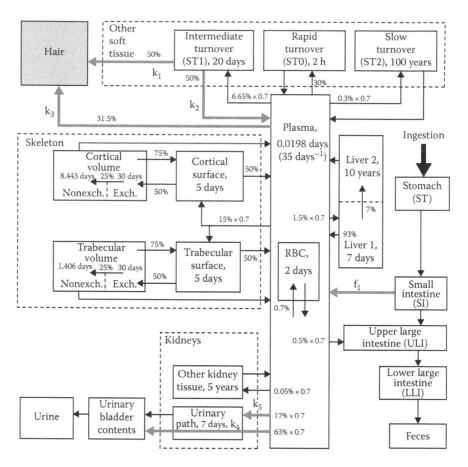

FIGURE 4.7 The compartmental model of ingested uranium, based on ICRP69 but including hair excretion pathway. (From Li, W.B. et al., *Health Phys.*, 96, 636, 2009. With permission.) The time to reach equilibrium is denoted as ST0, ST1 and ST2 for short (2 h), intermediate (20 days) and long (100 years), respectively, the transfer rates are shown as k and the absorption factors as f.

when the dose given to the rats was increased from 0.03 to 45 mg per kg body weight (La Touche et al. 1987). In another study of rats and rabbits that had free access for 91 days to water containing 600 mg L^{-1} of uranyl nitrate, the absorption ratio was 0.06% (Tracy et al. 1992). Thus, the common conclusion of all these studies is that only a small fraction of the ingested uranium is actually taken up by the body. The ratio between the intake (the amount of ingested uranium) and the uptake (the amount transferred to the bloodstream) is called the absorption factor and usually marked in the biokinetic models as f_1 and its value is typically 1%–2%, as mentioned earlier. In the International Commission on Radiological Protection (ICRP) model dealing with ingested uranium (ICRP69 1995), the absorption factor is given a value of 5% that is considered a conservative upper limit for exposure assessment.

The foundations for the model (ICRP69 1995) that describes the biokinetic behavior of ingested uranium in the human body are based on several studies, some of which were mentioned earlier, that were carried out in the 1980s and mainly on the publications of Leggett (1994) that summarized all available data at the time. An updated, detailed report on the biokinetic model for uranium discussing control of uranium intake at the workplace was published by Leggett and his group at Oak Ridge National Laboratory (ORNL) (Leggett et al. 2012). That model is based on the assumption that after ingestion, uranium rapidly appears in the bloodstream (La Touche et al. 1987) and is associated primarily with the red blood cells (Fisenne and Perry 1985) seen as RBC in Figures 4.2 and 4.7. It is assumed that there is equilibrium between the uranyl–albumin complex and ionic uranyl hydrogen carbonate complex ($UO_2HCO_3^+$) in the blood plasma (Moss et al. 1983). These uranyl ions have a high affinity for phosphate, carboxyl, and hydroxyl groups and can combine with proteins and nucleotides to form stable complexes (Moss et al. 1983). One fraction of the uranium in the bloodstream is removed rapidly by the kidneys and accumulates in the kidneys, the skeleton, and other soft tissues, whereas little is found in the liver (La Touche et al. 1987). It has been shown that uranyl ions can replace calcium in the hydroxyapatite complex of bone crystals (Moss et al. 1983) making the skeleton a site of uranium accumulation (Wrenn et al. 1985).

The results of laboratory animal studies and of people who are chronically exposed to uranium in their drinking water show that the amount of soluble uranium accumulated internally is proportional to the intake from ingestion. The total body burden of uranium in humans is estimated to be 40 µg, with approximately 40% of this being present in the muscles, 20% in the skeleton, and 10%, 4%, 1%, and 0.3% in the blood, lungs, liver, and kidneys, respectively (Igarashi et al. 1987).

Once equilibrium is attained between uranium in the blood and the other organs (skeleton and soft tissues), it is gradually excreted in the urine and feces. As mentioned earlier, excretion of ingested uranium is mainly (around 98%) through the feces, but removal of the uranium fraction that has entered the bloodstream is distributed between urine, feces, hair, nails, and perspiration (Figure 4.1). The rate of removal of uranium through urine depends in part on the pH of tubular urine. The uranyl hydrogen carbonate complex is stable under alkaline conditions and is excreted in the urine but low pH values would induce dissociation of this complex and the uranyl ion may then bind to cellular proteins in the tubular wall, which may then impair tubular function (Berlin 1986).

The biokinetic models estimate the half-life of uranium in the different compartments. Thus, for (rat) kidney the half-life has been estimated to be approximately 15 days, while for the skeleton half-lives of 300 and 5000 days have been projected, based on a two-compartment model (Wrenn et al. 1985). In another study using a 10-compartment model, half-lives of 5–11 days were estimated for the (rat) kidney and 93–165 for clearance from the skeleton (Sontag 1986). The overall elimination half-life of uranium under conditions of normal daily intake has been estimated to be between 180 and 360 days in humans (Berlin 1986).

After the uranium enters the bloodstream, regardless of the pathway through which it penetrated the body originally, it becomes distributed among the different organs. The ICRP 1995 biokinetic models describe this process for ingested (ICRP69 1995) and inhaled (ICRP71 1995) radionuclides and divide the body into schematic compartments as shown in Figures 4.2 and 4.7. The ICRP model of uranium assumes that some of the ingested uranium is transferred from the gastrointestinal (GI) tract, mainly from the small intestine, into the blood. After the discovery that a significant quantity of the uranium that enters the bloodstream is removed from the body through the hair (Karpas et al. 2005b), the ICRP model was modified (Li et al. 2009) to account for this additional excretion route. A minor quantity is also removed through toe nails and fingernails, but this was not included in the model. According to this model as shown in Figure 4.7, about 30% of the uranium content in the blood, regardless of its origin or entrance pathway, is transferred to the rapid turnover tissues compartments (ST0) while the remainder 70% is transferred to other compartments where the turnover is slow (ST2) or intermediate (ST1) or to other organs (compartments) like the liver, skeleton, large intestine, and kidneys. The blood is the main source (63%) for the excretion of uranium in urine and the urinary path contributes another 12%. Thus, about half (75% of the 70% of the uranium in the blood) is removed through the urine bladder or the hair. Clear evidence that the amount of uranium excreted through urine or through hair is proportional to the amount of uranium consumed through drinking water can be seen in Figure 4.8 (Karpas et al. 2005b) that is the summary of measurements of uranium content in water, urine, and hair of 205 residents of south Finland.

Although the effects of chronic ingestion of uranium in drinking water on its concentration in urine were widely studied, only a small number of studies on the excretion kinetics of ingested uranium were published. Studies of this sort require the ingestion of a controlled quantity of a dissolved uranium compound and periodical measurements of the amount of uranium excreted in urine. Regardless of the amount consumed, this is defined as *acute* ingestion as opposed to continuous or chronic ingestion. In one such study (Karpas et al. 1998), five volunteers drank a glass of grapefruit juice that contained 100 µg of dissolved uranium in the form of uranyl nitrate hexahydrate. The isotopic composition of the uranium was 0.4% ^{235}U, a fact that made it possible to follow the exchange kinetics of the ingested uranium with the uranium (of natural isotopic composition) stored in the body. The uranium and creatinine concentrations in each urine voiding were determined. The results were in line with several other studies that reported an uptake factor of 1%–2% (the ratio between the amount of uranium that was ingested and excreted in urine). The results also showed that the maximum concentration of uranium in urine was found 5–15 h after ingestion and that the position and value of this maximum varied slightly among the five volunteers. It was also found that within 3 days after ingestion the uranium concentration in urine returned to its normal value for these participants. For example, the change in the uranium concentration in samples of urine from one of the participants in the study as a function of the elapsed time after consumption, adapted from the original study (Karpas et al. 1998), is seen in Figure 4.9.

Finally, the use of isotopic composition measurements showed that the mechanism for exchange of freshly ingested uranium with the uranium stored in the body did not follow either the first in-first out (FIFO) model or the last in-first out (LIFO) model, but was a composite of the two (Karpas et al. 1998).

Highlights: Understanding the behavior of uranium in the body is a prerequisite for being able to assess the uranium body burden on the basis of a bioassay. The biokinetic models are the tool used to carry out this task. These models are based on

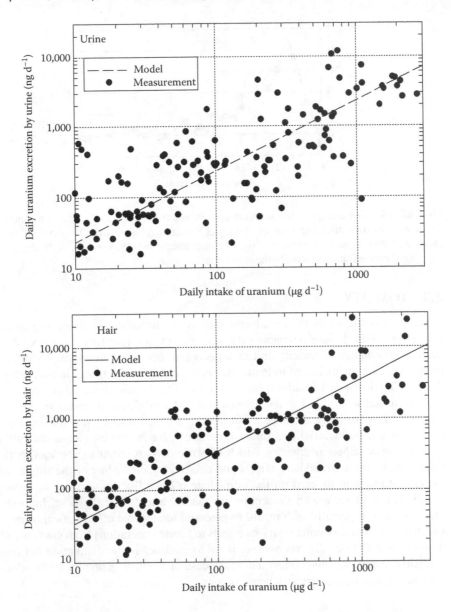

FIGURE 4.8 Daily uranium excretion by urine and hair as a function of daily uranium intake. (Measured concentrations are taken from Karpas, Z. et al., *Health Phys.* 88, 229, 2005b. With permission.)

studies that involved laboratory animals and people who were chronically or accidentally exposed to uranium. The models correlate the level of uranium found in a bioassay (usually urine) and the content in the body, regardless of the pathway through which it entered the bloodstream. Examples of the analytical procedures deployed in the bioassays are discussed in Section 4.4.

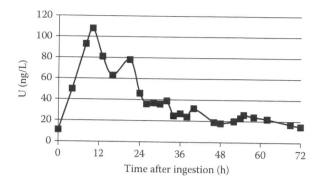

FIGURE 4.9 The change in the uranium concentration in samples of urine as a function of the elapsed time after ingestion of 100 µg of uranium, adapted from the original study (Karpas, 1998). Time zero represents the U concentration before ingestion. Note that after 72 h the U concentration returns to the zero value.

4.3 TOXICITY

Toxicology is the study of the adverse effects of chemicals on living organisms (Klaassen 1995). Before describing the specifics of uranium toxicity, we wish to discuss the concept of toxicity and its aspects that pertain to uranium. The afore-mentioned citation attributed to Paracelsus may be used to put things in proportion. There is no doubt that uranium, like many (if not all) xenobiotic substances, can be detrimental to health, but the amount that is harmful, the mechanism through which it enters the body, the organs that may be affected, and the possible injurious effects on genetic material and future generations are of interest (Klaassen 1995). One important aspect of toxicity, which is particularly relevant to uranium, is the difference between immediate effects and delayed toxicity. Most of the studies on toxicity of uranium focused on the immediate effects, like impairment of kidney functions that in view with Paracelsus are mainly a matter of the dose, but more recent studies, especially with regard to potential teratogenic effects of exposure to DU, have attributed extreme injurious effects to future generations from contact with uranium (see Frame 4.2). Yet another point to consider is the distinction between reversible and irreversible effects that in the case of uranium generally follow along the lines of immediate versus delayed toxicity.

4.3.1 TOXICITY OF URANIUM

The three main exposure pathways to uranium compounds are through the diet, inhalation, or injury, as described earlier. The uptake of uranium after ingestion is affected mainly by soluble uranium compounds (insoluble compounds are excreted through feces with little or no hazardous health effects) as discussed earlier. However, for inhaled compounds, the chemical and physical characteristics determine how long the uranium will be retained in the lungs and what fraction will enter the bloodstream during a given time period. The common practice is to divide these compounds into

FRAME 4.2   THE DISPUTE ABOUT THE HEALTH EFFECTS OF EXPOSURE TO DEPLETED URANIUM

Depleted uranium is probably one of the chemical substances with the worst public relations record, as a very cursory Google search shows (all accessed on March 8, 2014). Some of the titles: *Depleted uranium as malicious as Syrian chemical weapons, Depleted uranium contamination is still spreading in Iraq, Depleted uranium—The real dirty bombs*, and *Depleted uranium—far worse than 9/11: Depleted uranium dust—public health disaster for the people of Iraq and Afghanistan*, are just a few examples. However, the evidence-based scientific literature lends little support to these gloomy headlines.

A special issue of the *Journal of Environmental Radioactivity* in 2003 (Vol. 64, issues 2–3, pp. 87–259), compiled shortly before the Second Gulf War, was dedicated to the issues of DU munitions. Some articles related to the controversy on the health effects of exposure to DU munitions, mainly after the war in Kosovo and the 1991 Gulf War in which DU munitions were deployed. In this special issue, among other topics, the relevance of the ICRP models was noted in a letter to the editor (Valentin and Fry 2003), an overview on DU was presented (Bleise et al. 2003), the biokinetics of uranium from embedded DU fragments (Leggett and Pellmar 2003a), and genomic effects (Miller et al. 2003) were discussed. The work of Miller et al., one of the leading experts on DU health effects (Miller 2007), is especially interesting as it focuses on the effects that are manifested in delayed reproductive death and micronuclei formation (Miller et al. 2003). It was shown that exposure of cell lines and cultures to DU, nickel, and gamma radiation, as well as tungsten alloys, resulted in genomic instability of progeny cells. The effects were somewhat greater for DU exposure but were not unique for DU and, although not specifically stated in the article, similar effects are expected for exposure to natural uranium.

There is practically no controversy regarding the short-term health effects of exposure to depleted uranium that are accepted as being comparable to those related to exposure to natural uranium or other heavy metals and concern kidney functions (see, e.g., Kurttio et al. 2002; Sztajnkrycer and Otten 2004; Selden et al. 2009; Arzuaga et al. 2010; Kathren 2011). Monitoring several markers of kidney functions in cohorts of Gulf War (mostly the 1991 war) veterans who were exposed to DU, experimental animal studies and populations that are chronically exposed to elevated levels of uranium in their drinking water show similar results. To quote the conclusion of one of these articles: "There were no clear signs of nephrotoxicity from uranium in drinking water at levels recorded in this study, but some indications of an effect were observed using uranium in urine as a measure of overall uranium exposure. The clinical relevance of these findings remains unclear (Selden et al. 2009)." There appears to be an agreement that uranium (natural or depleted) kidney toxicity and short-term health effects are no worse than that of other heavy metals.

Therefore, the heated dispute about the teratogenic effects of depleted uranium is not anticipated. Most of the literature on these effects is related to the Fallujah area in Iraq where depleted uranium and phosphorous munitions were used in the 2003 Gulf War. According to some researchers, the outcome of exposure to DU munitions or inhalation of DU aerosols can be expressed in an increase in the incidence rate of abortions, premature births, stilted embryo development, congenital anomalies, and delivery of deformed babies (Hindin et al. 2005; Alaani et al. 2011a,b; Takato 2013). The toxicological data on birth malformation due to natural and depleted uranium were reviewed by Hindin et al. (2005), where it was stated that although no pathway has been found by which environmental DU can reach the reproductive cells the evidence supports plausibility of "increased risk of birth defects in offspring of persons exposed to DU."

The concentration of uranium and 51 other elements in hair samples collected from a cohort of 26 mothers and fathers of children diagnosed with congenital anomalies in the Fallujah region of Iraq was determined by ICPMS (Alaani et al. 2011b). The only element that was found to deviate from the *normal* range of literature values was uranium. It should be noted that the reported concentrations of uranium in hair of this Iraqi study were considerably below the values found in other studies, like the Finnish population that consume elevated uranium levels in their drinking water (Karpas et al. 2005b). Furthermore, the Iraqi study did not include a control group from the same area that makes it extremely difficult to distinguish this group from the regular population. Data on the uranium background level in soil samples and water (tap and river) were also presented. The uranium level in the six soil samples was quite low (below 1 mg kg^{-1} in five samples and ~1.5 mg kg^{-1} in the sixth sample) and the reported $^{238}U/^{235}U$ ratio in all six soil samples was below the accepted ratio in natural uranium, indicating exposure to slightly *enriched* uranium—a substance not used in DU ammunition!

The report compiled in 2013 by Human Rights Now (HRN), a nongovernment organization (NGO), focused on the effects of DU exposure on Iraqi civilians after the 2003 Gulf War (Takato 2013). Based on data from the Fallujah General Hospital, the report states that around 15% of infants born since 2003 had some congenital birth defects. A caveat in the summary of this report, "However, without sufficient disclosure of detailed information related to toxic weapons used during the conflict, the cause of problem has not yet been identified," somewhat falls short of attributing these defects to DU alone. A variety of different types of birth defects were observed by the HRN team that also interviewed doctors and families but did not perform any bioassays to determine exposure to uranium. The report quotes the study on the hair bioassays, mentioned earlier, for the content of uranium as evidence for involvement of DU in birth defects (Alaani et al. 2011b) but also states that lead and mercury that are also associated with munitions may have been involved.

It is interesting to note that in a different study of congenital birth defects in Iraqi cities the 17-fold increase in Basra hospitals is attributed to lead (Al-Sabbak et al. 2012).

In a comprehensive report on the toxicological profile for uranium exposure, the panel of experts addressed the question of birth defects (ATSDR 2013). The report states that no specific human study has identified uranium as responsible for birth defects and therefore cannot conclude that such a correlation exists. They continue by saying that "Some studies in animals exposed to high levels of uranium during pregnancy which caused toxicity in the mothers, have resulted in early deaths and birth defects in the young." The report continues, "In some rat studies, enriched uranium exposure during pregnancy caused changes in brain function in the offspring. Similar studies found changes in the ovaries of female offspring." The report also critically evaluated the studies on developmental effects and teratogenicity of DU aerosol exposure. It states that the review mentioned earlier (Hindin et al. 2005) is based mainly on data that were published in the media or presented at Iraqi conferences and not in peer-reviewed journals. It also iterates that some of the facts (like hydrocephalus birth rates) are implausible and that some other studies (Basra residents) are incomplete. Another drawback of the Hindin review, according to the ATSDR report, is that comparisons were not based on exposed versus unexposed specific persons but between residents in areas in which DU was deployed and other areas where it was not deployed. The report also questions some of the findings of the Alaani et al. paper (Alaani et al. 2011b) regarding the infant mortality and the declared alterations in male:female sex ratio in Fallujah. Other criticism concerns the reported birth defect that were not confirmed by medical records and the fact that there was no control group and wonder why only results of data from 5 years were analyzed while the study period covered 10 years (ATSDR 2013).

three categories or classes: those that are excreted rapidly within days (D for days or F for fast), those that are excreted within years or very slowly (Y for years or S for slow) and those that fall in between these two classes (W for weeks or M for medium or intermediate excretion time). The older terminology used D, W, and Y to describe the residence times of different uranium compounds in lung fluid (experiments were usually carried out with simulated lung fluid) while the more modern classification uses F, M, and S for the same purpose. Typical common class F compounds include UF_6, UCl_4, UO_2F_2, and $UO_2(NO_3)_2 \cdot 6H_2O$; class M compounds are UF_4 and UO_3 while UO_2 and U_3O_8 belong to class S (Craft et al. 2004). Exposure could also consist of a mixture of different classes (different compounds or formally similar compounds with different properties depending on their preparation) complicating exposure dose assessment (Kravchik et al. 2008).

A study on the effects of repeated inhalation of UO_2 by rats was conducted and the main finding was that the predictions of the biokinetic model were consistent with the results of these experiments (Monleau et al. 2006). Exposure to uranium may also concern the effects of the radioactive progeny, especially radium and polonium, in the lungs (see also Frame 4.1 for the effect on the Wismut miners).

As far as injury from DU munitions is concerned, it is assumed that the fragments belong to the class that will be dissolved slowly in the body (Leggett and Pellmar 2003b). The biokinetics of embedded DU was found to be similar to the commonly studied forms of uranium with regard to long-term accumulation in body organs. Any discussion of the health effects and toxicity of uranium must take into account these characteristics.

In some specific cases, which are very rarely documented, uranium may also pose a health hazard from an external source through emission of neutrons and gamma radiation as a result of a criticality incident, as occurred in Tokai Mura, Japan, in 1999. This could occur if a sufficient amount of fissile nuclides (^{235}U) accumulates under certain geometric and chemical conditions (briefly discussed later) and in more detail in Chapter 1.

Finally, another aspect of uranium toxicity must also be addressed, which is the teratogenic effect, that is, disturbing the proper growth and development of an embryo or fetus. This is a point of contention between the opponents of the use of DU munitions that attribute horrendous deformities in children whose parents were exposed to DU fragments or dust (HRN 2013) and other scientists who dispute these conclusions (Sztajnkrycer and Otten 2004). The dispute about the health effects of exposure to DU is discussed in Frame 4.2.

4.3.2 CHEMICAL TOXICITY

Uranium, like other heavy metals, such as lead, cadmium, and mercury, has detrimental effects on the human body, particularly on organs such as the kidneys. These nonessential elements are toxic at very low doses with a long biological half-life so that exposure to them is potentially harmful. The kidney can absorb and accumulate divalent metal ions and is the first target organ of heavy metal toxicity (Barbier et al. 2005). The normal operation of the nephrons in the kidney is to remove toxic substances from the blood into the bladder from which they are excreted in urine. Renal damage depends on the type of heavy metal, on the amount reaching the kidney and the duration of exposure. Both acute and chronic toxicity have been shown to cause nephropathies, and the severity can range from tubular dysfunctions to renal failure that may even lead to death. The uptake of heavy metals depends on the form (free or bound) of the metal and the part of the nephron where absorption of the metal occurs:

> While in the bone, a very small amount of radiation is emitted but the radiation is very diffuse, so the bone marrow is not effectively irradiated. The uranyl ion does not readily interfere with any major biochemical process except for depositing in the tubules of kidney. Animal experiments have shown that, in heavy doses, uranium can cause damage to kidneys. (56/04 2001)

There are somewhat contradictory reports and studies on the general health effects of uranium and particularly on the effects on kidney functions. In the comprehensive study of 325 people of a Finnish population that regularly consume large amounts of uranium in their drinking water (Kurttio et al. 2002), it was shown that continuous uranium intake from drinking water, even at relatively high exposures, did not have cytotoxic effects on kidneys in humans. In that study, the median uranium concentration in drinking water was 28 μg L^{-1} (the maximum was 1920 μg L^{-1}), resulting in the median daily uranium intake of 39 μg day^{-1}. The urine and serum concentrations of calcium, phosphate, glucose, albumin, creatinine, and β-2-microglobulin were measured to evaluate possible renal effects. Uranium concentration in urine was statistically associated with increased fractional excretion of calcium and phosphate. The main conclusions of the study were that uranium exposure is weakly associated with altered proximal tubulus function and despite chronic intake of water with high uranium concentration no effect on glomerular function was observed. In a follow-up study (Kurttio et al. 2006), kidney functions were determined in a group of 95 men and 98 women aged 18–81 years who had used drinking water from drilled wells for an average of 16 years. The median uranium concentration in drinking water was 25 μg L^{-1} (maximum 1500 μg L^{-1}). Ten indicators of kidney function (urinary N-acetyl-gamma-D-glucosaminidase, alkaline phosphatase, lactate dehydrogenase, gamma-glutamyltransferase, and glutathione-S-transferase; serum cystatin C; and urinary and serum calcium, phosphate, glucose, and creatinine) were measured as well as supine blood pressure. Indicators of cytotoxicity and kidney function did not show evidence of renal damage and no statistically significant associations with uranium in urine, water, hair, or toenails were found. However, uranium exposure was associated with slightly greater diastolic and systolic blood pressures, and cumulative uranium intake was associated with increased glucose excretion in urine. The conclusion was that continuous uranium intake from drinking water, even at relatively high exposures, was not found to have cytotoxic effects on kidneys in humans (Kurttio et al. 2006).

The chemical form of the absorbed uranium can also play a role. The most common species of uranium likely to be absorbed by the intestine through consuming food and water or through inhalation of soluble uranium compounds is the uranyl, UO_2^{+2}, ion. However, the anion counterpoint may also play a role in affecting the physiology. For example, if UF_6 vapors are inhaled, the hydrolysis in lung fluid would lead to production of HF, which would affect the normal operation of the lungs as it causes lesions and scarred tissues.

Toxicity studies of uranium compounds were carried out mainly on laboratory animals like rabbits, dogs, and rats, but in some cases humans that were inadvertently exposed to uranium compounds through inhalation or injury by DU munitions were also subject to several studies. In addition, populations that are known to consume high concentrations of uranium in their diet, like residents of southern Finland mentioned earlier and of Nova Scotia in Canada, were tested for renal damage and other pathological conditions. These later studies encompassed a much larger number of subjects than the other studies of human exposure. Undoubtedly the most extreme reported case of exposure to uranium through ingestion is of a person who advertently consumed 15 g of uranyl acetate (about 8 g of uranium) in an attempt to harm himself, in addition to abusing several other drugs (Pavlakis et al. 1996). This person suffered from several medical problems, including acute kidney failure, and had to undergo dialysis for 2 weeks. Six months after the incident, his kidney functions were not fully restored, but he did not suffer from fatal kidney damage despite exposure to this high level of uranium (Pavlakis et al. 1996). In fact, it has been claimed that there is no documented case of fatality caused by chemical effects of exposure to uranium (Kathren 2011).

The two most severe cases of catastrophic failure of nuclear power plants, Chernobyl in 1986 and Fukushima in 2011, are not good case studies for exposure to uranium because of the other highly toxic radionuclides (fission products and plutonium, mainly) that were released into the atmosphere so the relative risk from uranium was negligible.

A particular group that has been the subject of many toxicological studies is the cohort of US troops that were exposed to DU munitions during the first Gulf War in 1991. These included troops that were hit by DU fragments that remained embedded in their bodies and were regularly examined by medical teams to assess their health and determine the uranium content in their urine (and sometime also in blood) (McDiarmid et al. 2009). A 16-year follow-up study reported on a group of 35 members of the 77 member cohort who were victims of DU munitions (by friendly fire). It was found that the cohort continued to excrete elevated levels of uranium in urine but few clinically significant health effects related to uranium were found. Two years later, similar results were observed (McDiarmid et al. 2011). No information regarding birth defects of the offspring of these people was reported, so we assume that no such effect was observed (see Frame 4.2).

A large-scale study including 2499 exposed and nonexposed firefighters, police officers, and hangar workers was carried out after an airliner with DU ballast crashed in a suburb of Amsterdam (Bijlsma et al. 2008). Urine samples were collected 8.5 years after the accident and the uranium concentrations were determined by sector field inductively coupled plasma mass spectrometry. Exposed personnel were compared with their nonexposed colleagues. The median uranium concentrations were low, around 2 ngU g^{-1} creatinine, and median values of albumin–creatinine ratio and fractional excretion of beta(2)-microglobulin were well below the level for microalbuminuria and for tubular damage, respectively. No statistically significant differences between exposed and nonexposed workers were found, and it was concluded that no disturbed kidney function parameters were observed.

An interesting review (Arzuaga et al. 2010) of the kidney toxicity effects of long-term exposure to natural uranium and DU states that "The kidney was observed to be a target of uranium toxicity following oral and implantation exposure routes in several animal species. The interpretation and importance of the observed changes in biomarkers of proximal tubule function are important questions that indicate the need for additional clinical, epidemiological, and experimental research."

In a study of mice that consumed drinking water containing uranium, estrogen-like effects were observed. It was reported that exposure to uranium that is an endocrine-disrupting chemical may increase the risk of fertility problems and reproductive system cancer (Raymond-Whish et al. 2007). There are no such studies of these effects on humans.

In a 2011 presentation, the different estimations of the lethal dose (LD50) from the consumption of uranium were summarized (Kathren 2011). "Recommended Provisional Acute LD50 Doses: For acute oral intake of soluble U: 5 g; Acute inhalation intake of soluble U: 1 g; Above are most restrictive cases; LD50 for insoluble U compounds is much higher."

Highlights: Almost all studies of chronic or accidental exposure to uranium did not find significant chemical toxicological effects as far as kidney functions are concerned. Thus, the effect of exposure to uranium is probably the same, or even diminished, as exposure to other heavy metals (*lead bullets are probably more dangerous than uranium bullets*—Prof. Otto G. Raabe). However, this pertains to the immediate or short-term effects of the chemical toxicity of uranium, while the delayed effects, particularly on the fertility of exposed populations and on the proper development of future generations, are still the subject of controversy (Frame 4.2). An objective comprehensive study on the delayed effects, in addition to carcinogenic outcome, of populations that were occupationally or chronically exposed to uranium, is warranted. This study should also focus on birth defects and developmental problems among the offspring of these populations.

4.3.3 RADIOLOGICAL TOXICITY

Due to the low specific radioactivity of natural uranium (about 1 Ci per 3 metric tons or 81 ng U-238 per Bq; Frame 4.3), the main health effects are through its chemical toxicity. However, in some cases, the radiological effects may be dominant, like in the case of exposure to enriched uranium or when insoluble uranium compounds enter the body through inhalation or injury and are retained in the body for an extended period. The fact that the decay products of uranium (progeny) are also radioactive may enhance the radiological health effects.

In that case, the local damage to the cellular material close to the embedded compound could be severe. In addition, the effects could be delayed and affect the genetic material (teratogenic effect) so will only be expressed in the offspring of the exposed person, as claimed for populations living in zones where DU munitions were used (like Iraq).

A special case of radiological effects of uranium is when a criticality incident occurs. A criticality accident, or an uncontrolled nuclear chain reaction, may inadvertently occur if a sufficient amount of ^{235}U accumulates under certain conditions. The criteria for criticality control are known by the acronym MAGIC MERV for Mass, Absorption, Geometry, Interaction, Concentration, Moderation, Enrichment, Reflection, and Volume (for more details, see Frame 1.3 in Chapter 1). The result would be the emission of neutrons and gamma radiation, as occurred in Tokai-Mura, Japan, in 1999, and resulted in the death of two of the plant employees (WNA 2007). A brief discussion on the units for measuring radiation and exposure of the public is presented in Frame 4.3.

4.3.4 OTHER HEALTH EFFECTS

The use of DU munitions in the Gulf Wars (1991 and 2003) and in Kosovo (in the 1990s) by allied forces has given rise to accusations of serious health concerns not only for the people directly exposed but also for their offspring. There is quite a clear division between the group that contends that DU munitions are harmful, whether fragments embedded in soft tissues or uranium oxide dust particles that were inhaled, that are mostly residents of the areas afflicted by DU munitions and other, mainly Western, researchers. The case of Iraqi residents who were allegedly exposed to DU munitions has gained publicity among the groups opposed to the use of these weapons (Alaani et al. 2011a,b). In a review of the literature regarding the teratogenicity of

FRAME 4.3   UNITS FOR MEASURING RADIATION
AND EXPOSURE OF THE PUBLIC

Radiation is energy that travels through space after being emitted from a source. Depending on the energy, radiation can be nonionizing like visible light, radio waves and microwaves, or ionizing radiation if the energy is sufficient to remove an electron from the irradiated object and produce charged particles (ions). The radiation source can be artificial (electronic) or natural (radioactive nuclides or cosmic radiation). Radioactive atoms are unstable atoms that emit their excess energy in the form of electromagnetic radiation (gamma or x-rays) or as discrete particles (alpha or beta particles, neutrons, etc.).

We are immersed in *background* radiation from natural and artificial sources, as seen in Table 4.6. Exposure to radiation can be from an external source irradiating the whole body or given organs or tissues resulting in an *external radiation dose*. Internally deposited radioactive material may cause an *internal radiation dose* that is usually focused on a specific organ or tissue.

In the old system of units, exposure was described as *radiation absorbed dose (rad)*, *dose equivalent (rem)*, *roentgen (R)*, or *Curie (Ci)*. The official system of measurement, the International System of Units (SI), uses the *Gray* (Gy) and *Sievert* (Sv), where 1 Gy = 100 rad and 1 Sv = 100 rem. 1 Gy corresponds to 1 J of energy deposition in 1 kg of material. It should be noted that the Gray (and rad) are purely physical quantities reflecting the amount of absorbed radiation, while the Sievert (and rem) reflects the biological effect of the absorbed radiation and depends on the radiation type and energy.

Thus, the Gray expresses the amount of radiation absorbed by the body and the Sievert takes into account the effect of the radiation on the human body. For electromagnetic radiation (gamma and x-rays), these are considered as equal, that is, a dose of 1 Gy is equivalent to 1 Sv. However, for energetic particles absorbed by the body, the biological effects are more severe, so that a dose of one Gy of radiation by alpha particles would be equivalent to 20 Sv.

The Becquerel (Bq) is defined as the number of disintegrations per second (dps) and is proportion to the older unit 1 Ci = $3.7 * 10^{10}$ Bq. Radioactive transformation events are expressed in units of dps but because instruments are not 100%

TABLE 4.6
Radiation Exposure from Various Sources

Source	Exposure
External background radiation	60 mrem/year, US average
Natural K-40 and other radioactivity in body	40 mrem/year
Air travel round trip (NY-LA)	5 mrem
Chest x-ray effective dose	10 mrem per film
Radon in the home	200 mrem/year (variable)
Man-made (medical x-rays, etc.)	60 mrem/year (average)

TABLE 4.7

Annual Radiation Dose Limits for Various Populations According to Different Agencies

Annual Radiation Dose Limits	Agency
Radiation worker—5000 mrem	(NRC, *occupationally* exposed)
General public—100 mrem	(NRC, member of the public)
General public—25 mrem	(NRC, D&D all pathways)
General public—10 mrem	(EPA, air pathway)
General public—4 mrem	(EPA, drinking water pathway)

Source: Based on file:///D:/Uranium%20book/Chapter%204/Radiation%20Basics.htm.

efficient the actual measurement is *counts per second* (cps). The same radioactive source (with given dps) could yield different count rates (cps) in different instruments, depending on the efficiency of the measurement device. Thus, calibration of the efficiency of the instrument is required for each measured energy range.

One roentgen deposits 0.00877 Gy (0.877 rad) of absorbed dose in dry air, or 0.0096 Gy (0.96 rad) in soft tissue. Specifically for uranium-238, the specific activity is as follows:

$$1 \text{ Bq }^{238}\text{U/kg} = 81 \text{ ppb U } (81 * 10^{-9} \text{ gU/g})$$

Regulatory dose limits are set by international, federal, and state agencies for the general public and occupational exposure to limit cancer risk (Table 4.7). Other radiation dose limits are applied to potential specific biological effects with workers' skin and lens of the eye that will not be elaborated here.

exposure to DU aerosols (Hindin et al. 2005), the authors conclude that the "human epidemiological evidence is consistent with increased risk of birth defects in offspring of persons exposed to DU," although the causal pathway between parental DU exposure and birth of offspring with defects is yet to be established (Frame 4.2).

Highlights: Compared with the chemical toxicity of uranium, the immediate radiological hazards of natural or DU are negligible. The exceptions include enriched uranium, especially highly enriched uranium that may pose a long-term risk for the development of cancer. Exposure to uranium (even DU dust) may also cause possible teratogenic effects on the development of embryos and birth defects—but, as mentioned earlier, this is not universally accepted.

4.4 BIOASSAYS FOR URANIUM EXPOSURE

The methods of assessing exposure to uranium are derived from the biokinetic models that were discussed earlier. Analytical procedures for the determination of the uranium content in excreta: mainly urine and sometimes also feces, blood (serum or plasma), discarded tissues like hair and nails are deployed. The association of the uranium content in these bioassays to the body burden is gained through these models. A method based on the emanation of gamma rays from uranium compounds that are present in the lungs, uranium lung burden (ULB), is used to assess inhaled uranium compounds as shown on a commercial website of one of the manufacturers of the measurement device (Canberra n.d.). In specific cases, where the uranium-containing material is on external surfaces, like open wounds or in the nostrils, swipe samples of these areas may be collected and analyzed. Postmortem analysis of the uranium content in bone or soft tissues may also be carried out but is very rarely done.

PROCORAD is an international organization that carries out annual interlaboratory intercomparisons for the determination of radioactive substances in urine and feces (Berard et al. 2003). During the last years, participation typically included over 60 laboratories from more than 20 different countries and the annual meetings (held in different European countries) created opportunities for a frank exchange of analytical techniques and discussion of analytical problems and their solution (http://www.procorad.org/en). (For proper disclosure, the author had participated in several of these meetings.)

It should be emphasized that the presence of uranium in drinking water and food, as well as the possibility of inhalation of uranium compounds, means that every single individual will be exposed to uranium, and thus, traces of uranium are expected to be found in all bioassays, if the analytical sensitivity is sufficient. The assessment of exposure to DU has received special attention due to the controversy surrounding the possible health effects. Therefore, analytical methods that can also supply accurate and sensitive information on the isotopic composition of uranium in bioassays are preferred over the analytical methods that can only quantify the uranium. Namely, methods based on mass spectrometry can provide a full account on the isotopic composition of uranium (^{234}U/^{238}U, ^{235}U/^{238}U, and ^{236}U/^{238}U) while alpha spectrometric techniques can be used to determine the ^{234}U/^{238}U ratio.

Table 4.8 summarizes the findings of some of the major studies that were carried out in Sweden by Rodushkin et al., where reference values for the concentration of uranium in urine, whole blood, serum, hair, and nails are given for populations that are not occupationally exposed to uranium compounds.

As these reference values are quite low and can be accurately determined only with modern analytical techniques, the older reports should be regarded with some degree of caution. The values shown in Table 4.8 are regarded by the author as unofficial reference values and do not include populations that may be exposed to elevated levels of uranium through food and drinking water. Several studies have focused on special cases of populations thus exposed or that have been involved in conflict areas where DU munitions were used or are employed by the uranium mining or nuclear industries. The reference values presented here have not taken these studies into account. In the opinion of the author, whenever the study population of unexposed individuals involves a large number of subjects, the statistical analysis of the mean should exclude the top values (above the 95th percentile) because even members of the population that are not considered as *exposed* may be inadvertently and unknowingly be exposed to uranium through their diet or the environment.

It should also be noted that many publications focus on the development of analytical methods and determining their sensitivity, specificity, accuracy, and validating them

TABLE 4.8
Summary of Reference Values of Trace Elements in Human Biological Materials

Bioassay	Median	Range
Whole blood[a]	13 ng L^{-1}	8–35 ng L^{-1}
Whole blood[b]		1.1–32 ng L^{-1}
Serum[b]		0.5–19 ng L^{-1}
Serum[c]		1.4–15 ng L^{-1}
Urine[d]		<1.4–17.4 ng L^{-1}
Urine[b]		0.7–19 ng L^{-1}
Urine[c]		12–16 ng L^{-1}
Hair	36 ng g^{-1}	6–436 ng g^{-1}
Nails	8 ng g^{-1}	2–47 ng g^{-1}

Note: Reference values for uranium in bioassays (based on the studies by Rodushkin et al.), http://www.alsglobal.se/website/var/assets/media-se/pdf/reference_data_biomonitoring_120710.pdf.

[a] Rodushkin et al. *Fresensius J. Anal. Chem.* 364 (1999) 338–346.
[b] Rodushkin et al. *Recent Res. Dev. Pure Appl. Chem.* 5 (2001) 51–66.
[c] Rodushkin et al. *Anal. Bioanal. Chem.* 380 (2004) 247–257.
[d] Rodushkin et al. *J. Trace Elem. Med. Biol.* 14 (2001) 241–247.
[e] Rodushkin et al. *Sci. Total Environ.* 262 (2000b) 21–36.

from synthetic urine samples or spiked urine samples (standard addition methods) and may sometimes include a small number of real urine samples, but do not provide a broad dataset of real populations.

Urine: The values presented in Table 4.8 are based, as mentioned earlier, mainly on the publications of Rodushkin and coworkers in Sweden, but are in agreement with several other studies. For example, uranium concentrations up to 40 ng L^{-1} have been reported for populations in Israel (Lorber et al. 1996), up to 34.5 ng L^{-1} in the United States for the 95th percentile (Ting et al. 1999), up to 7.8 ng L^{-1} in Japan (Tolmachyov et al. 2004). The average daily excretion of uranium in urine of omnivorous 119 women and 119 men in Germany was estimated as being 0.18 ± 0.26 and 0.25 ± 0.16 µg day^{-1}, respectively (Anke et al. 2009). These relatively high values, when considered in conjunction with the reported ingested uranium values of the same study, lead to apparent absorption factors of 6.5% for women and 5.6% for men. Note that the assessment of the daily consumption of uranium through drinking water assumes that the average amount of drinking water consumed is 1.2 and 1.6 L day^{-1} for females and males.

Blood: There are fewer studies on the uranium concentration in blood than in urine. In Japan for five subjects, the range was 7.2–12.2 ng L^{-1} (Tolmachyov et al. 2004), for whole blood samples collected in a glass tube a similar value of 12 ng L^{-1} was reported, but for samples collected in plastic tube the uranium concentration was only 2 ng L^{-1}.

Feces: The feces have to be dry-ashed before measurement of the uranium concentration and the results are reported on the basis of the ash. The assay is usually performed only after exposure and not on a routine basis. Procorad organization sends out samples of fecal ashes for determination uranium as part of the laboratory intercomparison program. Uranium levels are usually in the range of 0.2–0.8 μg U g^{-1} fecal ash. The levels of uranium in fecal ash samples of miners in a nickel production facility that were not exposed to uranium were below ~0.3 μg U g^{-1} fecal ash for 25 out of 29 samples and the highest concentration was just below 1 μg U g^{-1} fecal ash (Azeredo et al. 2000). The activity of uranium and plutonium in 119 routine fecal ash samples of employees at AWE was determined by alpha spectrometry and the mean activity for uranium was 48.8 mBq day^{-1} (range 0.14–266 mBq day^{-1}) (Cockerill et al. 2006). Higher values were found in a study of the average daily excretion of uranium in feces of omnivorous women and men in Germany was estimated as being 2.41 ± 2.72 and 2.69 ± 3.25 μg day^{-1}, respectively (Anke et al. 2009).

Hair: The concentration of uranium in hair samples of 114 subjects from northeast Sweden shown in Table 4.8 was determined with ICPMS after acid digestion and a range of 6–436 ng g^{-1} with a mean of 57 ± 65 ng g^{-1} and a median value of 36 ng g^{-1} was reported (Rodushkin and Axelsson 2000b). Quite similar results were found in a study of 99 people residing in the south of Israel where the reported mean was 62 ± 42 ng g^{-1} and the median value was 50 ng g^{-1} for the entire study population (Gonnen et al. 2000). The statistical analysis, according to nonparametric distribution, of the uranium content in hair samples found no significant difference between female and male subjects or between smokers and nonsmokers, but younger people (under 45 years) had a slightly higher level of uranium in hair than older participants in the study.

The concentration of uranium in hair samples of the select population living in the south of Finland that obtain their drinking water from drilled wells, often containing elevated levels of uranium, was studied (Karpas et al. 2005b). As expected, the range of concentrations varied widely, in accordance with the concentration in drinking water, so the 5th and 95th percentiles had 26 and 28,400 ng U g^{-1} hair, respectively, and the concentration in the 50th percentile was 730 ng U g^{-1}.

Nails: The concentration of uranium in fingernails samples of 96 Swedish subjects, shown in Table 4.8, was determined with ICPMS after acid digestion and a range of 2–47 ng g^{-1} with a mean of 17 ± 37 ng g^{-1} and a median value of 8 ng g^{-1} was reported (Rodushkin and Axelsson 2000b). The concentration of uranium in toenail samples of the residents of south Finland that consume elevated levels of uranium in their drinking water showed a broad range from 7 ng g^{-1} for the 5th percentile to 6278 ng g^{-1} for the 95th percentile and the concentration in the 50th percentile was 141 ng g^{-1} (Karpas et al. 2005b). Once again, this broad range of concentrations reflected the amount of uranium consumed daily from drinking water.

Highlights: The level of uranium concentration in all bioassays for populations that are not exposed to uranium is low, typically ranging from parts-per-trillion (ng L^{-1}) in urine to parts-per-billion (ng g^{-1}) for hair samples. Therefore, sensitive analytical instrumentation is required for accurate determination of uranium at these levels. If the isotopic composition is also required to distinguish between exposure sources, then preconcentration may be required, as described in some of the examples given later.

4.4.1 ANALYTICAL METHODS FOR THE DETERMINATION OF URANIUM IN URINE

Urinalysis is the most common method for assessment of internal exposure to uranium and estimating its content in the body (i.e., the body burden). However, even this straightforward approach has several variations regarding sample collection, sample treatment, and analysis, and there is no globally accepted *official*, definitive method. In the following section, some of the sample collection and treatment strategies, as well as some case studies of the analytical methods used for urinalysis, will be presented. For detailed procedures, interested readers can follow the articles and reports that are cited and briefly discussed here, bearing in mind that each method has its advantages and limitations.

Sample collection: There is quite extensive and ambiguous literature concerning the urine sample collection. The *standard* approach is a collection of all the urine excreted by an individual during a 24 h period. Compliance with this collection method is not very convenient for active people who either have to carry a collection vessel with them during their daily activities or need to stay close to the collection vessel. This is more practical for bedridden patients connected to a catheter. In addition, studies have shown that there are quite large variations between the volumes of urine collected from the same person on different days (Marco et al. 2008). One way to overcome these variations is to normalize the urine sample to some internal parameter like the creatinine (2-amino-1-methyl-1*H*-imidazol-4-ol) concentration in the urine, the urine salts content (like total dissolved solids (TDS)), or the urine density (specific gravity). Even with normalization there are variations in samples collected on different days from the same individual, although not as large as observed without normalization. Due to the difficulties in verifying full compliance to the 24 h sample collection regime, other sampling strategies are used. One of those is the *simulated 24 h* collection where all the urine excreted during 8 h (a typical work day) is collected and the result multiplied by three, or pooling the urine collected for 8 h on three different days. Another common method is to collect a *spot sample* from a single voiding. In each case, normalization is required in order to correct for the difference in urine salt content (and uranium concentration) due to the state of hydration of the individual. Models have been developed to estimate the amount of creatinine in urine that is expected to be excreted diurnally by an individual on the basis of weight, height, gender, and age (Boeniger et al. 1993). Thus, once the concentrations of uranium and creatinine in urine have been determined, the diurnal amount can be calculated even from a *spot sample* and the body burden can be estimated.

Sample treatment: The sensitivity and specificity of the different analytical procedures that are used to determine the uranium content in urine dictate the treatment methodology. Some methods take the raw urine sample and introduce it directly into the analytical device, like flow injection-ICPMS (Lorber et al. 1996) or after performing a simple dilution (Ting et al. 1996). Other methods require total destruction of the organic matter before using laser-induced fluorescence (Karpas et al. 1996), yet some other methods include a very tedious sample preparation procedure involving destruction of organic matter, co-precipitation of the uranium, ion exchange separation and deposition of the purified preconcentrated uranium before alpha spectrometry (IAEA 2000). Other aspects of sample treatment include the use of an internal standard (either an isotope of uranium or a spike with a different element) to estimate the recovery efficiency and compensate for matrix effects, or carrying out a series of standard additions for improved accuracy. These sample preparation procedures will be discussed in more detail in the appropriate sections.

Stability and preservation of urine samples: In many cases, the urine samples cannot be analyzed immediately after collection, and therefore they need to be stored temporarily until the analysis can be carried out. The most common strategy is to store the urine samples in a refrigerator at ~4°C, sometimes they are stored in a deep-freeze at approximately −18°C that causes the urine to solidify (and later need to be defrosted). The samples should be acidified with nitric acid before storage due to degradation of urea that leads to production of ammonia and raises the pH of the urine. This may lead to the formation of precipitates and co-precipitation of part of the uranium, which could lead to erroneous results. The question of stability of urine samples has been addressed in several studies. For example, the effect of storage conditions on the stability of the urine matrix was investigated (Krystek and Ritsema 2002). Three *real life* (unspiked) urine samples and three spiked samples were stored under different conditions (acidification with 5% HNO_3, storage at room temperature and in a refrigerator). The main conclusion was that storage at room temperature without acidification was not recommended because the recovery of uranium was reduced by about 20% while with either acidification or refrigeration the loss was below 10% (Krystek and Ritsema 2002).

Analytical methods: A large variety of analytical methods have been deployed to determine the uranium content in urine, as described later. Nowadays, the dominant method is the use of inductively coupled plasma mass spectrometry (ICPMS) that is sensitive, fast, requires little sample preparation, and provides information on the isotopic composition as well as quantitative data. In a 2003 overview of the evolution of techniques for determination of uranium in urine, the authors, who are the organizers of the Procorad intercomparison program, clearly state, "The fact that more and more of the laboratories taking part in the Procorad intercomparison exercises are using ICP-MS demonstrates the suitability of this technique for mass analysis of uranium in urine in radiation protection monitoring in normal and accidental situations. Compared with widely used techniques such as classic fluorimetry, laser fluorimetry or phosphorescence, ICP-MS can provide useful quantitative information in a very simple way (Berard et al. 2003)." Since this was written, the trend toward the extensive use of ICPMS for urinalysis has gained even more momentum and is the method of choice in leading radio-toxicological laboratories, as evident from the examples presented in the following sections.

   Despite all these caveats, urine analysis is still by far the most commonly used bioassay for the assessment of exposure to uranium. For the general public, not involved in the nuclear industry or processes with uranium, the concentration of uranium in urine depends on many factors, but mainly on the amount of ingested uranium in food and drinking water. Thus, several studies in different countries found values for the uranium content in urine in the range of <2–20 ng L^{-1}, or taking the *reference* female and male diurnal urine excretion volumes of 1.2 and 1.6 L day^{-1}, respectively, the values are in the range of <2.5 to 32 ng day^{-1}, respectively. A survey of 500 US residents was carried out in order to establish the reference range concentrations of uranium in urine and the mean concentration was 11.0 ng L^{-1} (the 5th to 95th percentile range was 1.42–34.5 ng L^{-1}) (Ting et al. 1999). Consumption of elevated levels of uranium in drinking water would lead to higher values that in extreme cases may be two or three orders of magnitude higher (Karpas et al. 2005b).

As mentioned earlier, there are several analytical methods that have been developed for the analysis of uranium in urine. Some of the older methods required separation and preconcentration of the uranium from the urine sample, but modern methods, based mainly on ICPMS, allow direct determination of the uranium in raw urine or after simple dilution. In the following section, some of the older analytical methods for determining uranium in urine will be described. This section is based in part on a review that was carried out at Oak Ridge National Laboratory (Bogard 1996).

Alpha spectrometry can be used to determine the concentration of uranium in urine samples, and a detailed description of one of these procedures is presented here in order to demonstrate its complexity and the amount of labor involved (IAEA 2000). The uranium in the urine sample must be separated, purified, and preconcentrated by ion exchange before being deposited on a thin metal disk for alpha spectrometry. The IAEA method uses a spike of ^{232}U as a tracer to determine the chemical yield. The schematic outline of the procedure is shown in Figure 4.10.

First, a spike of ^{232}U was added to a 500 mL urine sample. Concentrated nitric acid and hydrogen peroxide were used to destroy the organic matter and to convert all uranium species into uranyl (UO_2^{+2}) ions and the sample was then boiled to dryness. Two more aliquots of nitric acid were added and this step was repeated twice to ensure that no organic matter was left in the sample. Then hydrochloric acid was added to leach the uranium from the dry precipitate. After the supernatant solution was cooled, the uranium was co-precipitated with iron hydroxide produced in situ by addition of ammonia to the iron carrier. The precipitate was dissolved in hydrochloric acid and passed through an anion exchange column. The uranium was eluted from the column with 100 mL of 1 M HCl. Then sodium bisulfate was added and the sample was evaporated to dryness. After that two aliquots of concentrated nitric acid were added and evaporation to dryness was carried out again. The precipitate was finally dissolved in 20 mL of 0.5 M HNO_3. After cooling, the uranium was once again co-precipitated, this time with a small amount of lanthanum fluoride (micro-co-precipitation). The sample was filtered and dried with an infrared (IR) lamp and then the filter was placed with quick drying glue on a stainless steel disk and finally counted by alpha spectrometry. The rather tedious procedure is presented here to demonstrate the efforts that were involved in determining uranium in urine at a low level of activity (MDA = 0.5 mBq) (IAEA 2000).

One study that used alpha spectrometry to determine background levels of uranium in human urine found a mean of 23 ng L^{-1} in 12 subjects, while a value of 3.4 ng L^{-1} was reported for analysis by thermal ionization mass spectrometry (TIMS) of a

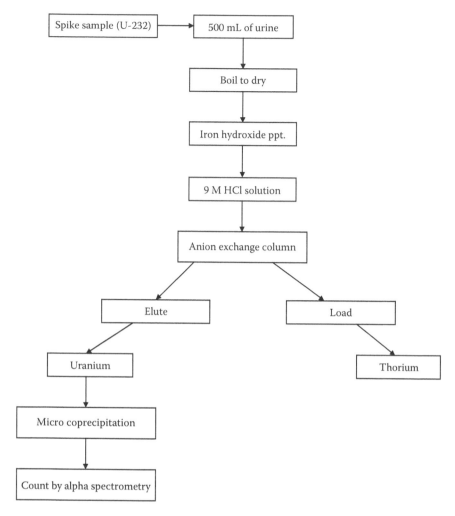

FIGURE 4.10 Outline of radiochemical separation processes used to prepare uranium bioassay samples for alpha spectroscopy. (Adapted from training manual IAEA, *Determination
of Uranium in Urine by Alpha Spectroscopy*, IAEA, Vienna, Austria, 2000.)

single urine sample (Wrenn et al. 1992). One of the ASTM standard test methods for
screening possible exposures to uranium is based on radiochemical determination of
uranium isotopes in urine by alpha spectrometry (C1473 2011). Some parts are similar to the procedure given earlier, but there are some differences. The 50 mL urine
samples were spiked with ^{232}U as a tracer, wet-ashed with nitric acid and hydrogen
peroxide to destroy organic matter. The uranium was separated and preconcentrated
on an anion exchange column and then electro-deposited on a metal disk or co-
precipitated with neodymium fluoride for alpha spectrometry. The method is used
mainly for the determination of ^{238}U and ^{234}U due to their prominence in the alpha
spectrum and the limit of detection depends on the parameters of the analytical system and on the counting time (C1473 2011).

Liquid scintillation spectroscopy is a variation of alpha spectrometry that has a higher counting efficiency and a much easier sample preparation procedure. The Photon-Electron Rejecting Alpha Liquid Scintillation (PERALS) device uses pulse shape discrimination techniques for rejecting interferences from β-particles and gamma radiation, thus reducing the background level and improving the limit of detection. Liquid–liquid extraction into an organic phase can be used to selectively extract the complex containing uranium. This is combined with efficient light collection and electronic signal processing, and the method has been deployed to improve resolution and enhance sensitivity (Dazhu et al. 1991).

UV-visible spectrometry is a classical analytical method that utilizes the fact that uranyl ions form colored complexes with chelating agents. Arsenazo-III is one of the reagents used for spectrophotometric determination of uranium as it forms a complex with high molar absorbance at 653 nm. The uranium in the urine sample is oxidized by nitric acid and then separated on an anion exchange column prior to forming a complex with Arsenazo-III. A detection limit of 0.29 µg L^{-1} was reported for an aliquot of 100 mL seawater (that is essentially similar to urine as far as matrix effects are concerned) using a dual-beam spectrophotometer (Nakashima et al. 1992). This method, or a variant of it, can also be used to determine uranium in urine although its minimum detectable level is evidently too high for monitoring the uranium concentration in urine for unexposed populations.

Solid pellet fluorometry (or fluorimetry) is one of the classic older methods that were widely used to determine the uranium content in urine (Centanni et al. 1956). The urine sample is added to solid NaF or NaF/LiF that is fused by heating so that water and volatile organic and inorganic compounds are evaporated. The sample is then excited by UV radiation at 320–370 nm and the fluorescence at 530–570 nm is measured (perpendicular to the incident beam). The sensitivity is about 30 µg L^{-1} for a 0.1 mL sample, but after preconcentration by ion exchange detection limits of 0.1 ± 0.1 µg L^{-1} for a 10 mL have been reported (Dupzyk and Dupzyk 1979). Even this improved MDL is insufficiently sensitive for monitoring unexposed populations where the expected concentration of uranium in urine is below 0.02 µg L^{-1} (20 ng L^{-1}).

Laser-induced fluorescence (LIF) utilizes the property of uranyl ions to emit light at a specific wavelength (green light at 534 nm) when excited by energetic photons, typically UV light at 330 nm from a pulsed nitrogen laser (Decambox et al. 1991). The organic materials in the urine sample that may contribute to the fluorescence are destroyed by the use of concentrated nitric acid, hydrogen peroxide, and thiosulfate. This step is repeated until a clear solution is obtained. The laser excitation pulses are followed by temporal resolution so that fluorescence from any remaining traces of organic material will decay before the fluorescence from the uranyl ions will reach the photo-detector. In effect there are three levels of resolution to improve the specificity for uranyl ions: the selection of the exciting wavelength, the choice of the emitted wavelength, and the temporal resolution (Decambox et al. 1991). This method replaced the older, and less sensitive, solid pellet fluorescence described earlier that was in wide use in the past. The sensitivity of the technique was sufficient for the action limit threshold at the time, namely about 1.5 µg L^{-1} for a 2 mL urine sample (Karpas et al. 1996) but not sensitive enough for monitoring unexposed populations.

Kinetic phosphorescence analysis (KPA) is another optical method that utilizes the fact that uranyl ions emit light after excitation by an energetic photon. The use of a pulsed laser for irradiating the sample affords time resolution to differentiate between the decay of uranyl ions (that follow first-order kinetics) and other compounds that may be present in the urine sample. Reportedly, the method has a limit of detection (LOD) of 10 ng L^{-1} uranium in urine that is almost sufficient for measurement of background levels of unexposed populations (Moore and Williams 1992).

FTA has also been used to determine the concentration of uranium in water and urine (Sawant et al. 2011). The sample was placed in a polythene tube and a Lexan detector was also inserted in the tube so that it was submerged in the liquid sample. The tube was heat-sealed but a space of a few millimeters was intentionally left above the liquid. The urine samples and the calibration tubes were placed in a capsule that was then irradiated for 1 min with a flux of ~5 * 10^{13} n cm^{-2} s^{-1}. The Lexan detectors were removed from the capsule, etched with 6 M NaOH at 60°C for 1 h. The fission tracks in each detector were counted in 300–350 fields under a microscope at 400× magnification. The reported MDA was 3 ng per sample (Sawant et al. 2011), giving an MDL of 3 μg L^{-1} for a 1 mL urine sample.

Neutron activation analysis (NAA) and *delayed neutron analysis (DNA)* have also been used for bioassays of uranium. NAA is based on n-γ reactions that ^{238}U nuclides undergo when irradiated by neutrons. The unstable ^{239}U produced by neutron capture rapidly decays by β emission to ^{239}Np that subsequently emits another β particle (with a half life of 2.34 days) to produce ^{239}Pu. These β emissions are accompanied by γ-rays, that is, electromagnetic radiation that is measured. DNA focuses on ^{235}U with its large cross section for fission by thermal neutrons and some of the fission product nuclei undergo emission of delayed neutrons. The delayed neutron flux is proportional to the amount of ^{235}U in the sample. The three nuclear-based methods, FTA, NAA, and DNA, require access to a neutron source with a high flux. However, unlike FTA where the development of the Lexan detector and track counting can be performed any time after irradiation, with NAA and DNA the sample must be analyzed shortly after irradiation. It was reported that 1 μg of uranium can be detected in a 25 mL urine sample by DNA or that detection limits of 0.5 μg L^{-1} can be achieved for a 100 mL aliquot after preconcentration on an ion exchange resin (Bogard 1996).

The review of Bogard presents a table in which the different methods are compared and an abbreviated adaptation is shown in Table 4.9.

Thermal ionization mass spectrometry (TIMS) is a sensitive mass spectrometric technique that has been deployed in some cases to measure trace amounts of uranium in urine and its isotopic composition (Kelly et al. 1987). The authors report measurement of one freeze-dried urine standard sample (SRM 2670) and two actual urine samples collected from children. For TIMS measurements, chemical separation has to be performed prior to the analysis. In an earlier work by the same author (Kelly and Fassett 1983), a spike of ^{233}U was used to implement isotope dilution measurements of picogram quantities of uranium in biological tissues. As mentioned earlier, a single urine sample tested by TIMS gave 3.4 ng L^{-1} (Wrenn et al. 1992).

Inductively coupled plasma-mass spectrometry (ICPMS) is currently the method of choice in many laboratories for determination of uranium in urine and other

TABLE 4.9
Reported Uranium Measurement Levels in Excreta and Tissues for Several Analytical Techniques

Technique	Measurement Level (ng L⁻¹)	Notes
UV-Vis spectrophotometry	5,000–66,000	Anion exchange followed by the formation of colored complex using Arsenazo III
Fluorometry	100 ± 100	Standard deviation for U in water is given by ASTM expression
UV laser	1000–7000	Detection of 0.01 ng L⁻¹ U in aqueous solution following CaF_2 precipitation and measurement of phosphorescence in fused precipitate
KPA	15–30	Estimated detection limit ~10 ng L⁻¹
PERALS		No bio-assay reported, but attainable levels estimated like alpha-spectrometry
Alpha spectrometry	40 ^{238}U 8 ^{235}U 0.002 ^{234}U	Approximate, based on 24 h counting time, 30% CV and typical conditions
Neutron irradiation	NAA 0.001–5000 DNA 7 (^{235}U) FTA 0.1–0.7 (^{235}U)	Neutron flux $3*10^{13}$ cm^{-2} s^{-1} Neutron flux $3*10^{13}$ cm^{-2} s^{-1}, 25 mL Neutron flux $3*10^{17}$ cm^{-2}, 0.05 mL
Mass spectrometry	RIS 1000 ICPMS 74–8000	Capable of isotopic analysis

Source: Adapted from Bogard, J.S., *Review of Uranium Bioassay Techniques*, Oak Ridge National Laboratory, Oak Ridge, TN, 1996 (ORNL 6857).

matrices as noted by the Procorad intercomparison (Berard et al. 2003). Several experimental procedures have been described in the literature since the 1990s. Some of these methods that deployed a quadrupole-type device (ICP-QMS) used a flow injection system (FIAS) for direct analysis of raw urine (Karpas et al. 1996), or simple dilution of the urine sample in order to reduce the matrix effect (Ting et al. 1996; Zamora et al. 2002). In some cases, the urine was acidified upon arrival in the laboratory (to prevent precipitation due to the rise in pH when urea is degraded to ammonia) and a series of standard addition samples was tested (Ough et al. 2002). The use of magnetic sector ICPMS for more accurate isotope composition measurement (Karpas et al. 1998) is also quite common as summarized in a report of an interlaboratory comparison (Parrish et al. 2006) or several other publications (Pappas et al. 2002).

Preconcentration of uranium on a solid phase extraction (SPE) column combined with magnetic sector ICPMS has been used for accurate isotope ratio determination in small urine volumes with microwave digestion (Pappas et al. 2003) or without it (Pappas and Paschal 2006). The urine sample may be spiked with an internal standard such as europium (Allain et al. 1991), iridium (Ting et al. 1996), bismuth

(Baglan et al. 1998), thallium (Karpas et al. 2005b), or an isotope of uranium. Considerations for the selection of the element used as an internal standard should include its absence in urine, its mass (not too far from that of uranium), its isotopic composition, its cost and availability in pure form, and lack of polyatomic atoms that may be isobars of the uranium isotopes. Further variations in the deployment of ICPMS for the determination of uranium in urine can be found with the use of different sample introduction techniques and nebulizers. Due to the importance of uranium urinalysis by ICPMS, some detailed examples are presented in this section. In one study, isotope dilution ICP-MS was compared with alpha spectrometry for the determination of uranium in slightly diluted urine samples spiked with ^{233}U (Haldimann et al. 2001).

A flow injection analysis system (FIAS) for direct analysis of raw urine was mentioned earlier (Lorber et al. 1997). No sample preparation was carried out so the probability of cross contamination was negligible. The response profile of the ICPMS detector is typical of the FIAS sample introduction method, as shown for a calibration solution containing 100 ng L^{-1} uranium (top frame of Figure 4.11) and an actual raw urine sample containing 3.8 ng L^{-1} (bottom frame).

A short time (~8 s in this case, but the actual time depends on the sample flow rate and length of tubing) after the sample is carried from the sample loop a sharp increase in the uranium signal is observed reaching a maximum value after about 15 s, and then a gradual decrease is observed until the signal intensity returns to its background value at the termination of the measurements (50 s). The limit of detection was estimated to be 1.5 ng L^{-1} and the sample throughput was above 100 samples per shift. The matrix effect due to the total dissolved solids in the urine sample is somewhat alleviated by the fact that the raw urine gets diluted (about five-fold in this set-up) until it reaches the plasma torch. Studies have shown that the difference in signal intensity between a given uranium concentration in an aqueous solution and in raw urine is typically 20%–40%, depending on the dissolved solids content in the urine sample. An internal standard (e.g., Ir, Rh, Bi, Tl) can be used to correct this effect on a sample by sample basis, or the use of a pooled urine sample that is spiked with a known concentration of uranium for deriving an average batch correction. If instead of a limited urine sample contained in the FIAS loop (typically 100 μL) continuous flow is used for raw urine, then accumulation of solid deposits on the sampling and skimmer cones would eventually result in clogging their orifices. Dilution of the raw urine or injection of a small size sample by the FIAS system, overcome this drawback.

In the comprehensive study of the uranium and thorium levels in urine of 500 US residents, a magnetic sector ICPMS with a microconcentric nebulizer was used (Ting et al. 1999). The sample handling was carried out in a clean laboratory and all reagents were of high purity in order to avoid cross contamination. The samples were spiked with iridium that served as an internal standard and the creatinine concentration was also determined in order to normalize the results. The distribution of the uranium concentration, normalized to creatinine is shown in the top frame of Figure 4.12 while the bottom frame shows the frequency distribution of uranium in 350 urine samples in Israel (Karpas et al. 1996).

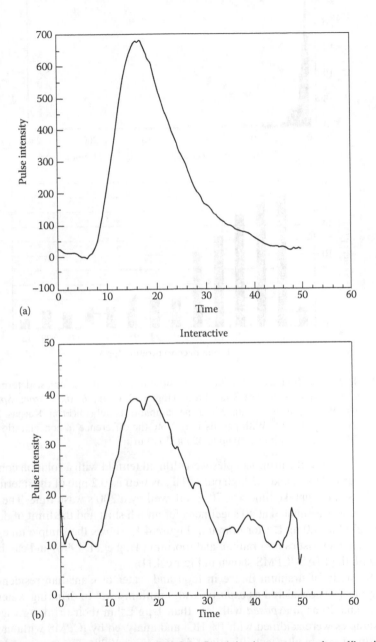

(a)

(b)

Interactive

FIGURE 4.11 (a) Typical response (cps) of the flow injection system to the calibration solution containing 100 ng L^{-1} uranium as aqueous uranyl nitrate. The x-axis is the time is seconds. (b) The response for a raw urine sample containing 3.8 ng L^{-1} uranium. (From Lorber, A. et al., *Anal. Chim Acta*, 334, 295, 1996. With permission.)

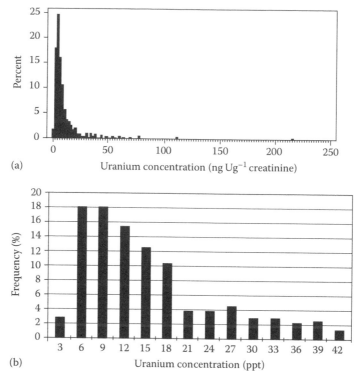

(a)

(b)

FIGURE 4.12 The frequency distribution of uranium in urine samples, determined by ICPMS for 500 residents in the US (a) (From Ting, B.G. et al., *J. Anal. Atom. Spectrosc.*, 11, 339, 1996. With permission.) and 350 urine sample in Israel (b) (From Karpas, Z. et al., *Health Phys.*, 71, 879, 1996. With permission). Note the difference in concentration scales: (a) up to 250 ng U g^{-1} creatinine; (b) up to 42 ng U L^{-1} urine.

In the US study, the urine samples were diluted tenfold with a solution containing 5% high purity HNO_3 and 5% high purity HF, as well as 0.2 ppb of the internal standard. The sample uptake time was 70 s, followed by a 200 s wash time. The sample throughput was estimated at 40 specimens for an 8 h shift and the limit of detection was reported as 1.0 ng L^{-1} for uranium. Figure 4.13 shows the sample take-up and washout characteristics for uranium and thorium (Ting et al. 1996) and can be compared with the FIAS-ICPMS shown in Figure 4.11.

In the study of uranium intake in food and water of Canadian residents from two regions (30 considered as *exposed* with uranium level in drinking water above 1 μg L^{-1} and 20 as *unexposed* with less than 1 μg L^{-1} in their drinking water), the urine samples were acidified with 1% HCl and analyzed by ICPMS without further pretreatment except ultracentrifugation (Zamora et al. 1998). The lowest and highest intake values of uranium covered a broad range of 0.34–570 μg day^{-1} but the corresponding uranium levels in urine were not given (although they can be calculated with an assumption on the amount of ingested water).

In a study that included 72 children and 87 adults in Germany, the concentration of 30 trace elements in urine was determined by ICPMS equipped with and octopole

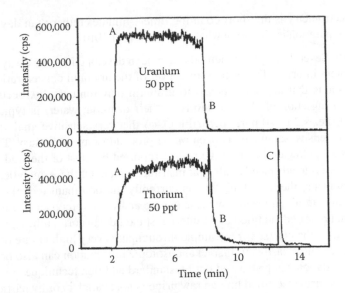

FIGURE 4.13   The sample take-up and washout characteristic response profile for uranium and thorium. (From Ting, B.G. et al., *J. Anal. Atom. Spectrosc.*, 11, 339, 1996. With permission.)

collision/reaction cell (Heitland and Koster 2006). The mean values for the concentration of uranium in the urine for the children and adults were 4 and 5 ng U L^{-1}, respectively, and the corresponding uranium normalized to creatinine value for both groups was 3 ng U g CRE^{-1}.

The level of uranium in urine and blood plasma was determined by ICPMS in a group of 20 healthy French residents of Paris in one of the earliest published studies using this technology (Allain et al. 1991). The samples were diluted threefold by 1% nitric acid and europium (25 μg L^{-1}) was used as an internal standard. The time for analysis of a sample was 2 min and a conservative limit of quantification (5σ) was reported as 35 ng L^{-1} for uranium in urine and plasma.

Regional variations of the daily urinary excretion of uranium of nonexposed subjects in different regions of Germany and different countries were studied by an international team using high-resolution ICPMS (Hollriegl et al. 2002). The daily excretion among healthy adults in 13 federal states in Germany varied from 8.8 to 30.2 ng U day^{-1}. Comparing the daily urinary excretion of uranium between populations in Germany (n = 285), Italy (n = 12), Great Britain (n = 5), and the Southern Urals (n = 9) showed quite a narrow range of 21.5 ± 19.4, 33.6 ± 3.7, 4.9 ± 3.6, and 5.4 ± 1.4 ng day^{-1}, while for 62 subjects in Jordan the mean was 322 ± 78 ng day^{-1} (Hollriegl et al. 2002).

A standard ASTM method for the determination of [237]Np, [232]Th, [235]U, and [238]U in urine is based on ICPMS measurement of these radionuclides after the sample is wet-ashed and the nuclides are separated and preconcentrated on a chromatographic resin column (C1614 2010). It is stated that due to isobaric interferences and limits on sensitivity some of the other actinides of interest ([239]Pu, [241]Am, and [234]U) cannot be assayed by this ICPMS procedure. In this case, the main advantage of ICPMS over alpha spectrometry is the short counting time—minutes rather than days—and the sample preparation is also somewhat simpler (C1614 2010).

Highlights: Several analytical methods have been developed for the assay of the uranium content in urine. The background level of the uranium concentration in urine of populations that are not exposed to uranium (environmentally, occupationally, or through ingestion of elevated levels in their drinking water) is typically <2 to 20 ng L^{-1} (i.e., <2 to 20 parts-per-trillion), so that very sensitive analytical instruments or extensive sample preconcentration procedures are necessary. This finding also casts a shadow of doubt on the results reported by most of the studies that did not have access to advanced analytical instrumentation. Of these analytical methods and instruments, the use of ICPMS is currently the dominant technique. Sample preparation is minimal (even raw urine can be analyzed), preconcentration or separation is not required, addition of an internal standard can correct the signal attenuation caused by the *matrix effect*, sample throughput is very high (a rate of dozens of samples per hour is easily attainable) and isotopic information can also be acquired. Method validation can be carried out by standard addition techniques by the use of reconstituted urine (certified human raw urine is not available) or by participation in interlaboratory intercomparison programs.

4.4.2 Urinalysis for Special Populations

The concentration of uranium in urine of the general population that is not occupationally, environmentally, or through dietary habits (mainly through elevated uranium levels in drinking water) exposed to uranium usually does not exceed 40 ng L^{-1}, as found in most of the modern studies described earlier. However, there are a few special population groups that are considered as being potentially exposed to uranium. These include employees of uranium mining operations and other uranium processing facilities, military personnel who were in conflict areas where DU munitions were used and inhabitants who live in those areas, as well as residents that consume elevated uranium levels in their drinking water. In principle, the same methods described earlier that are used for bioassays of the general population that are exposed only to relatively low-level natural uranium are also suitable for these special populations. However, the isotopic composition of anthropogenic uranium may be an important factor, indicating the source of exposure and is thus of interest. In this section, some examples of deployment of ICPMS-based methods for measurement of the isotopic composition of uranium in urine samples will be described. The full procedures can be seen in the cited articles and here only an outline of the main features of each method is presented.

In order to distinguish between exposure to uranium from natural sources and anthropogenic origins, the isotopic composition of uranium must be determined. Such a method using sector field ICPMS for measurement of $^{234}U/^{238}U$, $^{235}U/^{238}U$, and $^{236}U/^{238}U$ in small-volume urine samples was developed and validated (Arnason et al. 2013). The work placed special emphasis on the cleanliness of all reaction vessels used in the process, on the purity of all reagents, and on the low background contribution from the chromatographic resins used for separation and preconcentration of the uranium. Pooled samples of urine were meticulously prepared and divided into several containers that were spiked with controlled amounts of DU from certified reference materials. The performance of two popular chromatographic resins (TRU and UTEVA, both from Eichrom) was compared with special attention given to the blank background level after several washings of the resin columns. The method detection limits were impressive for ^{234}U, ^{235}U, ^{236}U, and ^{238}U in 2 mL urine samples and were reported as $9*10^{-4}$, $3*10^{-3}$, $2*10^{-4}$, and 0.3 ng L^{-1}, respectively, based on three standard deviations of the mean of 27 replicates in 10 runs over a period of 6 months. Naturally, using a larger sample, say 8 mL, will improve the MDLs by about a factor of 10 for the minor isotopes and a factor of 3 or 4 for ^{235}U and ^{238}U. Recovery from UTEVA columns was close to 100% even when the uranium concentration was 1–5 ng L^{-1} and a 2 mL urine sample was tested (i.e., 2–10 pg uranium per sample). The operational conclusion of this work is that by meticulous control of the blank level the ^{235}U/^{238}U ratio can be accurately determined (RSD ~ 1%) from a 1 mL urine sample when the uranium concentration is 50 ng L^{-1} (i.e., minimal occupational exposure level). Thus, the method can easily distinguish between exposure to DU, natural uranium or enriched uranium (Arnason et al. 2013).

An isotope dilution method with an ICPQMS equipped with direct reaction cell (DRC) to remove interferences from polyatomic ions was described (Kinman et al. 2009). A nebulizer suitable for high solids introduction was used for measurement of urine samples that were diluted fivefold with 10% nitric acid. Isotope ratios were precisely determined for urine sample that contained >54 ng L^{-1}. A three-stage rinsing procedure was run between samples to minimize salt deposition, memory effects, and instrumental drift.

Accurate and sensitive determination of the ^{236}U/^{238}U and ^{235}U/^{238}U ratios in human urine using a sector field-ICPMS was described (Gray et al. 2012). This procedure involved separation of uranium from the urine matrix and preconcentration by co-precipitation with calcium and magnesium after addition of high-purity ammonia at pH = 9. After settling for 24 h, the sample was centrifuged, the supernatant fluid was discarded, and the precipitate was dissolved in nitric acid. For optimal analytical conditions, the uranium concentration should be in the range of 5–200 ng L^{-1}, and the alkaline earths (combined calcium and magnesium) should be below 500 mg L^{-1}, so some of the processed samples had to be diluted to comply with these criteria. According to the authors, the advantage of co-precipitation of the uranium with calcium and magnesium is that the uranium concentration in the processed sample is relatively higher by a factor of 16 than by simply diluting urine to ensure that the level of total dissolved solids (TDS) is below 500 mg L^{-1}. When measuring the ^{236}U content by ICPMS a correction for the contribution of the uranium hydride ion, ^{235}UH$^+$ that has a nominal mass of 236 Da, is required, so sufficient standards and blanks have to be measured with every batch of samples. The use of a nebulizer that also does desolvation reduces this ^{235}UH$^+$ contribution. When the ^{236}U/^{238}U ratio is plotted as a function of the ^{235}U/^{238}U, as shown in Figure 4.14, a clear distinction could be made between the urine samples of the subjects with DU fragments and people who have uranium of natural isotopic abundance (Gray et al. 2012). The study population included 12 samples collected from people with embedded DU fragments and about 150 samples of inhabitants with natural uranium only.

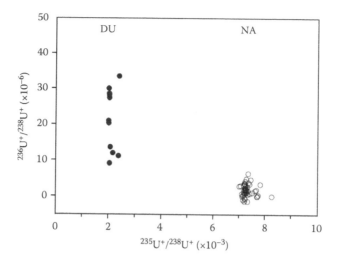

FIGURE 4.14 $^{236}U/^{238}U$ ratio expressed as a function of $^{235}U/^{238}U$ ratio for a subset of patient samples. Solid circles indicate patients with DU exposure and open circles patients with natural occurrence of uranium. NA, natural abundance. (From Gray, P.J. et al., *Microchem. J.*, 105, 94, 2012. With permission.)

A study to determine whether the uranium levels in urine samples of German peacekeeping personnel in the Balkans differed from the level in urine samples in a control group of civil servants and unexposed people was carried out (Oeh et al. 2007). Over 1300 urine samples were collected over a 7-year period. The samples from the personnel returning from the Balkans were collected within 1–12 months after returning to Germany and compared with a control group of residents of Germany who had not served abroad. The participants were provided with acid-cleaned polyethylene bottles and were requested to carry out a 24 h urine collection. Sample preparation was based on an earlier method where UV-photolysis with hydrogen peroxide and nitric acid were deployed to digest the urine sample and destroy the organic matter (Schramel 2002). Analysis of the urine samples was by sector field-ICPMS and the ^{238}U isotope was measured (Oeh et al. 2007).

Workers in uranium mining operations constitute another special population that may be occupationally exposed to uranium and indeed several studies have focused of this population. In one such study, urine samples were collected from a group of 10 uranium miners (hewers) in the Czech Republic and compared with urine samples collected from 27 members of the general public (mostly employees of NPRI) and 11 family members (Malatova et al. 2011). Analysis of the control group was carried out on 24 h samples with a high-resolution ICPMS and the analysis of the 50–100 mL urine samples of the hewers was done by the same method. As expected, the uranium level in the group of hewers was significantly higher than in the control groups. For the general population, the arithmetic mean was 10.3 ± 11.8 ng L^{-1}, for family members of the hewers it was 22.8 ± 17.1 ng L^{-1}, and for the group of 10 miners it was 295 ± 152 ng L^{-1} (Malatova et al. 2011).

A number of studies on the uranium urine levels of workers in underground gold mines in South Africa were surveyed (Winde 2013). High uranium levels were found in urine samples of workers in the Deelkraal Gold Mine. This was attributed to illegal consumption of chilled service water that contained up to 4000 μg L^{-1} in the Deelkraal gold mine and 20,000 μg L^{-1} in another site (W-Dreifontein). Furthermore, in some cases, neighboring mining communities may have been supplied with poor quality water pumped by the mines.

Isotopic ratio measurements can be used to determine the origin of elevated uranium urine levels in populations that consume large amounts of uranium in their drinking water (Karpas et al. 2005a). The fact that the ^{234}U/^{238}U ratio shows great variation (called disequilibrium and discussed in Chapter 1) in water that is in contact with minerals that contain uranium deposits is reflected in the bioassays of the people who consume the elevated uranium levels in their drinking water. The water and urine samples were measured without pretreatment (except acidification and spiking with thallium) by a flow injection system (FIAS). Two types of ICPMS devices were used to measure the ^{234}U/^{238}U ratio: a quadrupole ICP-QMS and a multi-collector magnetic sector ICPMS. The agreement between the two devices was excellent and a clear correlation was found between the ^{234}U/^{238}U ratio in drinking water and in excreted urine for each individual. It should be noted that the variance of the ^{234}U/^{238}U ratio was due to natural processes and not the outcome of anthropogenic activities.

The use of a dynamic reaction cell to reduce interferences intended to identify the presence of DU in urine was described (Ejnik et al. 2005). The idea was to deploy a reagent gas (oxygen in this case) to convert the U$^+$ ions into UO$_2^+$ ions in order to shift the uranium isotopic peaks to masses that have no isobaric interferences from polyatomic species. The method did improve the detection limit for uranium in urine to 0.1 pg mL^{-1} and the ^{235}U/^{238}U ratio can be measured accurately with uranium concentrations of 3 pg mL^{-1}. After validation of the methods with urine samples of known uranium content and isotopic composition, the efficiency of the method was demonstrated for 21 urine samples from patients who had embedded DU fragments (Ejnik et al. 2005).

Highlights: The picture that emerges from the urinalysis studies described earlier for unexposed and exposed populations quite clearly represents the state of the art. Methods that require extensive sample pretreatment, involving destruction of all the organic matter in the urine matrix that is usually followed by ion-exchange to separate and preconcentrated the uranium, are labor intensive, slow, expensive, and usually do not yield full information on the isotopic composition of uranium. These are now almost universally replaced by a variety of simple sample preparation procedures that are suitable for ICPMS analysis. Some methods require simple dilution and the use of an internal standard and others even allow direct introduction of raw urine into the ICPMS (with flow injection to minimize salt burden). In addition to superb sensitivity, the ICPMS-based method readily supply detailed isotopic composition, sometimes also of the minor isotopes (^{234}U and ^{236}U) not only of the ^{235}U/^{238}U ratio. It should also be noted that the result of the urinalysis is usually used in the biokinetic models to assess the body burden of uranium and that these models contain quite a large degree of uncertainty. So that even if the urinalysis is accurate and the urine sample is a perfect representative sample there will still be uncertainty in the body burden and internal exposure estimation.

4.4.3 BLOOD AND FECES ANALYSIS

The concentration of uranium in blood—whole blood, serum, or plasma—is not the standard method for the assessment of the body burden or for estimating uranium exposure levels. This is due to the difficulty and inconvenience of collecting blood samples as well as to the medical requirements for handling the samples. Therefore, fewer methods were developed for analysis of uranium in blood, fewer studies were published, and the total size of the population for which uranium in blood was determined is much smaller than those on whom urinalysis was performed. The situation for determining uranium in feces is even more challenging because the level of the uranium content in feces mainly reflects the amount of ingested uranium (not the uptake) and the sampling, handling, processing, and treatment of feces samples are unpleasant, labor-intensive, and relatively complicated.

Blood: Neutron activation analysis was used to determine the level of uranium in samples of blood, urine, and hair of exposed people (workers in uranium production facilities in Slovenia) and unexposed population (Byrne and Benedik 1991). Blood samples were collected by venipuncture from the arm. The samples were freeze-dried and powdered and well mixed before encapsulation in polythene vials that were heat-sealed. The samples were irradiated with a $4 * 10^{12}$ n cm^{-2} s^{-1} neutron flux for up to 30 min and rapidly ashed with 50 mg of uranium that served as a carrier. The uranium was then extracted with TBP in toluene and after cleaning the organic phase by acid washing the samples were counted with a HPGe gamma detector for 20 min. The 74.7 keV ^{239}U peak formed by neutron irradiation and the 185.7 keV ^{235}U peak (from the carrier) serve for calculating the uranium concentration in the sample and the chemical yield, respectively. The measured concentration of uranium in blood was between 5 and 39 ng kg^{-1} for the employees in the store, maintenance, mining, and milling, but between 25 and 113 ng kg^{-1} for the personnel involved in yellow cake production. The level in the control group was below 5.3 ng kg^{-1} in all cases.

A different approach, based on flow injection-ICPMS measurement of full blood and blood plasma, was reported (Lorber et al. 1997). Sample preparation involved centrifugation to separate the plasma from red blood cells. The results for unexposed population were quite similar to those found for unexposed people in Slovenia (Byrne and Benedik 1991). ICPMS was also used to determine the uranium level in blood samples collected from 25 females in Bavaria (Chaudhri and Watling 2010). The blood samples were collected in the morning, after the participants had breakfast, were centrifuged, freeze-dried, and shipped from Germany to Australia for analysis. There the samples were dissolved under reflux at 180°C by high-purity nitric acid, and NIST standards of tomato, apple, and peach leaves were used as standards (CRMs for uranium in blood are not available). The concentration range that was found was large, between 7 and 543 ng L^{-1}, and the high values were attributed to the uranium content in the water supply (Chaudhri and Watling 2010).

ICPMS was also used for the determination of DU in whole blood samples (Todorov et al. 2009). The samples were digested using a closed vessel microwave oven at 180°C. Standards for quantitative calibration were prepared by adding a ^{233}U spike and digestion with HNO_3 and H_2O_2. The blood concentrations found for Gulf War I veterans were between 9 and 800 ng L^{-1}, and ^{235}U/^{238}U isotopic ratios were between 0.205% and 0.721%, with the expected general trend that deviation from the natural abundance increased concomitantly with the increase in uranium concentration (Todorov et al. 2009).

Biomonitoring of 20 trace elements, including uranium, in blood and urine samples of 50 workers (26 males and 24 females) that were occupationally exposed to beryllium in a beryllium processing plant was carried out by sector field ICPMS (Ivanenko et al. 2013). Sample collection was performed with special attention to prevent cross-contamination, and the samples were frozen and stored at −20°C. The samples were thawed in the laboratory. The blood samples were digested with concentrated nitric acid (0.5 mL blood with 2 mL acid) in a closed vessel in a microwave oven for 25 min at 190°C and then diluted with Suprapure water to a final volume of 25 mL. The results showed a mean, median, and maximum concentrations of 40, 34, and 120 ng L^{-1}, respectively. Interestingly, the uranium level in most (92.5%) of the corresponding urine samples were below the LOD. The authors quote a Canadian study where the concentration range in blood was from <3 to 6 ng L^{-1} (Clark et al. 2007) but the data could not be confirmed.

The uranium level in blood, urine, and hair of workers in phosphate mines in Syria was determined by fluorimetry (Othman 1993). The preparation of the blood samples included addition of an anticoagulant, nitric acid, and hydrogen peroxide, followed by centrifugation. The samples were then heated on a platinum plate and the flux mixture was added. Finally, the fluorescence of the samples was recorded. A correlation was found between the level of uranium in the bioassays and the number of years of employment in the mines (Othman 1993).

Feces: Procorad organization conducts interlaboratory comparisons for determining the uranium (and other radionuclides) content in fecal ash. The preparation of the pooled samples was carried out by the organizers and each participating laboratory developed their own analytical procedure. In general, digestion of fecal ash samples is not easy. For example, at the Nuclear Research Center, Negev (NRCN) acid digestion with concentrated nitric acid and hydrogen peroxide was carried out in a sealed vessel in a microwave oven and the estimation for recovery of uranium varied from 50% to 80%. The concentration of uranium in fecal ash was below 1 μg U g^{-1}, and typically in the range of 0.2–0.8 μg U g^{-1}. In the study of Brazilian workers in nickel mines, samples were collected from 68 nonoccupationally exposed females and males (Azeredo et al. 2000). The fecal samples were dry ashed at 400°C, homogenized, and then ashed again at 600°C. Subsamples of 10 mg ashes were digested with 10 mL of concentrated HNO_3, diluted, and analyzed by ICPMS with In or Tl as internal standards. As mentioned earlier, uranium content in 25 out of 29 fecal ash samples was below 0.3 μg U g^{-1} and above 0.5 μg U g^{-1} for four samples (Azeredo et al. 2000). One of the interesting features of the British study of the uranium content in feces of 119 AWE (Atomic Weapons Establishment in Aldermaston, UK) employees is the broad range of uranium activity (0.14–226 mBq day^{-1}) that is partly due to the variability in the dry weight of the fecal ash that was between 0.3 and 9.7 g (the dry weight was 2.0%–8.8% of the wet weight) (Cockerill et al. 2006). [As shown in Frame 4.3, 1 Bq corresponds to 81 * 10^{-9} g of ^{238}U, so assuming secular equilibrium then 1 mBq would be equivalent to ~39 pg of uranium.] The samples were frozen until radiochemical analysis was performed: dry ashed at 550°C overnight, and then a tracer was added, and using microwave digestion in nitric and hydrofluoric acids. This was followed by separation on an anion-exchange, electro-deposition on a stainless steel planchette and counting the alpha spectrum for 168 h, giving a detection level of ~1 mBq per sample. The ratio between ^{234}U and ^{238}U showed disequilibrium in many samples so that depleted, natural, or enriched uranium could be differentiated (Cockerill et al. 2006).

Highlights: Analytical procedures for bioassays of uranium in blood and feces have been developed and used in a limited number of studies. Compared to urinalysis, the collection and handling of blood and feces are complicated and entail special procedures. The preparation of fecal samples involves dry-ashing at an elevated temperature and then acid digestion. For alpha spectrometry, separation, purification, and deposition precede counting (that could last several days per sample). Other analytical techniques described in the literature include neutron activation analysis, but ICPMS is currently more widely used.

4.4.4 URANIUM CONTENT ANALYSIS IN HAIR AND NAILS

Hair is especially attractive as a bioassay for the determination of uranium exposure for a number of reasons. Hair is an efficient bio-concentrator of uranium as reflected by the concentration ratio of uranium in hair that is three orders of magnitude higher than its level in blood (see Table 4.8). Sampling of hair is painless and safe and no special equipment is required. Sample preservation, storage, and transport are simple and inexpensive. Hair analysis can truly reflect chronic exposure and so can be used to assess the average over a long period of time (depending on the length of the hair strands). In addition, the hair samples can be collected long after an accidental exposure incident and still reflect the extent of exposure at this incident.

The two main drawbacks of hair analysis are the need to remove external contamination if the body's internal hold-up is to be assessed and the need to digest the hair samples for most analytical methods (laser ablation and neutron activation techniques, discussed later, are notable exceptions). A simplified methodology that does not require microwave digestion, used to overcome these two factors was developed and described in detail later (Gonnen et al. 2000), but this is by no means the only method used for the treatment of hair samples.

Hair: The procedure to remove external contamination used by Rodushkin and Axelsson (2000a) included stirring of the sample with acetone, deionized water and a Triton X-100 aqueous solution in an ultrasonic bath and shaker, followed by rinsing with ultrapure water and drying in a class 100 clean room. Microwave digestion was accomplished with 1:1 H_2O_2:HNO_3 in closed PFA or PTFE vessels for 30 min at 325 W power setting. The vessels were carefully vented to release the gases formed in the digestion process and the samples were diluted to 10 mL with purified water. In cases that the samples were not completely digested, a second digestion step was carried out with HF at 70°C, followed by evaporation to dryness and re-dissolving in nitric acid. Analysis of the samples was carried out by ICPMS with external calibration (Rodushkin and Axelsson 2000). The results of this study are described in Section 4.4 and are shown in Table 4.8.

A somewhat different approach to digestion of hair samples, after rinsing, included placing the samples in a clean high-pressure PTFE vessel with 2 mL of concentrated super-pure nitric acid overnight (D'Ilio et al. 2010). This is reportedly done in order to prevent strong chemical reactions that produce gases that lead to high pressures when the vessel is closed. Thus, hydrogen peroxide was added only before the vessels were placed in the microwave oven that was programmed to ensure efficient digestion.

A popular method (Gonnen et al. 2000) that is easy to implement used a similar rinsing procedure but did not require the use of a microwave oven and special digestion vessels described earlier (see Section 4.4 for a summary of the results of this study). The method involved removal of external contamination first by rinsing with deionized water, then with addition of an aqueous detergent (like 0.1% Triton X-100 in deionized water). This was followed by rinsing with an organic solvent (e.g., acetone), once again with deionized water, and finally, after repeating these stages, by purified water (18 MΩ cm^{-1}). After air-drying and weighing, the hair samples (50–100 mg of hair per sample) were placed in 15 mL polystyrene test tubes with screw caps. The digestion process was carried out with a 1:2 mixture of high-purity hydrogen peroxide (30% H_2O_2) and Suprapure concentrated nitric acid. Digestion at room temperature was usually complete within a few hours, and this could be expedited by heating the samples in a hot water bath. Caution must be practiced to prevent build up of gases produced during the digestion process and the simple solution was to allow excess gases to escape from the digestion vessel by keeping the screw cap loosely closed (Gonnen et al. 2000). The sample was then diluted with purified water and a spike of an internal standard (usually Tl, Ir, Bi, or Rh) was added. Analysis can be carried out by ICPMS or almost with any of the analytical methods described earlier. A schematic diagram of the sample preparation procedure for hair samples is shown in Figure 4.15. A similar sample preparation technique was used in other studies (Karpas et al. 2005b; Alaani et al. 2011a).

The uranium concentration in hair samples of Brazilian residents (18 young females and 4 males) was determined by ENAA (Akamine et al. 2007). Hair samples were collected from the occipital part of the head, cut into 2 mm sections, rinsed to remove external contamination, and placed on filter paper for drying. The hair samples, and samples spiked with uranium standards, were placed in thin cadmium capsules and irradiated for 16 h. After 4 days, to allow for decay of interfering radionuclides, the gamma activity of the samples was measured for 50,000 s, and the uranium content was determined from the intensity of the 106 and 278 keV peaks of [239]Np. Method validation was based on measurement of CRM (NIST 1575—pine needles). The uranium concentration was 2.1–49.8 ngU g^{-1} hair, with mean and median values of 15.4 and 10.7 ng g^{-1}, respectively.

Laser ablation coupled to ICPMS has been used to determine the concentration of uranium in hair samples (Rodushkin and Axelsson 2003) and even along a

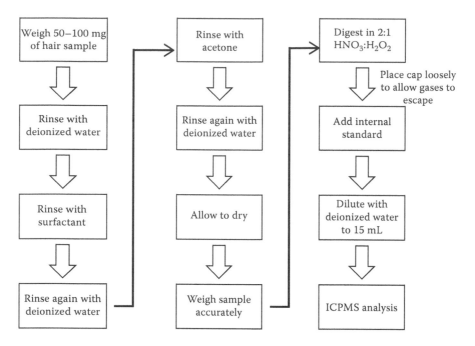

FIGURE 4.15 Schematic diagram of the procedure for preparing hair samples for analysis without an ultrasonic bath. (Based on Gonnen, R. et al., *J. Radioanal. Nucl. Chem.*, 243, 559, 2000.)

single hair strand (Elish et al. 2007; Sela et al. 2007). In this study (Rodushkin and Axelsson 2003), laser ablation was introduced for direct measurement of 55 elements (including uranium) in hair samples that were retained from the previous study (Rodushkin and Axelsson 2000b). The authors demonstrated that the concentrations of a number of elements can be monitored along a single hair strand in steps of 2.5 mm between sampling points and noted that smaller step sizes can be used if a single element, or small number of elements, is to be monitored. This indeed was shown later (Elish et al. 2007) when single pulses of a laser for ablation of small sectors along a single hair strand could pinpoint an exposure incident. The concentration of uranium was shown to reach a maximum value of ~2300 ng g^{-1} but after a few days returned to the level before the incident (assuming that the rate of hair growth is about 1 cm per month).

In another demonstration of the information that can be derived from hair samples by LA-ICP-SF-MS analysis (the system is presented schematically in Figure 4.16), it was shown that the uranium level in drinking water is indeed reflected in the uranium concentration in hair (Sela et al. 2007). A person who lived in an area in Israel where the uranium concentration in the communal drinking water supply was ~2 μg L^{-1} relocated to Germany where the uranium in drinking water was ~0.03 μg L^{-1}. Samples of hair were collected 8 months after moving to Germany and analysis of the uranium concentration was performed along a single strand of hair. The uranium level in the distal part of the hair that grew while the person was still in Israel was about 200 ng g^{-1}, while in the proximal part of the hair that grew after several

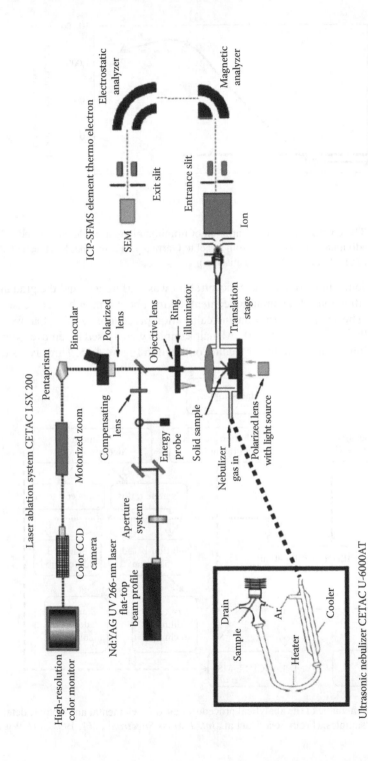

FIGURE 4.16 Schematic of the laser ablation–ICP-SFMS: coupling a Cetac LSX-200 with a double focusing sector field ICP-MS (Thermo Electron, Bremen). (From Sela, H. et al., *Int. J. Mass Spectrom.*, 261, 199, 2007. With permission.)

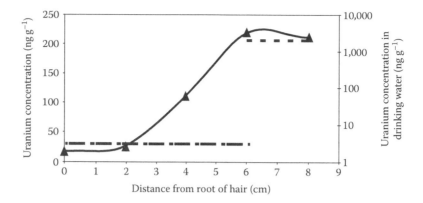

FIGURE 4.17   The changes in the concentration of uranium along a single hair strand collected from an individual that relocated from Israel to Germany. (From Sela, H. et al., *Int. J. Mass Spectrom.*, 261, 199, 2007. With permission.)

months of residence in Germany the concentration was ~20 ng g^{-1}, and the gradual decrease (detoxification) of the uranium content, taking about 4 months, can be seen in Figure 4.17. The sample preparation and calibration procedures used in that work are schematically shown in Figure 4.18. All hair samples were rinsed to remove external contamination according to the procedure outlined in Figure 4.15. Three types of

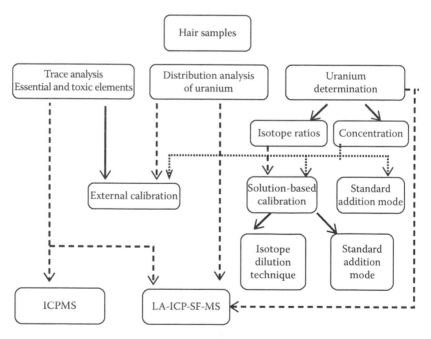

FIGURE 4.18   Schematic of the analytical procedures used for elemental and isotopic determination in hair samples. (From Sela, H. et al., *Int. J. Mass Spectrom.*, 261, 199, 2007. With permission.)

hair samples were prepared: digested samples according to the procedures outlined earlier, single other hair samples were fixed on a carbon stub, and other hair samples were powdered. The analytical procedures are outlined in Figure 4.18 and include isotope dilution and standard addition modes for accurate determination of the isotopic composition of uranium and its quantitative measurement.

Alpha spectrometry was deployed to determine the concentration of uranium in hair samples collected from 10 Greek monks (Kehagia et al. 2010). The samples were collected from the nape of the neck, rinsed with water and acetone, and air-dried prior to digestion in a microwave oven. The uranium was purified and preconcentrated on an anion exchange resin column, and finally electrodeposited on a stainless steel plate for alpha spectrometry. The detection limit in hair was reported as 5 ng g^{-1} for 7200 min counting time. The measured uranium concentration among these subjects varied from 12.1 to 170 ng U g^{-1} hair compared to that measured for staff members that was around 30 ng g^{-1}.

A study of hair samples collected from 22 Balkan residents showed that the concentration of most of the studied elements (Mn, Ni, Cu, Zn, Sr, and Cd) was in line with worldwide reported values, but the uranium concentration showed an exceptionally wide range of 0.9–449 ng g^{-1} (Zunic et al. 2012). The isotopic composition of uranium showed a natural $^{235}U/^{238}U$ ratio, although the samples were collected in conflict zones where DU munitions could have been used. Sample treatment involved rinsing followed by microwave digestion with HNO_3 and $HClO_4$. Analysis was by ICPMS with rhodium as an internal standard.

Nails: Nails, fingernails, and toenails have also been used for the assessment of exposure to uranium. The digestion procedures employed for hair (Gonnen et al. 2000; Rodushkin and Axelsson 2000a) are suitable for treatment of nails, and the same ICPMS-based analytical methods can be deployed (see Section 4.4 for the results of these studies). However, as bioassays for uranium nails are inferior to hair in two aspects, the accumulated time period is generally shorter and the absolute concentration is lower, as demonstrated in a number of comprehensive studies (Rodushkin and Axelsson 2000b; Karpas et al. 2005b). Figure 4.19 shows a log-log plot of the daily uranium excretion by nails as a function of the excretion through hair (both in units of ng day^{-1}) for participants in the Finnish study that ingested 10 μgU day^{-1} or more in their drinking water for females (triangles) or males (circles). The solid line is the regression line and the two dashed lines indicate the lower and upper trend lines that encompass 90% of the points (Karpas et al. 2005b).

Evidently, there is a clear correlation between the amount of uranium ingested through drinking water (see also Figure 4.8) and the amount excreted through hair and nails. The use of laser ablation ICPMS for the analysis of nails (Rodushkin and Axelsson 2003) can yield information that is not available by other techniques, namely, differentiating between the concentration of uranium (and 54 other elements) on the nails' surface as opposed to the internal parts of the nail. The results are truly surprising as the level of uranium (and several other elements) is much higher on the surface—by a factor of 30 and 50 for the two samples that were thus investigated. This result may have implications on the representativeness of the fully digested nails samples.

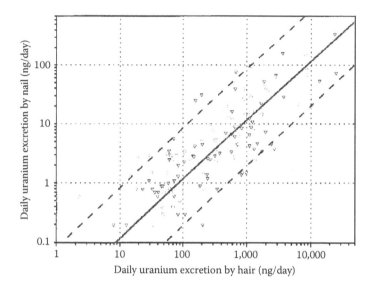

FIGURE 4.19 A log-log plot of the daily uranium excretion by nails as a function of the excretion through hair (both in units of ng day^{-1}) for individuals that ingested 10 µg U day^{-1} or more in their drinking water for females (triangles) or males (circles). The solid line is the regression line and the two dashed lines indicate the lower and upper trend lines that encompass 90% of the points. (From Karpas, Z. et al., *Health Phys.* 88, 229, 2005b. With permission.)

Highlights: Bioassays of hair for the assessment of uranium exposure are especially attractive due the fact that hair is an efficient bio-concentrator of uranium, in addition to the advantages of collecting, transporting, storing, and handling of hair samples. Furthermore, assays of hair samples provide a comprehensive picture of the history of exposure to uranium. Several digestion procedures have been developed for hair analysis and the use of ICPMS for measurement of the uranium content and isotopic composition is readily done. The background level of uranium in hair for people that are not exposed occupationally or environmentally to uranium, just like the case for urine, strongly depends on their dietary habits, mainly the consumption of uranium through drinking water. The *normal* level would thus be in the range of 5–100 ng U g^{-1} hair. This is three orders of magnitude higher than uranium in urine, but considering that 100 mg of hair are digested in 10 mL of solution leads to quite similar concentrations of uranium in both liquid sample types. The concentration of uranium in toenails or fingernails is two to three orders of magnitude lower than in hair (of the same subjects), and the finding of Rodushkin that the nail surface contains much higher levels than the internal parts of the nails makes this bioassay method less attractive than urinalysis or hair analysis (Rodushkin and Axelsson 2003). The fact that the distribution of the uranium along a single hair strand can readily be determined by laser ablation ICPMS provides a powerful tool for studying the exposure history of the subject and makes it possible to assess the extent of exposure a few weeks after the incident (depending on the frequency of having a hair cut).

4.4.5 URANIUM CONTENT ANALYSIS IN OTHER BIOASSAYS

Bones: The reference values for the uranium content in adult human bones, calculated on the basis of the ash-weight, were reported for the mean value, median, and range of 12 ± 14, 6.8, and 0.8–100 µg U kg⁻¹ ashes, respectively (Tandon et al. 1998). The authors note that different approaches were adopted for this type of calculation so they used ashed bones as the basis for their estimate and comparison between different countries, and the median values for several countries are shown in Figure 4.20.

The variation between the results of the different studies can be due to diet, age, and geographic locations but gender and bone type are not believed to be major factors.

Ultralow levels of uranium and plutonium isotopes in bone samples were measured by TIMS (Efurd et al. 2006). Sample preparation involved drying and ashing of the bone samples in a quartz crucible and dissolution of the ash in ultrapure acids. The uranium and plutonium in the sample were purified and concentrated with ion-exchange chromatography and measured by TIMS, with detection limits of 74 pg for ^{238}U and 8 fg for ^{239}Pu in 100 mg of ashes (this large difference of four orders of magnitude in the detection limit is mainly caused by blank levels that were 29 ± 15 pg and 2 ± 2 fg for ^{238}U and ^{239}Pu, respectively). Due to the low levels and limited amounts of bone material, the sample preparation should be carried out in a clean-room environment with very high-purity reagents and super-clean laboratory equipment. The uranium content in bones and other body organs of rats that were implanted with DU fragments was determined by ICPMS (Zhu et al. 2009). Sample preparation was done by adding aliquots of HNO_3 and H_2O_2 and heating to 150°C initially, followed by muffle furnace heating to 300°C for 2 h and then to 600°C for 4 h. The incinerated residues were dissolved in nitric acid and then diluted by purified water before analysis by ICPMS. As expected, uranium accumulated mainly in the bones and kidneys of the rats, and a positive correlation was found between the size of the implanted fragment and the uranium concentration (Zhu et al. 2009).

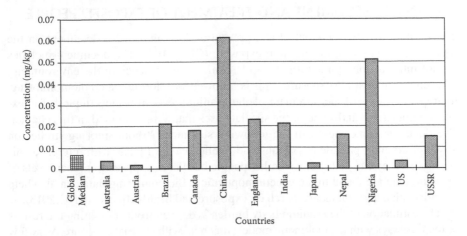

FIGURE 4.20 The median concentration of uranium in human bones, calculated on the basis of ashed bones. (From Tandon, L. et al., *Appl. Radiat. Isot.*, 49, 903, 1998. With permission.)

Teeth: A case study in which the uranium level in teeth was determined as a novel bioassay method was published (Prado et al. 2008). Forty-one teeth samples were collected from residents of a region in Brazil (city of Caetite) with large uranium deposits and 50 teeth from children (up to 10 years old) who attended a pediatric dental clinic in a uranium-free area. As no reference values for uranium content in teeth were available, it was assumed that the concentration in teeth should be similar to the concentration in bones (the worldwide average was estimated to be 0.5 ng g^{-1}). The sample treatment involved removal of the soft tissues by sonication in hydrogen peroxide, followed by calcinations at 800°C. A sample of the ashes was completely digested by nitric acid and indium was used as an internal standard prior to analysis by ICPMS. The finding of an average value of 2 ng U g^{-1} tooth in the control group was viewed as four times higher than worldwide average for bones, while the median for teeth of residents in of Caetite was significantly higher at 16 ng U g^{-1} (Prado et al. 2008).

Semen: In one rather unique study, ICPMS was used to determine the level of uranium in semen samples collected from a cohort of six veterans with embedded DU shrapnel fragments and a control group (Todorov et al. 2013). The average uranium concentration in the control group samples was 2.9 pg U g^{-1} semen while the level in the veterans ranged from below the detection limit (0.8 pg g^{-1}) to 3350 pg g^{-1}, which were consistent with their current uranium body burden. The potential effects, if any, on the offspring of the men with high uranium levels in semen were not reported.

Highlights: These bioassays (bones, teeth, and semen) are of limited value for routine measurements for assessment of exposure to uranium but should be viewed as novel media that may be used in special cases. The sample preparation procedures had to be customized for the bone and teeth bioassays, but once the samples were digested, the same analytical techniques deployed in the other assays could be used.

4.5 DOSE ASSESSMENT AND TREATMENT OF EXPOSED PEOPLE

The Centers for Disease Control and prevention has published a document on the topic of biomonitoring for uranium exposure (CDC 2013). The document includes a brief introduction on the properties of uranium, its presence in the environment, and sources of possible exposure. This is followed by information on biomonitoring, mainly the results of measurement of the uranium level in urine samples of unexposed populations and some of the special cases that were discussed in this chapter. The document concludes with the following statement: "Biomonitoring studies on levels of uranium provide physicians and public health officials with reference values so that they can determine whether people have been exposed to higher levels of uranium than are found in the general population. Biomonitoring data can also help scientists plan and conduct research on exposure and health effects" (CDC 2013).

The estimation of the uranium body burden is derived from combining the results of the bioassays with the biokinetic models that deal with internal exposure. Very few estimates of the total body burden and the distribution among the different body organs in humans have been published. As mentioned earlier, one publication estimated the total uranium content in a human body as being 40 μg, with approximately 40% in the muscles, 20% in the skeleton, and 10%, 4%, 1%, and 0.3% in the blood, lungs, liver, and kidneys, respectively (Igarashi et al. 1987). In a study on rats, the uranium content in blood, kidney, liver, and bone was determined by delayed neutron activation analysis (DNAA) and the conclusion was that following ingestion the uranium localized rapidly and about equally in the kidney and bone (La Touche et al. 1987). The study found that the uranium level in blood rose rapidly after ingestion but started to quickly decline after about 30 min.

The uranium content in several organs of two individuals that donated their bodies to science was determined by kinetic phosphorescence analysis (KPA) (Marshall et al. 1995). The total skeletal burdens of these two people were calculated as 41.5 and 55.9 μg (4.84 ± 2.79 and 5.75 ± 3.19 ng U g^{-1} tissue). There were large variations in the uranium content in the soft tissues with 0.699 ± 2.959 and 0.259 ± 6.390 ng U g^{-1} whereby the concentration in the lungs was significantly higher than this average value. The concentration of uranium in the tracheobronchial lymph nodes was 22.8 ng g^{-1} and considerably higher than the ICRP 23 (ICRP23 1975) value (updated as ICRP89 2002), but in the kidneys was 0.94 and 0.97 ng g^{-1} for the two individuals and thus much lower than the 22 ng g^{-1} assigned in ICRP 23. Overall, the sum of the soft tissues burden was 28 and 16.6 μg, and the total uranium body burden was calculated as 69.5 and 72.5 μg for these two people, which after normalization to their body weight were in quite good agreement with the reference man value of 90 μg.

The health effects of uranium are often compared with those of lead, so as a final conclusion of this chapter we bring the following quote:

> Metallic lead has considerably higher toxicity than metallic uranium... Lead within the body affects the nervous system and several biochemical processes, while the uranyl ion does not readily interfere with any major biochemical process except for depositing in the tubules and glomeruli of kidney, where damage may occur at high concentration.

> **Prof. Otto G. Raabe PhD, CHP**
> *Institute of Toxicology & Environmental Health*
> *University of California*

4.6   SUMMARY

The issues regarding exposure to uranium and consequent health effects have received considerable public attention. The main topics were concerned with the effects of the atom bombs, the release of radioactive materials following major accidents in nuclear power plants, the use of DU munitions, the effects of the progeny of natural uranium (mostly radon), and the environmental impact of pollution. Uranium is undoubtedly at the core of all these items but due to its relatively low specific activity is hardly the main culprit of the health-related issues. Nonetheless, highly sensitive and sophisticated analytical methods have been developed to monitor the presence of uranium in the environment (Chapter 3), its content in food products and drinking water, and its concentration in a whole suite of biological assays. Reference values in all these media for unexposed populations are well established (as shown throughout this chapter), and although not officially adopted by a regulating body, can serve as a starting point to assess the impact of exposure on the general public. There appears to be a consensus among most researchers in the field that uranium may pose a minor risk as a heavy metal (e.g., although less so than mercury or lead), and with regard to its radioactivity the risk from natural uranium is negligible. There are a few exceptions: emanation of radon from building materials or from types of soil that contain high concentrations of uranium, proximity to mine tails that hold uranium as a by-product (phosphates, gold, etc.), consumption of elevated levels of uranium in drinking water may also have a minor effect on kidney function (although none proven). Specific populations that may be exposed to uranium and its progeny are employees of the uranium industry (miners, millers, workers in uranium processing plants, nuclear power plants and people dealing with irradiated fuel disposal), and, still much in controversy, people residing in areas where DU munitions were widely deployed.

The bioassays developed for assessment of the internal uranium contamination (body burden) are capable of regularly determining the uranium content in urine, feces, hair, nails, bones, and teeth samples. Biokinetic models use the results of these measurements to evaluate the body burden, but it should be kept in mind that these general models may not apply exactly to each individual, so there is quite a large uncertainty that is built-in to the assessment of internal exposure. We have presented the analytical methods and procedures used for the measurements and have tried to demonstrate the variability and complexity of the methods.

REFERENCES

56/04, F. (2001). Uranium-238 in the 2001 total diet study. London, U.K.: UK Government.

Akamine, A.U., Silva, M.A.D., Saiki, M. et al. (2007). Determination of uranium in human head hair of a Brazilian populational group by epithermal neutron activation analysis, *J. Radioanal. Nucl. Chem.* 271, 607–609.

Akhter, P., Khaleeq-Ur-Rahman, M., Shiraishi, K. et al. (2003). Uranium concentration in typical Pakistani diet, *J. Radiat. Res.* 44, 289–293.

Alaani, S., Savabieasfahani, M., Tafash, M. et al. (2011a). Four polygamous families with congenital birth defects from Fallujah, Iraq, *Int. J. Environ. Res. Public Health* 8, 89–96.

Alaani, S., Tafash, M., Busby, C. et al. (2011b). Uranium and other contaminants in hair from the parents of children with congenital anomalies in Fallujah, Iraq, *Confl. Health* 5, 15.

Allain, P., Berre, S., Premel-Cabic, A. et al. (1991). Investigation of the direct determination of uranium in plasma and urine by ICPMS, *Anal. Chim. Acta* 251, 183–185.

Al-Sabbak, M., Ali, S.S., Savabi, O. et al. (2012). Metal contamination and the epidemic of congenital birth defects in Iraqi cities, *Bull. Environ. Contam. Toxicol.* 89(5), 937–944.

Anke, M., Seeber, O., Muller, R. et al. (2009). Uranium transfer in the food chain from soil to plants, animals and man, *Chemie der Erde* 69(Suppl. 2), 75–90.

Arnason, J.G., Pellegri, C.N., Parsons, P.J. et al. (2013). Determination of uranium isotope ratios in human urine by sector field inductively coupled plasma mass spectrometry for use in occupational and biomonitoring studies, *J. Anal. Atom. Spectrom.* 28, 1410–1419.

Arzuaga, X., Rieth, S.H., Bathija, A. et al. (2010). Renal effects of exposure to natural and depleted uranium: A review of the epidemiologic and experimental data, *J. Toxicol. Environ. Health B, Crit. Rev.* 13, 527–545.

ATSDR. (2013). Toxicological profile for uranium. Atlanta, GA: US Agency for Toxic Substances and Disease Registry.

Azeredo, A.M.G.F., Lipsztein, J.L., and Miekeley, N.T. (2000). Simultaneous determination of Ni, Nb, U and Th in urine and fecal ash samples using ICP-MS. In *10th International Radiation Protection Association* (p. P-3a-147). Hiroshima, Japan: IRPA.

Baglan, N., Berard, P., Trompier, F. et al. (1998). How to reduce the uncertainties of committed dose derived from urinary uranium measurements: Investigation of new protocols using ICP-MS, *Radiat. Prot. Dosimetry* 79, 477–480.

Barbier, O., Jacquillet, G., Tauc, M. et al. (2005). Effect of heavy metals on, and handling by, the kidney, *Nephron Physiol.* 99, 105–110.

Barratta, E.J. and Mackill, P. (2001). Determination of isotopic uranium in food and water, *J. Radioanal. Nucl. Chem.* 248, 473–475.

Becker, A., Tobias, H., and Mandler, D. (2009). Electrochemical determination of uranyl ions using a self-assembled monolayer, *Anal. Chem.* 81, 8627–8631.

Berard, P., Monteguw, A., Briot, F. et al. (2003). Procorad's international intercomparisons highlight the evolution of techniques used to determine uranium in urine, *Radiat. Prot. Dosimetry* 105, 447–450.

Berlin, M. (1986). Uranium. In Friberg, L., Nordberg, G.F., and Vouk, V.B. (eds.), *Handbook on the Toxicology of Metals* (pp. 623–637). Amsterdam: Elsevier/North Holland Biomedical Press.

Bijlsma, J.A., Slottje, P., Huizink, A.C. et al. (2008). Urinary uranium and kidney function parameters in professional assistance workers in the Epidemiological Study Air Disaster in Amsterdam (ESADA), *Nephrol. Dial. Transplant.* 23, 249–455.

Bijwaard, H., Dekkers, F., and van Dillen, T. (2011). Modelling lung cancer due to radon and smoking in Wismut miners: Preliminary study, *Radiat. Prot. Dosimetry* 143, 380–383.

Birke, M., Rauch, U., Lorenz, H. et al. (2010). Distribution of uranium in German bottled and tap water, *J. Geochem. Explor.* 107, 272–282.

Bleise, A., Danesi, P.R., and Burkart, W. (2003). Properties, use and health effects of depleted uranium (DU): A general overview, *J. Environ. Radioact.* 64, 93–122.

Boeniger, M.F., Lowry, L.K., and Rosenberg, J. (1993). Interpretation of urine results used to assess exposure with emphasis on creatinine adjustments: A review, *Am. Ind. Hyg. Assoc. J.* 54, 615–627.

Bogard, J.S. (1996). Review of uranium bioassay techniques, ORNL 6857. Oak Ridge, TN: Oak Ridge National Laboratory.

Boice, J.D., Mumma, M.T., and Blot, W.J. (2010). Cancer incidence and mortality in populations living near uranium milling and mining operations in grants, New Mexico, 1950–2004, *Radiat. Res.* 174, 624–636.

Byrne, A.R. and Benedik, L. (1991). Uranium content of blood, urine and hair of exposed and non-exposed persons determined by radiochemical neutron activation analysis, with emphasis on quality control, *Sci. Total Environ.* 107, 143–157.

C1473, A. (2011). Standard test method for radiochemical determination of uranium isotopes in urine by alpha spectrometry. West Conshohocken, PA: ASTM.

C1614, A. (2010). Standard practice for determination of ^{237}Np, ^{232}Th, ^{235}U and ^{238}U in urine by inductively coupled plasma mass spectrometry and gamma-ray spectrometry. West Conshohocken, PA: ASTM.

Canada. (2000). Uranium in Nova Scotia's drinking water, Nova Scotia, Canada; http://www.novascotia.ca/nse/water/uranium.asp (accessed August 3, 2014).

Canberra. (n.d.). Actinide (uranium/plutonium) lung counter model 2270; http://www.canberra.com/products/hp_radioprotection/model2270-lung-counter.asp (accessed August 3, 2014).

CDC. (2013). Biomonitoring summary. Retrieved February 27, 2014, from www.cdc.gov/biomonitoring/Uranium BiomonitoringSummary.html.

Centanni, F.A., Ross, A.M., and De Sesa, M.A. (1956). Fluorometric determination of uranium, *Anal. Chem.* 28, 1651–1657.

Chaudhri, M.A. and Watling, R.J. (2010). Measurement of uranium in blood-plasma of Bavarian females. In *10th Radiation Physics and Protection Conference* (pp. 111–113), Nasr City, Egypt.

Clark, N.A., Teschke, K., Rideout, K. et al. (2007). Trace element levels in adults from the west coast of Canada and associations with age, gender, diet, activities and levels of other trace elements, *Chemosphere* 70, 155–164.

Cockerill, R.J., Breese, R., Tallon, M. et al. (2006) A review of the faecal sampling programme at AWE. In *Symposium on the Second European IRPA Congress* (pp. TA-3). Paris, France: IRPA.

Coons, T. (2011). Uranium toxicology and epidemiology. Danville, VA: Institute for Advanced Learning and Research, http://www.geos.vt.edu/events/uranium/pdf/1345-1430_Coons_U_Toxicology_&_Epidemiology.pdf (accessed August 3, 2014).

Craft, E.S., Abu-Qare, A.W., Flaherty, M.M. et al. (2004). Depleted and natural uranium: Chemistry and toxicological effects, *J. Toxicol. Environ. Health B* 7, 297–317.

D5174, A. (2013). Standard test method for trace uranium in water by pulsed laser phosphorimetry. West Conshohochen, PA: ASTM.

D6239, A. (2009). Standard test method for uranium in drinking water by high resolution alpha-liquid scintillation spectrometry. West Conshohocken, PA: ASTM.

Dazhu, Y., Yongjun, Z., and Mobius, S. (1991). Rapid method for alpha counting with extractive scintillator and pulse shape analysis, *J. Radioanal. Nucl. Chem.* 147, 177–189.

Daneshvar, G., Jabbari, A., Yamini, Y. et al. (2009). Determination of uranium and thorium in natural waters by ICP-OES after on-line solid phase extraction and preconcentration in the presence of 2,3-dihydro-9,10-dihydroxy-1,4-anthracenedion, *J. Anal. Chem.* 64, 602–609.

Decambox, P., Mauchien, P., and Moulin, C. (1991). Direct and fast determination of uranium in human urine samples by laser-induced time-resolved spectrofluorometry, *Appl. Spectrosc.* 45, 116–118.

D'Ilio, S., Violante, N., Senofonte, O. et al. (2010). Determination of depleted uranium in human hair by quadrupole inductively coupled plasma mass spectrometry: Method development and validation, *Anal. Methods* 2, 1184–1190.

Dupzyk, I.A. and Dupzyk, R.J. (1979). Separation of uranium from urine for measurement by fluorometry or isotope dilution mass spectrometry, *Health Phys.* 36, 526–529.

Efurd, D.W., Steiner, R.E., LaMont, S.P. et al. (2006). Processing bone samples for the determination of ultra low-levels of uranium and plutonium, *J. Radioanal. Nucl. Chem.* 269, 679–682.

Elish, E., Karpas, Z., and Lorber, A. (2007). Determination of uranium concentration in a single hair strand by LAICPMS applying continuous and single pulse ablation, *J. Anal. Atom. Spectrom.* 22, 540–546.

Enderle, G.J. and Friedrich, K. (1999). Uranium mining in East Germany ("Wismut"): Health consequences, occupational medical care and workers' compensation, *Int. Arch. Occup. Environ. Health* 72(Suppl. 3), M42–M49.

Ejnik, J.W., Todorov, T.I., Mullick, F.G. et al. (2005). Uranium analysis in urine by inductively coupled plasma dynamic reaction cell mass spectrometry, *Anal. Bioanal. Chem.* 382, 73–79.

EPA. (2006). Uranium–radiation protection. http://www.epa.gov/radiation/radionuclides/uranium.html (accessed August 3, 2014).

Fisenne, I.M. and Perry, P.M. (1985). Isotopic U concentration in human blood from New York City donors, *Health Phys.* 49, 1272–1275.

Fisenne, I.M., Perry, P.M., Decker, K.M. et al. (1987). The daily intake of [235,235,238]U, [228,230,232]Th and [226,228]Ra by New York City residents, *Health Phys.* 53, 357–363.

Fisenne, I.M. and Welford, G.A. (1986). Natural U concentration in soft tissues and bone of New York City residents, *Health Phys.* 50, 739–746.

Garcia, F., Barioni, A., Arruda-Neto, J.D.T. et al. (2006). Uranium levels in the diet of Sao Paolo City residents, *Environ. Int.* 32, 697–703.

Gonnen, R., Kol, R., Laichter, Y. et al. (2000). Determination of uranium in human hair by acid digestion and FIAS-ICPMS, *J. Radioanal. Nucl. Chem.* 243, 559–562.

Gray, P.J., Zhang, L., Xu, H. et al. (2012). Determination of the $^{236}U/^{238}U$ and $^{235}U/^{238}U$ ratios in human urine by inductively coupled plasma mass spectrometry, *Microchem. J.* 105, 94–100.

Grosche, B., Kreuzer, M., Kreisheimer, M. et al. (2006). Lung cancer risk among German male uranium miners: A cohort study, 1946–1998, *Br. J. Cancer* 95, 1280–1287.

Haldimann, M., Baduraux, M., Eastgate, A. et al. (2001). Determining picogram quantities of uranium in urine by isotope dilution inductively coupled plasma mass spectrometry. Comparison with alpha spectrometry, *J. Anal. Atom. Spectrom.* 16, 1364–1369.

Halicz, L., Segal, I., Gavrieli, I. et al. (2000). Determination of the $^{234}U/^{238}U$ ratio in water samples by inductively coupled plasma mass spectrometry, *Anal. Chim. Acta* 422, 203–208.

Heitland, P. and Koster, H.D. (2006). Biomonitoring of 30 trace elements in urine of children and adults by ICP-MS, *Clin. Anal. Acta* 365, 310–318.

Hindin, R., Brugge, D., and Panikkar, B. (2005). Review: Teratogenicity of depleted uranium aerosols: A review from an epidemiological perspective, *Environ. Health* 4(17), 1–19.

Hirose, K. and Sugimura, Y. (1981). Concentration of uranium and the activity ratio of $^{234}U/^{238}U$ in surface air: Effect of atmospheric burn-up of Cosmos-954, *Meteorol. Geophys.* 32, 317.

Hollriegl, V., Roth, P., Werner, E. et al. (2002). Regional variation of urinary excretion of uranium in non-exposed subjects. In *European IRPA Congress* (p. 44). Florence, Italy: IRPA.

HRN. (2013). Report of a fact finding mission on congenital birth defects in Fallujah, Iraq in 2013. Tokyo, Japan: Human Rights Now.

IAEA. (2000). Determination of uranium in urine by alpha spectroscopy. Vienna, Austria: IAEA.

ICRP23. (1975). Report of the task group on reference man: A report. [2006.] Oxford, Tarrytown, NY: Pergamon Press. http://www.worldcat.org/title/report-of-the-task-group-on-reference-man-a-report/oclc/664140856?referer=di&ht=edition (accessed August 3, 2014).

ICRP69. (1995). Age dependent doses to members of the public from intake of radionuclides—Part 3: Ingestion dose coefficients. *Ann. ICRP* 25, 1.

ICRP71. (1995). Age dependent doses to members of the public from intake of radionuclides. Part 4: Inhalation dose coefficient. *Ann ICRP* 25, 3.

ICRP89. (2002). Basic anatomical and physiological data for use in radiological protection reference values. *Ann. ICRP* 32, ICRP Publication 89.

Igarashi, Y., Yamakawa, A., and Ikeda, N. (1987). Plutonium and uranium in Japanese human tissues, *Radioisotopes* 36, 433–439.

Ivanenko, N.B., Ivanenko, A.A., Solovyev, N.D. et al. (2013). Biomonitoring of 20 trace elements in blood and urine of occupationally exposed workers by sector field inductively coupled plasma mass spectrometry, *Talanta* 116, 264–269.

Kahlos, H. and Asikainen, M. (1980). Internal radiation doses from radioactivity of drinking water in Finland, *Health Phys.* 39, 108–111.

Kapsimalis, R., Landsberger, S., and Ahmed, Y.A. (2009). The determination of uranium in food samples by Compton epithermal neutron activation analysis, *Appl. Radiat. Isot.* 67, 2097–2099.

Karpas, Z., Halicz, L., Roiz, J. et al. (1996). Inductively coupled plasma mass spectrometry as a simple, rapid, and inexpensive method for determination of uranium in urine and fresh water: Comparison with LIF, *Health Phys.* 71, 879–885.

Karpas, Z., Lorber, A., Elish, E. et al. (1998). The uptake of ingested uranium after low "acute intake", *Health Phys.* 74, 337–345.

Karpas, Z., Lorber, A., Sela, H. et al. (2005a). Determination of $^{234}U/^{238}U$ ratio: Comparison of multi-collector ICPMS and ICP-QMS for water, hair and nails samples and comparison with alpha spectrometry for water samples, *Radiat. Prot. Dosimetry* 118, 106–110.

Karpas, Z., Paz-Tal, O., Lorber, A. et al. (2005b). Urine, hair, and nails as indicators for ingestion of uranium in drinking water, *Health Phys.* 88, 229–242.

Kathren, R.L. (2011, March 24). Toxicity of uranium: A brief review with special reference to man. Retrieved November 6, 2013, from http://www.ucdenver.edu/academics/colleges/PublicHealth/research/centers/maperc/training/healthphysics/Documents/Kathren.pdf.

Kehagia, K., Bratakos, S., Kolovou, M. et al. (2010). Hair analysis as an indicator of exposure to uranium, *Radiat. Prot. Dosimetry* 144, 423–426.

Kelly, W.R. and Fassett, J.D. (1983). Determination of picogram quantities of uranium in biological tissues by isotope dilution thermal ionization mass spectrometry with ion counting, *Anal. Chem.* 55, 1040–1044.

Kelly, W.R., Fassett, J.D., and Hotes, S.A. (1987). Determining picogram quantities of U in human urine by thermal ionization mass spectrometry, *Health Phys.* 52, 331–336.

Kinman, W.S., LaMont, S.P., and Steiner, R.E. (2009). A rapid isotope dilution ICPMS procedure for uranium bioassay, *J. Radioanal. Nucl. Chem.* 282, 1027–1030.

Klaassen, C. (1995). *Casarett and Doull's Toxicology: The Basic Science of Poisons.* New York: McGraw-Hill.

Kravchik, T., Oved, S., PazTal-Levy, O. et al. (2008). Determination of the solubility and size distribution of radioactive aerosols in the uranium processing plant at NRCN, *Radiat. Prot. Dosimetry* 131, 418–424.

Krystek, P. and Ritsema, R. (2002). Determination of uranium in urine—Measurement of isotope ratios and quantification by use of inductively coupled plasma mass spectrometry, *Anal. Bioanal. Chem.* 374, 226–229.

Kurttio, P., Auvinen, A., Salonen, L. et al. (2002). Renal effects of uranium in drinking water, *Environ. Health Perspect.* 110, 337–342.

Kurttio, P., Harmoinen, A., Saha, H. et al. (2006). Kidney toxicity of ingested uranium from drinking water, *Am. J. Kidney Dis.* 47, 972–982.

La Touche, Y.D., Willis, D.L., and Dawydiak, O.I. (1987). Absorption and biokinetics of U in rats following anoral administration of uranyl nitrate solution, *Health Phys.* 53, 147–162.

Leggett, R.W. (1994). Basis for the ICRP's age-specific biokinetic model for uranium, *Health Phys.* 67, 589–610.

Leggett, R.W., Eckerman, K.F., and McGinn, C.W. (2012). Controlling intake of uranium in the workplace: Applications of biokinetic modeling and occupational monitoring data, ORNL/TM-2012/14. Oak Ridge, TN: Oak Ridge National Laboratory.

Leggett, R.W. and Pellmar, T.C. (2003a). The biokinetics of uranium migrating from embedded DU fragments, *J. Environ. Radioact.* 64, 205–225.

Leggett, R.W. and Pellmar, T.C. (2003b). The biokinetics of uranium migrating from embedded DU fragments, *J. Environ. Radioact.* 64, 205–225.

Li, W.B., Karpas, Z., Salonen, L. et al. (2009). A compartmental model of uranium in human hair for protracted ingestion of natural uranium in drinking water, *Health Phys.* 96, 636–645.

Lorber, A., Halicz, K., and Karpas, Z. (1997). Uranium in urine and serum of "normal" populations: A FIAS-ICPMS study. In Tanner, H.A. (ed.), *Plasma Source Mass Spectrometry: Developments and Applications.* London, U.K.: Royal Society of Chemistry.

Lorber, A., Karpas, Z., and Halicz, L. (1996). Flow injection method for determination of uranium in urine and serum by ICPMS, *Anal. Chim Acta* 334, 295–301.

Lucas, H.F. and Markun, F. (1970). Thorium and uranium in blood, urine and cigarettes. Argonne National Laboratory Radiation Physics Division annual report, Part 2 (pp. 47–52), ANL-7760. Argonne, IL: Argonne National Laboratory.

Malatova, I., Beckova, V., Tomasek, L. et al. (2011). Content of uranium in urine of uranium miners as a tool for estimation of intakes of long-lived alpha radionuclides, *Radiat. Prot. Dosimetry* 147, 593–599.

Marco, R., Katorza, E., Gonnen, R. et al. (2008). Normalization of spot urine samples to 24-h collection for assessment of exposure to uranium, *Radiat. Prot. Dosimetry* 130, 213–223.

Marshall, E.T., Toohey, R.E., Cossairt, J.D. et al. (1995). Distribution of uranium in two whole body donors, USTUR-0034. Richland, WA: WSU.

McDiarmid, M.A., Englejardt, S.M., Dosrey, C.D. et al. (2009). Surveillance results of depleted uranium-exposed Gulf War I veterans: Sixteen years of follow-up, *J. Toxicol. Environ. Health A* 72, 14–29.

McDiarmid, M.A., Englejardt, S.M., Dosrey, C.D. et al. (2011). Longitudinal health surveillance in a cohort of Gulf War veterans 18 years after first exposure to depleted uranium, *J. Toxicol. Environ. Health A* 74, 678–691.

Medinsky, K. (1996). Toxicokinetics. In Kurtiss, C.D. (ed.), *Casarett and Doull's Toxicology: The Basic Science of Poisons*, Chapter 8. New York: McGraw Hill.

Miller, A.C. (2007). *Depleted Uranium: Properties, Uses and Health Consequences*. Boca Raton, FL: Taylor & Francis/CRC Press.

Miller, A.C., Brooks, K., Stewart, M. et al. (2003). Genomic instability in human osteoblast cells after exposure to depleted uranium: Delayed lethality and micronuclei formation, *J. Environ. Radioact.* 64, 247–259.

Misdaq, M.A. and Bourzik, W. (2004). Evaluation of annual committed effective dose to members of the public in Morocco due to ^{238}U and ^{232}Th in various food materials, *J. Radiol. Prot.* 24, 391–399.

Monleau, M., Bussy, C., Lestaevel, P. et al. (2006). The effect of repeated inhalation on the distribution of uranium in rats, *J. Toxicol. Environ. Health A* 69, 1629–1649.

Moore, L.L. and Williams, R.L. (1992). A rapid method for determining nanogram quantities of uranium in urine using the kinetic phosphorescence analyzer, *J. Radioanal. Nucl. Chem.* 156, 223–233.

Moss, A.M., McCurdy, R.F., Dooley, K.C. et al. (1983). Uranium in drinking water—Report on clinical studies in Nova Scotia. In Brown, S.S. and Savory, J. (eds.), *Chemical Toxicology and Clinical Chemistry of Metals* (pp. 149–152). London, U.K.: Academic Press.

Nakashima, T., Yoshimura, K., and Taketatsu, T. (1992). Determination of uranium (VI) in seawater by ion-exchange phase absorptiometry in aresnazo III, *Talanta* 39, 523–527.

Neves, O. and Abreu, M.M. (2009). Are uranium-contaminated soil and irrigation water a risk for human vegetable consumers? A study case with *Solanum tuberosum, Phaseolus vulgaris* L. and *Lactuca sativa* L., *Ecotoxicology* 18, 1130–1136.

Oeh, U., Priest, N.D., Roth, P. et al. (2007). Daily uranium excretion in German peacekeeping personnel serving in the Balkans compared to ICRP model prediction, *Radiat. Prot. Dosimetry* 127, 329–332.

OMEE. (1996). Monitoring data for uranium—1990–1995. Toronto, Canada: Ontario Ministry of Environment and Energy, Ontario Drinking Water Surveillance Program.

Othman, I. (1993). The relationship between uranium in blood and the number of working years in the Syrian phosphate mines, *J. Environ. Radioact.* 18, 151–161.

Ough, E.A., Lewis, B.J., Andrews, W.S. et al. (2002). An examination of uranium levels in Canadian forces personnel who served in the Gulf War and Kosovo, *Health Phys.* 82, 527–532.

Pappas, R.S. and Paschal, D.C. (2006). Simple changes improve sample throughput for determination of low concentration uranium isotope ratios in small volumes of urine, *J. Anal. Atom. Spectrom.* 21, 360–361.

Pappas, R.S., Ting, B.G., Jarret, J.M. et al. (2002). Determination of uranium-235, uranium-238 and thorium-232 in urine by magnetic sector inductively coupled plasma mass spectrometry, *J. Anal. Atom. Spectrosc.* 17, 131–134.

Pappas, R.S., Ting, B.G., and Paschal, D.C. (2003). A practical approach to determination of low concentration uranium isotope ratios in small volumes of urine, *J. Anal. Atom. Spectrom.* 18, 1289–1292.

Parrish, R.R., Thirlwall, M.F., Pickford, C. et al. (2006). Determination of $^{238}U/^{235}U$, $^{236}U/^{236}U$ and uranium concentration in urine using SF-ICP-MS and MC-ICP-MS: An interlaboratory comparison, *Health Phys.* 90, 127–138.

Pavlakis, N., Pollock, C.A., McLean, G. et al. (1996). Deliberate overdose of uranium: Toxicity and treatment, *Nephron* 72, 313–317.

Prado, G.R., Arruda-Neto, J.D.T., Sarkis, J.E.S. et al. (2008). Evaluation of uranium incorporation from contaminated areas using teeth as bioindicators—A case study, *Radiat. Prot. Dosimetry* 130, 249–252.

Rajusth, S., Sharma, P., Srilakshmi, C. et al. (2013). Voltammetric ultra trace determination of uranium: Method development, application and validation, *JEST-M* 2, 39–42.

Ratliff, A.E. (2008). Development of a colorimetric test for quantification of uranium in drinking water. In *234th ACS National Meeting* (pp. NUCL-047, NUCL-040). Boston, MA: ACS; http://tigerprints.clemson.edu/cgi/viewcontent.cgi?article=1384&context=all_theses (accessed August 3, 2014).

Raymond-Whish, S., Mayer, L.P., O'Neal, T. et al. (2007). Drinking water with uranium below the U.S. EPA water standard causes estrogen receptor–dependent responses in female mice, *Environ. Health Perspect.* 115, 1711–1716.

Roberts, K.D. (2008). A survey of naturally occurring uranium in groundwater in southwestern North Dakota. http://www.ndhealth.gov/wq/gw/pubs/uranium.htm (accessed August 3, 2014).

Rodushkin, I. and Axelsson, M.D. (2000a). Application of double focusing sector field ICP-MS for multielement characterization of human hair and nails. Part I. Analytical methodology, *Sci. Total Environ.* 250, 83–100.

Rodushkin, I. and Axelsson, M.D. (2000b). Application of double focusing sector field ICP-MS for multielemental characterization of human hair and nails. Part II. A study of the inhabitants of northern Sweden, *Sci. Total Environ.* 262, 21–36.

Rodushkin, I. and Axelsson, M.D. (2003). Application of double focusing sector field ICP-MS for multielement characterization of human hair and nails. Part III. Direct analysis by laser ablation, *Sci. Total Environ.* 305, 23–39.

Sahoo, S., Satpati, A.K., and Reddy, A.V.R. (2013). Stripping voltammetric determination of uranium in water samples using Hg-thin film modified multiwall nanotube incorporated carbon paste electrode, *Am. J. Anal. Chem.* 4, 141–147; https://www.itia.ntua.gr/hsj/redbooks/222/iahs_222_0071.pdf.

Salonen, L. (1994). 238U series radionuclides as a source of increased radioactivity in groundwater originating from Finnish bedrock. In *Future Groundwater Resources and Risk* (pp. 71–84). Helsinki, Finland: IAHS.

Sawant, P.D., Prabhu, S.P., and Kalsi, P.C. (2011). Application of fission track technique in solution media for analysis of uranium in bioassay samples, *Radiat. Prot. Environ.* 34, 74–76.

Schramel, P. (2002). Determination of 235U and 238U in urine samples using sector field inductively coupled plasma mass spectrometry, *J. Chromatogr. B* 778, 275–278.

Sela, H., Karpas, Z., Zoriy, M. et al. (2007). Biomonitoring of hair samples by laser ablation inductively coupled plasma mass spectrometry (LA-ICP-MS), *Int. J. Mass Spectrom.* 261, 199–207.

Selden, A.L., Lundholm, C., Edlund, B. et al. (2009). Nephrotoxicity of uranium in drinking water from private drilled wells, *Environ. Res.* 109, 486–494.

Shrivastava, A., Sharma, J., and Soni, V. (2014). Various electroanalytical methods for the determination of uranium in different matrices, *Bull. Faculty Pharm. Cairo Univ.* 51, 113–129.

Singh, N.P., Burleigh, D.P., Ruth, H.M. et al. (1990). Daily U intake in Utah residents from food and drinking water, *Health Phys.* 59, 333–337.

Sontag, W. (1986). Multicompartment kinetic models for the metabolism of americium, plutonium and uranium in rats, *Hum. Toxicol.* 5, 163–173.

Sugiyama, H. and Isomurac, K. (2007). Contents and daily intakes of gamma-ray emitting nuclides, ^{90}Sr and ^{238}U using market-basket studies in Japan, *J. Health Sci.* 53, 107–118.

Sullivan, M.F., Ruemmler, P.S., Ryan, J.L. et al. (1986). Influence of oxidizing or reducing agents on gastrointestinal absorption of U, Pu, Am, Cm and Pm by rats, *Health Phys.* 50, 223–232.

Sztajnkrycer, M.D. and Otten, E.J. (2004). Chemical and radiological toxicity of depleted uranium, *Milit. Med.* 169, 212–216.

Takato, N. (2013). 10 Years after the Iraq War: Innocent lives are still dying and suffering; A report of a fact finding mission on congenital birth defects in Fallujah, Iraq in 2013. Tokyo, Japan: Human Rights Now.

Tandon, L., Iyengar, G.V., and Parr, R.M. (1998). A review of radiologically important trace elements in human bones, *Appl. Radiat. Isot.* 49, 903–910.

Ting, B.G., Paschal, D.C., and Caldwell, K.L. (1996). Determination of thorium and uranium in urine with inductively coupled plasma mass spectrometry, *J. Anal. Atom. Spectrosc.* 11, 339–342.

Ting, B.G., Paschal, D.C., Jarrett, J.M. et al. (1999). Uranium and thorium in urine of United States residents: Reference range concentrations, *Environ. Res. Sect. A* 81, 45–51.

Todorov, T.I., Ejnik, J.W., Guandalini, G. et al. (2013). Uranium quantification in semen by inductively coupled plasma mass spectrometry, *J. Trace Elem. Med. Biol.* 27, 2–6.

Todorov, T.I., Xu, H., Ejnik, J.W. et al. (2009). Depleted uranium analysis in blood by inductively coupled plasma mass spectrometry, *J. Anal. Atom. Spectrom.* 24, 189–193.

Tolmachyov, S.Y., Kuwabara, J., and Noguchi, H. (2004). Flow injection extraction chromatography with ICP-MS for thorium and uranium determination in human body fluids, *J. Radioanal. Nucl. Chem.* 261, 125–131.

Tosheva, Z., Stoyanova, K., and Nikolchev, L. (2004). Comparison of different methods for uranium determination in water, *J. Radioanal. Nucl. Chem.* 72, 47–55.

Tracy, B.L., Quinn, J.M., Lahey, J. et al. (1992). Absorption and retention of uranium from drinking water by rats and rabbits, *Health Phys.* 62, 65–73.

Valentin, J. and Fry, F.A. (2003). What ICRP advice applies to DU?, *J. Environ. Radioact.* 64, 89–92.

Wang, J. and Setiadji, R. (1992). Selective determination of trace uranium by stripping voltammetry following adsorptive accumulation of the uranium cupferron complex, *Anal. Chim. Acta* 264, 205–211.

WHO (2004). Uranium in drinking water: Summary statement, WHO/SDE/WSH/03.04/118; http://www.who.int/water_sanitation_health/dwq/chemicals/uraniumsum.pdf (accessed August 3, 2014).

Winde, F. (2013, February 22). Uranium pollution of water—A global perspective on the situation in South Africa. Inaugural lecture. Vaal Triangle Campus, South Africa: North-West University. ISBN 978-1-86822-629-0.

Wise (2012, October 15). Uranium in soil and building material—Individual dose calculator. Retrieved March 2, 2014, from http://www.wise-uranium.org/rdcush.html.

WNA. (2007). *Tokaimura Criticality Accident.* London, U.K.: World Nuclear Association.

Wrenn, M.E., Durbin, P.W., Howard, B. et al. (1985). Metabolism of ingested U and Ra, *Health Phys.* 48, 601–633.

Wrenn, M.E., Ruth, H., Burleigh, D. et al. (1992). Background levels of uranium in human urine, *J. Radioanal. Nucl. Chem.* 156, 407–412.

Yousefi, S.R., Ahmadi, S.J., Shemirani, F. et al. (2009). Simultaneous extraction and preconcentration of uranium and thorium samples by new modified mesoporous silica prior to inductively coupled plasma optical emission spectrometry determination, *Talanta* 80, 212–217.

Zamora, M.L., Tracy, B.L., and Zielinski, J.M. (1998). Chronic ingestion of uranium in drinking water: A study of kidney bioeffects in humans, *Toxicol. Sci.* 43, 68–77.

Zamora, M.L., Zielinski, J.M., Meyerhof, D.P. et al. (2002). Gastrointestinal absorption of uranium in humans, *Health Phys.* 83, 35–45.

Zhu, G., Tan, M., Li, Y. et al. (2009). Accumulation and distribution of uranium in rats after implantation with depleted uranium fragments, *J. Radiat. Res.* 50, 183–192.

Zikovsky, L. (2006). Determination of uranium in food in Quebec by neutron activation analysis, *J. Radioanal. Nucl. Chem.* 267, 695–697.

Zoellner, T. (2010). *Uranium: War, Energy and the Rock that Shaped the World*. London, U.K.: Penguin Books.

Zunic, Z.S., Tokonami, S., Mishra, S. et al. (2012). Distribution of uranium and some selected trace metals in human scalp hair from Balkans, *Radiat. Prot. Dosimetry* 152, 220–223.

5 Trace Analysis and Nuclear Forensics

Absence of evidence is **not** evidence of absence. We need to report uncertain results and do it clearly.

Anderson (2004, with regard to HIV-1 transmission but this is also so true for nuclear forensics)

They have been saying for a long time that Iraq made an effort to import active uranium, and my colleague demonstrated the other day that they came to the conclusion that it was a fake document that everybody is relying upon.

Hans Blix

5.1 INTRODUCTION

5.1.1 NUCLEAR FORENSICS

Nuclear forensics (NF) is defined in Wikipedia as "the investigation of nuclear materials to find evidence for example the source, the trafficking and the enrichment of the material." A scientifically based definition is more specific: "Nuclear forensic analysis seeks to determine the physical, chemical, elemental and isotopic characteristics of nuclear [or radiological] material of unknown origin" (Glaser and Bielefeld 2008). A legalistic definition of NF states "scientific analysis of nuclear or other radioactive material, or of other evidence that is contaminated with radioactive material in the context of legal proceedings, including administrative, civil, criminal or international law" (quoted in Mayer et al. 2013). There are several other definitions, but whatever definition is used, the objectives are still the same: to characterize the radioactive material in order to trace its source and transport route and to verify that it conforms to its declared application.

Historically, an impressive early demonstration of the capability of nuclear forensics was the detection of the first Soviet atomic test (in 1948), which was achieved by the collection of airborne particles that were released in the test by a weather reconnaissance plane (see the declassified report and map in Glaser and Bielefeld 2008). The stated objectives of nuclear forensics are divided into several areas (adapted from Glaser and Bielefeld 2008): combating illicit trafficking of nuclear materials, environmental sampling to detect undeclared activities involving nuclear materials, implementing IAEA safeguards, and verifying compliance with treaties (mainly the Comprehensive Test Ban Treaty (CTBT), Fissile Materials Cut-off Treaty (FMCT), Non-Proliferation Treaty (NPT)), national security enhancement (gathering measurement intelligence on real and perceived enemies), and deterrence at early stages. Postdetonation activity—after explosion of an improvised nuclear device (IND) or a radioactive dispersion device (RDD)—nuclear forensics should be able to supply information on the type of device, its construction materials, its transport route, its origin and age, with the objective of identifying the perpetrators and their assistants (also known as attribution). With regard to RDD, there is a serious concern about the preparedness of the authorities and the public, but this is beyond the scope of this book (Sohier and Hardeman 2006).

Uranium plays a major part in any nuclear program and is the source, directly or indirectly, of most nuclear and radioactive materials. The definitions of nuclear forensics given earlier for characterizing the physical properties (e.g., morphology), chemical content (like the major uranium compounds), elemental features (what impurities are present), and isotopic composition (including minor isotopes) should also include age dating (chronology of the most recent time that uranium was purified from its progeny), last legal owner, and wherever possible the structure of the components accompanying the fissile material (like packaging or triggering devices, etc.).

The interest in nuclear forensics has increased considerably in recent years as seen in Figure 5.1 that depicts the number of publications with the term nuclear forensics between 1996 and 2014 (based on SciFinder). Note that the term actually gained popularity only in 2005, so this figure may be biased. There are numerous reasons for this increased interest. First, smuggling activities of nuclear materials increased quite dramatically after the Soviet Union disbanded into several independent states in the early 1990s. Some of these had stores of fissile materials that were not as closely guarded as before, and this was combined with a dire economic situation so that thefts of fissile materials became tempting and possible. A second factor was

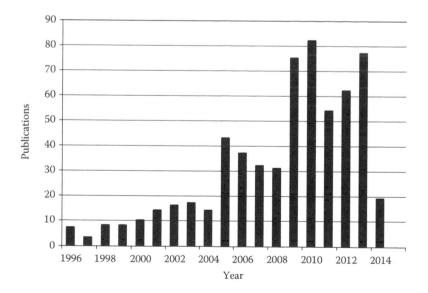

FIGURE 5.1 Annual number of publications on Nuclear Forensics from 1996 to 2014. (Based on SciFinder, accessed April 29, 2014.)

the discovery of a myriad of undeclared (and unsuspected) nuclear activities in Iraq after the First Gulf War in 1991 (Donohue and Zeisler 1993). The teams of IAEA inspectors were surprised by the breadth and depth of the clandestine efforts made by the Saddam regime to obtain a nuclear weapon (Donohue 2002). A third factor was the exposure of the *black market network* in nuclear technology and hardware that was run by Pakistani Abdul Qadeer Khan (A.Q. Khan). The fact that in 2003 Libya had disclosed that it purchased (but had not yet installed) a complete uranium isotope enrichment plant from the A.Q. Khan network surprised the world and raised concerns about spreading nuclear proliferation. Other alarm bells started ringing when India and Pakistan (who were not signees of the NPT) carried out underground nuclear tests in 1998 and when it was discovered that North Korea ignored the agreements and treaties it had signed and detonated its own nuclear device. Furthermore, intelligence sources attributed North Korean assistance to the fledgling plutogenic nuclear reactor that was destroyed in Syria in 2007. Finally, Iran's nuclear program, supposedly only for civilian purposes but with a long history and a body of evidence for consistently misleading the IAEA inspectors and running a clandestine military program, increased the need for efficient safeguards.

In parallel, technical advances in analytical instrumentation (see, e.g., Mayer 2013; Mayer et al. 2005, 2007, 2013; IAEA 2006; Piksaikin et al. 2006; Burger et al. 2009; Stanley et al. 2013), better understanding of sampling methods, improved remote sensing and photometric devices, and tighter control of sales of dual-use equipment and chemicals, have boosted the capability of the safeguards implemented to prevent nuclear proliferation (Mayer et al. 2011). The basic concept underlying nuclear forensics is that nuclear activities inevitably lead to the release of particulate matter or gases that can be collected and detected. Processing nuclear materials leads to emission of gases, effluents, and particles that are dispersed in and around the plant and the environment; these can be sampled and analyzed (Tamborini et al. 2004).

A partial list of the *tool-kit* available for nuclear forensics investigations is shown in Table 5.1 (IAEA 2006). A somewhat different list of the tools that are used to characterize seized nuclear materials (pellets or powder) is shown in Table 5.2 (Wallenius et al. 2006). One more point that is almost unique for nuclear forensics: when a smuggled radioactive material is seized or when undeclared nuclear activities are suspected, the law enforcement forces are expected to provide a preliminary characterization of the material within a short time (usually 24 h) and more detailed and accurate data after a few more days, as shown in Table 5.3 (IAEA 2006). These publications will be mentioned in context with the subject matter of this chapter.

Nuclear forensics has been the subject of many review articles, reports, presentations, and an excellent book dedicated to nuclear forensic analysis (Moody et al. 2005), as well as the focus of numerous international and national conferences and working groups sponsored by the IAEA and other organizations (IAEA 2002, 2014). For example, in July 2014, the "International conference on advances in nuclear forensics: countering the evolving threat of nuclear and other radioactive material out of regulatory control" is scheduled. In this chapter, we will try to demonstrate some of these advances, with focus on the role played by uranium in the field of nuclear forensics.

TABLE 5.1

Part of the Nuclear Forensics *Tool-Kit*

Measurement Goal	Technique	Type of Information	Typical Detection Limit	Spatial Resolution
Survey	High-resolution gamma spec.	Isotopic	ng–µg	
Elemental and isotopic bulk analysis	Chemical assay	Elemental	mg	
	Radiochemistry/ RA counting methods	Isotopic/elemental	fg–pg	
	TIMS	Isotopic/elemental	pg–ng	
	ICP-MS	Isotopic/elemental	pg–ng	
	GD-MS	Isotopic/elemental	0.1–10,000 ppb	
	XRF	Elemental	10 ppm	
	XRD	Molecular	~5%	
	GC–MS	Molecular	ppm	
	Infrared	Molecular	ppm	
Imaging	Visual inspection	Macroscopic		0.1 mm
	Optical microscopy	Microscopic		1 µm
	SEM	Structure		1.5 nm
	TEM	Structure		0.1 nm
Microanalysis	ICP-MS	Isotopic/elemental	pg–ng	
	TIMS	Isotopic	pg–ng	
	SIMS	Isotopic/elemental	0.1–10,000 ppb	0.2–1 µm
	SEM/EDS or WDS	Elemental	0.1–2 atom%	1 µm
	XRD	Molecular	~5 atom%	

Source: Adapted from International Atomic Energy Agency, Nuclear forensics support: Technical guidance reference manual, IAEA nuclear security series 2, IAEA, Vienna, Austria, 2006. With permission.

Notes: TIMS, thermal ionization mass spectrometry; ICP-MS, inductively coupled plasma mass spectrometry; GD-MS, glow discharge mass spectrometry; XRF, x-ray fluorescence; XRD, x-ray diffraction; GC–MS, gas chromatography–mass spectrometry; SEM, scanning electron microscope; TEM, transmission electron microscope; SIMS, secondary ion mass spectrometry; EDS, energy-dispersive sensor; WDS, wavelength-dispersive sensor.

5.1.2 ADAPTATION OF ANALYTICAL TECHNIQUES FOR TRACE ANALYSIS

In Chapter 1, the basic principles of the main analytical techniques that are used for the determination of uranium, its isotopic composition, and the impurities content were described. In addition, throughout the book, several examples of specific sample preparation procedures, including preconcentration and separation, and of measurement protocols were presented. However, trace analysis, ultratrace analysis, and particularly single particle analysis require a different approach. Many examples of sample handling and preparation procedures are presented later on a case-by-case

TABLE 5.2
Summary of the Information That Can Be Obtained from Nuclear (U, Pu) Material

Parameter	Information	Analytical Techniques
Appearance	Material type (e.g., powder, pellet)	Optical microscopy
Dimensions (pellet)	Reactor type	Database
U, Pu content	Chemical composition	Titration, HKED, IDMS
Isotopic composition	Enrichment → intended use; reactor type	HRGS, TIMS, ICPMS, SIMS
Impurities	Production process; geo-location	ICPMS, GDMS
Age	Production date	AS, TIMS, ICPMS
$^{18}O/^{16}O$ ratio	Geo-location	TIMS, SIMS
Surface roughness	Production plant	Profilometry
Microstructure	Production process	SEM, TEM

Source: Adapted from Wallenius, M. et al., *Forensic Sci. Int.*, 156, 55, 2006. With permission.

Notes: HKED, hydride K-edge densitometry; IDMS, isotope dilution mass spectrometry; HRGS, high-resolution gamma spectrometry; TIMS, thermal ionization mass spectrometry; ICPMS, inductively coupled plasma mass spectrometry; SIMS, secondary ion mass spectrometry; GDMS, glow discharge mass spectrometry; AS, alpha spectrometry; SEM, scanning electron microscopy; TEM, transmission electron microscopy.

TABLE 5.3
Suggested Sequence for Laboratory Techniques and Methods for Characterizing Seized or Suspected Nuclear Material

Technique/Method	24 h	1 Week	2 Months
Radiological	Estimated activity		
	Dose rate (α,β,γ,n)		
	Surface contamination		
Physical	Visual inspection	SEM/EDS	TEM (EDX)
	Radiography	XRD	
	Weight, density		
	Dimensions		
	Optical microscopy		
Traditional forensics	Fingerprints fibers		
Isotopic analysis	Gamma spectroscopy	Mass spectrometry	Radiochemical separations
	Alpha spectroscopy	SIMS, TIMS, ICP-MS	
Elemental/chemical		ICP-MS, XRF, assay	GC–MS
		(titration, IDMS)	

Source: Adapted from International Atomic Energy Agency, Nuclear forensics support: Technical guidance reference manual, IAEA nuclear security series 2, IAEA, Vienna, Austria, 2006. With permission.

Notes: SEM, scanning electron microscope; EDS, energy-dispersive sensor; TEM, transmission electron microscopy; SIMS, secondary ion mass spectrometry; TIMS, thermal ionization mass spectrometry; ICP-MS, inductively coupled plasma mass spectrometry; XRF, x-ray fluorescence; ID-MS, isotope dilution mass spectrometry; XRD, x-ray diffraction; GC–MS, gas chromatography–mass spectrometry.

basis, so in this section we shall focus on the adaptation of the three widespread mass spectrometric techniques (TIMS, SIMS, and ICPMS) that are currently in use for ultratrace and individual particle analysis. It should be noted that for single particle analysis by TIMS and ICPMS the relevant particles must first be located and transferred to the mass spectrometer, while SIMS and LA-ICPMS can in principle screen the sample on the swipe material, locate the uranium-bearing particles, and analyze them.

5.1.3 Thermal Ionization Mass Spectrometry

The basic principles of thermal ionization mass spectrometry (TIMS) operation were described in Chapter 1: a drop of the liquid sample is deposited on a filament, a low electric current heats the filament, and the solution is evaporated to dryness. The filament current (temperature) is then raised and atoms of the sample are emitted and ionized (either by the same filament or by a second electron emitting filament). The ions are accelerated by an electric field, pass through an electrostatic analyzer (ESA) that focuses the ion beam before it enters a magnetic field that deflects the ions into a curved pathway (in some devices, the ions enter the magnetic field before the ESA—referred to as reverse geometry). Heavy and light ions are deflected by the field at different curvatures that depend on their mass-to-charge ratio. A detector at the end of the ion path measures the ion current (or counts the ion pulses). There are many variations of ion sources, ion separation devices, and detectors that are used in TIMS instruments and specifically adapted for ultratrace or particle analysis.

Filaments: Filaments are usually made of rhenium, tungsten, or thoriated tungsten. It was found that treatment of the rhenium filament with benzene vapors forms a thin layer of carbonization that leads to higher ionization efficiencies. The efficiency is calculated as the fraction of ions formed from a given number of atoms, for uranium usually between 1:100 and 1:10,000, as shown in Table 5.4. Liquid samples can be deposited on this filament by the standard procedure described earlier or absorbed in a hollow bead that is placed (usually glued) on the filament. Single particles can also be placed directly on the filament for analysis. When the sample is heated, some fractionation occurs: at the beginning of the analytical cycle, the lighter isotopes are preferentially desorbed and the measurement is biased to overestimate the lighter isotopes. This gradually depletes the lighter components on the filament and for a certain period the evaporated material truly represents the original composition of the sample. Finally, the fraction of the heavy components on the filament increases by such an extent that causes a reverse mass bias. This process has been demonstrated for a standard CRM010 containing 1% ^{235}U (Mathew et al. 2013). The use of a set of standards or a double-spike containing a known $^{233}U/^{236}U$ ratio can be used to derive a correction factor for the $^{235}U/^{238}U$ ratio measurement. An alternative is to carry out a *total evaporation* (TE) procedure where the entire sample is evaporated and analyzed and each of the ion currents is integrated so that mass biases are eliminated (Mathew et al. 2013).

Ion separation: In TIMS devices, the separation of the ions according to their mass-to-charge ratio is based on the deflection of the ion beam in a magnetic field.

TABLE 5.4

Total Efficiency (Atoms Loaded to Ions Detected) for Thermal Ionization of Uranium Using Various Sample Preparation and Loading Techniques

Analytical Method	Element and Sample Size (pg)	Ion Source	Sample Form/ Additive	Total Efficiency %
TIMS	U	Re filament	Liquid load	10^{-2}–10^{-1}
	U	Re filament/ carbon	Liquid load/ carbon	0.6–1.2[a]
	U/10^4	Re filament	Electrodeposition	0.015
	U/100–500	Re cavity	Resin bead/ carbon	5.8[s]
	U/10^6	W cavity	Liquid, liquid + carbon	~3.5[a], 8.5[a]
	U	Re, Ta, W cavity		39[a,b], 4.5[a,b], 15[a,b]
	U	W cavity	Carrier, 0.5–2 mg	15.1[a,b]
	U/7–605	Re filament	Cation Resin bead/carbon	0.58[b]
	U/7–71	Ta filament	Resin bead/ carbon	0.02[b]
	U/50–75	Re cavity	Liquid, carbon	0.021[b]
	U	Ta cavity	Liquid, carbon	<0.025
RIMS	U/10^4			10^{-2}
AMS	U			10^{-2}
(MC)-SF-ICPMS	U	MicroMist or Meninhardt	Nebulizer	0.004–0.03
	U	Nanovolume flow injection	Nebulizer	~0.1
	U	Cetac Aridus	Nebulizer	~0.1
	U	Apex	Nebulizer	~0.2
	U	Cetac Aridus	Nebulizer	~1

Source: Adapted from Burger, S. et al., *Int. J. Mass Spectrom.*, 286, 70, 2009. With permission.

[a] Only ionization efficiency.

[b] Average or median value.

The mass scan can be performed by changing the magnetic field strength or the voltage in the electrostatic analyzer. The performance of the device is assessed mainly by the resolution (ability to differentiate between ions of close mass-to-charge ratios) and the sensitivity (ability to measure low signals and distinguish them from the background). Other operational parameters are the *abundance sensitivity* (tailing of a large signal that may affect neighboring peaks), the stability of a signal, the reproducibility of a measurement (that depends on several parameters), the time required for a measurement, etc. Some TIMS instruments can be operated in different modes like high-, medium-, and low-resolution concomitant with a trade-off between resolution and sensitivity that can affect the accuracy of the measurements.

Detectors: The two common types of detectors for measuring ion currents are the Faraday cup (or plate) and the electron multiplier (or multi-ion-counting (MIC)). The former is less sensitive but more stable than the latter, and in some cases, both types are used in a single device. Modern TIMS instruments are usually equipped with an array of detectors positioned so that each of the relevant ion beams impinges on a separate detector. The high-intensity beams are directed to Faraday detectors while the weaker signals are measured by ion counting detectors. Multicollector devices (MC–TIMS) are advantageous especially for small samples in which several ions need to be measured (e.g., ^{234}U, ^{235}U, ^{236}U, and ^{238}U) as all ions are measured all the time while with a single detector the ions must be measured sequentially in *peak hopping* mode (or in mass scan mode) that means that while one ion mass is measured the others are not counted and wasted. In addition, a single detector used for measuring ion currents that differ in intensity by several orders of magnitude (e.g., $^{236}U/^{238}U$ ratio) may suffer from limited linearity issues.

5.1.4 SECONDARY ION MASS SPECTROMETRY

The principle of operation of secondary ion mass spectrometry (SIMS) devices was described in Chapter 1: an energetic ion beam (primary beam) impinges on the sample causing secondary ions and neutrals to be desorbed from the surface and enter a mass spectrometer where these ions are analyzed. A schematic of a SIMS device is shown in Figure 5.2 (Betti 2005). This device includes two ion sources (Duoplasmatron and cesium), a sample chamber with a transfer rod,

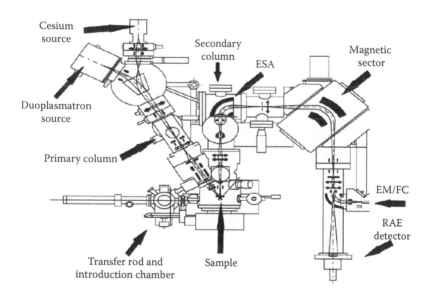

FIGURE 5.2 A schematic of a SIMS device that includes two ion sources (Duoplasmatron and cesium), a sample chamber with a transfer rod, a magnetic sector mass spectrometer with a Faraday cup (FC), resistive anode encoder (RAE), and electron multiplier (EM) detectors. (From Betti, M., *Int. J. Mass Spectrom.*, 242, 169, 2005. With permission.)

a magnetic sector mass spectrometer with a Faraday cup (FC), resistive anode encoder (RAE), and electron multiplier (EM) detectors. More advanced devices also have an optical microscope, an electron flood gun, a total ion count detector, and several ion detectors (five or seven, depending on the configuration). SIMS devices use different primary ion beams that typically consist of O_2^+, Ga^+, Cs^+, Ar^+, O^-, and O_2^- ions (some other types of ion sources are also available) and the energy of these ions is usually 10–15 keV (and −10 to −15 keV for negative ions). The secondary ions are also accelerated (4–10 keV, typically) in order to be effectively introduced into the mass spectrometer. A large geometry device (LG-SIMS) was developed to enhance the sensitivity for particle detection and characterization, as described later (Ranebo et al. 2009). The ion separation and detection can be carried out by a magnetic sector mass spectrometer, a time-of-flight (TOF-MS) (Hocking et al. 2012) or a quadrupole (QMS) (see Chapter 1 for the principle of operation of these three devices). In nuclear forensics applications, SIMS is an invaluable tool: it can be used for screening a sample in order to detect and locate uranium containing particles, it can map the coordinates of these particles and can create an image of the distribution of elements or isotopes in the sample. In addition, SIMS can be applied to characterize the particles' shape and morphology, elemental constituents, and the uranium isotopic composition. The main limitation is the rate at which a sample can be analyzed and the possible interferences from polyatomic ions that have the same mass as the uranium isotopes (see Table 4 in Ranebo et al., 2009).

5.1.5   INDUCTIVELY COUPLED PLASMA MASS SPECTROMETRY

The general principle of operation of inductively coupled plasma mass spectrometry (ICPMS) was explained in Chapter 1: a liquid sample is sprayed by a nebulizer as an aerosol into a chamber where large droplets are drained and fine droplets are transported to a plasma torch in which the effective temperature is 7,000–10,000 K. The droplets evaporate, the dry aerosol particles dissociate to molecules that are atomized and ionized, and the ions are introduced through differential pumping and a set of ion lenses into a mass spectrometer where they are analyzed. There are several types of nebulizers that can be selected according to the application on hand. Those that are particularly suitable for trace analysis comprise high ionization efficiency (~0.1% or 1:1000) and low sample consumption rate (see Table 5.4). Laser-ablation (LA-ICPMS) can be used for direct introduction of solid samples, including individual particles and for screening samples, as described earlier (Sela et al. 2007). This is particularly effective for quick analysis as sample preparation is minimal and single particle measurement times are short. The mass spectrometer used in the system can be the relatively inexpensive and simple quadrupole (ICP-QMS) or a magnetic sector device while ICP-TOF-MS devices are less common. There are two main types of magnetic sector devices used in combination with the ICP: the ICP-sector field-MS that has a single detector and uses peak hopping to measure different ions and the multicollector (MC-ICPMS) that uses an array of detectors to simultaneously measure several ions. The resolution of magnetic sector ICPMS instruments can also be altered, from R = 300 to R = 4000

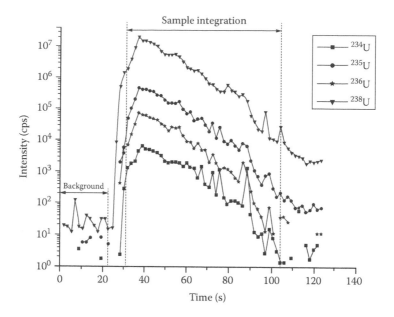

FIGURE 5.3 Typical laser ablation signal for ^{234}U, ^{235}U, ^{236}U, and ^{238}U from a low-enriched uranium oxide particle (21 μm × 25 μm) with a low energy setting of laser power so that only ~20% of the particle was ablated. (From Varga, Z., *Anal. Chim. Acta*, 625, 1, 2008. With permission.)

(e.g., Glaser and Bielefeld 2008; Varga 2008). For example, Figure 5.3 shows a typical laser ablation signal for ^{234}U, ^{235}U, ^{236}U, and ^{238}U from a low-enriched uranium oxide particle (21 μm × 25 μm) with a low-energy setting of laser power so that only ~20% of the particle was ablated (Varga 2008).

The advantages and limitations of these mass spectrometers are the same as described earlier. ICPMS has a very high throughput compared to TIMS, and the MC-ICPMS has quite similar accuracy in isotopic ratio measurements and sensitivity as the SIMS and TIMS methods.

5.2 NUCLEAR FINGERPRINTS

Nuclear materials, uranium in particular, may have distinct fingerprints in all stages of the nuclear fuel cycle (NFC). These are indicative of the source of the uranium (minerals and ores), the industrial processes used to separate it from waste materials to produce uranium ore concentrates (UOC), the procedures to make purified nuclear grade uranium in the uranium conversion facility (UCF), the enrichment method and ^{235}U content, the type of nuclear fuel (oxide or other), the irradiation regime in the reactor (burn-up), the reprocessing (if plutonium was separated from the irradiated fuel), and the disposal. In addition, the chronometry or *age dating* that indicates the most recent separation and purification of the uranium from its progeny also provides important forensic information. *Classical* forensics methods are also helpful as they offer additional data on the origin and transport routes, but these aspects are

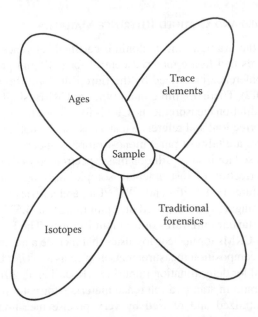

FIGURE 5.4 Diagram showing how individual forensic signatures define a subset of materials from which an interdicted sample may have originated. Multiple signatures can be collectively applied to minimize the set of possible origin materials. (Adapted from Kristo, M.J. and Turnet, S.J., *Nucl. Instrum. Methods Phys. Res. B*, 294, 656, 2013. With permission.)

beyond the scope of this book (for a recent review on microsample preparation for forensic applications, see Kabir et al. 2013). These can be summarized in a diagram that shows how combining data from classical forensics with isotopic composition analysis, age determination, and the content of elemental impurities can focus on the source of the sample (Figure 5.4).

Nuclear forensics, like all branches of analytical chemistry, is based on the ability to characterize, identify, and quantify the components of samples. In addition to analytical instrumentation, sample preparation techniques and proper methodology, standards and certified reference materials are required for calibration of instrumentation and for method validation. The urgent need for new radiological certified reference materials for nuclear safeguards and forensics has been reviewed by a large group of scientists, mainly from the US National Laboratories after a multiagency workshop (Inn et al. 2013). The ramifications and consequences of a nuclear forensics investigation following a nuclear event like detection of an improvised nuclear device (IND) or a radiological dispersion device (RDD) before it is detonated or, postdetonation, could lead to far-reaching political and military actions against the suspect perpetrators. Therefore, the requirements for accuracy and timeliness of the analysis are beyond those of any other routine measurement. In particular, an urgent need for three types of reference materials was mentioned by the participants of the workshop: nuclides for isotope dilution measurements (^{233}U, ^{236g}Np, ^{244}Pu, and ^{243}Am); CRMs for analysis of actinides and fission products in postdetonation debris, and CRMs for safeguards and environmental samples (Inn et al. 2013).

5.2.1 Standards and Certified Reference Materials

An overview of the uranium and plutonium certified reference materials that are used for safeguards and fissile material control for bulk and particle analysis was presented at a conference organized by the European Joint Research Centre (JRC) (Jakopic et al. 2012). There are three main types of CRMs: single isotope standards (^{233}U) for isotope dilution measurements; CRMs for isotope ratio calibration and mass discrimination correction; and reference materials for age dating. As far as uranium is concerned, there are (almost pure) mono-isotopic standards (^{233}U, ^{235}U, ^{236}U, and ^{238}U) that can be used for isotope dilution mass spectrometry (IDMS), dual standards (^{233}U/^{236}U) for correction of instrumental isotopic mass bias, uranium–plutonium double spike mixtures (^{233}U/^{242}Pu and ^{235}U/^{239}Pu), and a series standards with certified ^{235}U content ranging from depleted uranium to LEU at 4.52%. In addition, there are several other standard mixtures of uranium isotopes. The French nuclear industry has developed CRMs for nuclear forensics that include a number of standards for uranium isotopic composition measurements as well as a ^{233}U/^{236}U double-spike and standards certified for determination impurities (Roudil et al. 2012).

The ^{238}U/^{235}U ratio in standard reference materials for natural uranium has been meticulously scrutinized and revised by very precise measurements (Brennecka et al. 2010; Richter et al. 2010; Hiess et al. 2012). The NBS SRM 960 (NBL CRM 112a) served as the *consensus* value for ^{238}U/^{235}U of 137.88 but interlaboratory measurements with an accurate double-spike (^{233}U/^{236}U) gave a slightly different average of 137.837(15). An alternative standard (IRMM-184) with a certified ^{238}U/^{235}U value of 137.697(41) has been confirmed by several laboratories (Richter et al. 2010).

Interlaboratory exercises focusing on nuclear forensics have been initiated by the Joint Research Centre (JRC) of the European Community. The two most recent exercises were called NUSIMEP-6 (Aregbe et al. 2008) and NUSIMEP-7 (Truyens et al. 2013) (NUSIMEP—Nuclear Signatures Inter-laboratory Measurement Evaluation Programme). In these exercises, identical samples were sent to the participating laboratories and the predefined analytes (content and isotopic ratios) were measured and reported to the coordinators, who then compiled the results, compared them with the target value, and produced a report. The remainder of the samples that were not consumed in the measurements can subsequently serve as standards for future measurements. One other benefit of these intercomparisons is that the accuracy and effectiveness of the different analytical procedures and the instrumentation can be assessed. In addition, corrective measures can be implemented in order to improve the performance. Finally, the cooperation between the scientists and the laboratories can lead to the development of standard procedures and recommended methods. This is essential for nuclear forensics where the implications of the analytical results, as mentioned earlier, may evoke political and even military actions in some cases.

5.2.2 Fingerprinting of Uranium Ores

As mentioned in Chapters 1 and 2, uranium is present in a large variety of minerals and its commercial production is carried out from rich ores that contain several percent of uranium to poor minerals that may include only a few hundred parts-per-million (grams-per-ton) of uranium. Extraction of uranium from the latter is usually a by-product of other mining activities like phosphate manufacture, columbite ores, or gold mines. The uranium ores from different sources may contain slightly different isotopic compositions, different amounts of impurities and trace elements, and different mineral compositions. A brief summary may be found in an MSc thesis (Hinrichsen 2010).

One of the first studies that characterized fine differences in the uranium isotopic composition in ore samples from various locations used a quadrupole mass spectrometer with a gas inlet to measure the $^{235}U/^{238}U$ ratio in UF_6 and a TIMS device for the minor isotope ratios (Richter et al. 1999). In this study, the ore samples were collected from Gabon, Czech Republic, Canada, Namibia, France, Finland, and Australia and were processed in different milling facilities (usually at or near the mine). Each sample was split into two parts: one subsample was converted to gaseous UF_6 and the mass spectrometric measurement was calibrated against two isotopic reference materials. The other subsample was dissolved in nitric acid and purified by anion exchange columns and then deposited on a rhenium filament for TIMS analysis. The main objectives of the study were to provide accurate measurement of the variations in the isotopic composition of uranium and set the *best measurement from a single terrestrial source* and update the current IUPAC values. The suggested values of this study were 54.20 (42)% ppm for ^{234}U, 0.72041 (36)% for ^{235}U, and 99.27417 (36)% for ^{238}U (i.e., $^{238}U/^{235}U$ = 137.802, quite far from the value derived earlier; Richter et al. 2010). The suggested range of the natural variations was 54–51 ppm for ^{234}U, 0.7207%–0.7201% for ^{235}U, and 99.2748%–99.2739% for ^{238}U. In two of the ore samples, the presence of ^{236}U was established: in the sample from the Czech Republic, this was attributed to anthropogenic contamination, while in the sample from Gabon this was indicative of the famous Oklo natural reactor (see, e.g., Bentridi et al. 2011).

The slight variations in $^{238}U/^{235}U$ isotopic ratios have been attributed to different mechanisms (Brennecka et al. 2010; Hiess et al. 2012). However, unequivocally there are large variations in the $^{234}U/^{238}U$ ratio in minerals and water sources as described in detail in Chapter 1 (Frame 1.2). The exact measurement of the isotopic composition of uranium is therefore helpful in determining the origin of the ore. In order to obtain a high degree of accuracy, extensive sample treatment is required. In one study (Brennecka et al. 2010) of the exact $^{238}U/^{235}U$ ratio in *yellow cake*, the samples were dissolved in 4 M HNO_3 (~100 mg in 20 mL) and a 250 µL aliquot of the solution was treated with 4 mL of 3 M HNO_3 + 0.05 M HF. Uranium was separated from matrix on a UTEVA chromatographic resin column. A double isotopic spike $^{233}U/^{236}U$ was added and high precision isotope ratio measurements were carried out with a multicollector ICPMS instrument equipped with an Apex-Q nebulizer. Differences in the $^{238}U/^{235}U$ and $^{235}U/^{234}U$ isotopic ratios were found between uranium deposits that were formed in low-temperature and high-temperature redox conditions in sandstone and nonredox conditions in quartz pebble deposits. These very slight differences can serve to identify and verify the source of uranium *yellow cake* at the front end of the NFC (Brennecka et al. 2010). In a different publication, which was not concerned with nuclear forensics applications, a comprehensive compilation of the $^{238}U/^{235}U$ ratio in a large variety of geological samples was included, as well as samples of extraterrestrial origins. This can help distinguish between uranium ores from different types of deposits and of specific origins (Hiess et al. 2012).

The elemental composition of uranium ores, especially the distribution of the rare-earth elements (REEs), can provide a unique fingerprint of the source of the ore, as shown in the examples given later. In one study, the anion composition of the uranium ores was shown to also be helpful in differentiating between uranium ore concentrates (UOCs) of different origins (mainly phosphorite and quartz pebble conglomerate) from Canada (15 samples), Australia (7 samples), and 2 samples from the United States (Keegan et al. 2012). Samples were prepared for analysis by dissolution in 8 M HNO_3 and 0.1 M HF in a Teflon vial at 90°C. The content of uranium and 65 trace elements was determined by sector-field ICPMS with a single collector using rhodium as an internal standard to correct for matrix effects. In addition, aqueous leach tests for 24 h at room temperature were carried out to determine the anionic content by ion chromatography after filtering the leachate. Canonical analysis of principle coordinates (CAP) was used to analyze the data and the main impurities responsible for group distinctiveness were identified and shown in Table 5.5. It should be noted that the elemental and anionic content in the UOC can be due to the original mineral but could also be influenced by the method used in the ore processing. For example, the processing method affected the content of impurities: UOC from one facility had a high calcium content that was attributed to the lime precipitation step while high magnesium content in another facility was due to MgO precipitation procedure used there. This is also relevant for the chloride and sulfate content and the Cl^-/SO_4^{2-} ratio that could arise from the mineral or the process. Thus, from the nuclear forensics aspect, the content of select impurities in uranium ore concentrates can serve for identifying the origin of the material (Keegan et al. 2012).

The same sample treatment method (dissolution in 8 M HNO_3 and 0.1 M HF at 90°C) was used for the determination of REEs in uranium-bearing materials by ICP-SF-MS (Varga et al. 2010a). The concentration of the REEs was determined directly after dissolution and then after purification and preconcentration on a TRU™ chromatographic resin. The quantitative results were quite similar, but the limit of

---

TABLE 5.5

Summary of Key Parameters Responsible for Distinguishing Uranium Ore Concentrates Sample Groups Found to Have Characteristic Impurities

	UOC Sample Group	Key Parameters Responsible for Group's Distinctiveness
Geology	Phosphorite	Cd Zn F/Cl Os P V Lu Pd Yb Y Tl Tm Re
	Quartz Pebble conglomerates	Rb Th Cs Pb Ti S NO_3/Cl Sm Tl Sc (+Tl when phosphorite excluded)
	Unconformity related	Mo (when phosphorite excluded)
Region—Canada	Elliot lake	Bi S Cs Rb Tl Pb Th Zn In NO_3/Cl Sm Ti
	Bancroft	P Fe Ta Au Hf Sb

Source: Adapted from Keegan, E. et al., *Appl. Geochem.*, 27, 1600, 2012. With permission.

detection for most elements improved from several nanograms per gram for direct dissolution to about 0.1 ng g^{-1} for samples that were chromatographically separated. The method was validated with a certified reference material and three samples of *yellow cake* were shown to have different lanthanide profiles (also referred to as REE patterns), implying that the origin of each material has a unique fingerprint. The study also found that the Ba/La ratio varied from 0.2 to 16 and this too is indicative of the source of the material.

The use of REEs signature as a means to determine the origin of uranium deposits was also investigated in another study (Mercadier et al. 2011). Sixty-six samples of uranium oxides from 18 sources and different types of deposits were analyzed. First, areas within each sample were selected after scanning electron microscopy (SEM), electron microprobe (EMP), and SIMS for U/Pb dating. The REE signature was determined by SIMS and LA-ICPMS. The REE content reflects the conditions of mineral formation (temperature, hydrothermal, saline brines, presence of ligands, redox conditions, etc.). These differences can be used as markers to determine the origin of uranium deposits and as a nuclear forensics tool for attribution of illegal nuclear trafficking.

As lead propagates through the front end of the uranium production process, its isotopic composition can serve as an indicator of the source of uranium ores. Lead has four natural isotopes (^{208}Pb, ^{207}Pb, ^{206}Pb, and ^{204}Pb) and three of them also have a radiogenic source (^{208}Pb from decay of ^{232}Th, ^{207}Pb and ^{206}Pb from decay of ^{235}U and ^{238}U, respectively, as shown in Chapter 1). It was found that the lead content and isotopic composition varies widely from mine to mine so that it may be useful for nuclear forensic purposes (Svedkauskaite-Legore et al. 2007). In a later publication by the same group, strontium isotopes appeared to be better indicators of the source of UOC (Varga et al. 2009). The ratio between the two lead progenies of uranium (^{206}Pb/^{207}Pb) can help determine the age of the raw ore material, but due to the high variations of lead isotopic composition within the mine area and the possible effects of anthropogenic contamination the attribution of a UOC sample to a given mine, based on lead isotopic, may be uncertain. The digestion procedure of the UOC samples was according to the process described earlier (10 M HNO$_3$ heated to 90°C for 3 h in a Teflon vessel). The lead and strontium were then separated from the matrix on chromatographic resins and finally analyzed by a multicollector ICPMS. The method was validated by analysis of SRM-981 for the lead isotopes and SRM-987 for the strontium isotopes, and the total procedure was confirmed by samples of basaltic rock reference materials. As mentioned earlier, even within a given mine large variation were found in the ^{207}Pb/^{206}Pb ratio. For example, in four samples from the Beverley mine (Australia), the ^{207}Pb/^{206}Pb ratio was 0.2332 ± 0.0252, 0.14298 ± 0.00072, 0.1765 ± 0.0011 and 0.1379 ± 0.0029 and in two samples from Rossing mine (Australia) values of 0.16820 ± 0.00087 and 0.82390 ± 0.00014 were obtained. On the other hand, the results for the ^{87}Sr/^{86}Sr ratio were less prone to variability within the mining site (Varga et al. 2011). The combination of the two ratios increases the confidence of UOC source attribution (Varga et al. 2009).

As noted, uranium ore concentrates are a commodity that is an attractive material for diversion for proliferation purposes (Kristo and Turnet 2013). The origin of UOC can be traced by comparison of chemical and physical features to a library database like the one at Lawrence Livermore National Laboratory (LLNL). This database uses 30 parameters from over 1800 samples of UOC from all the production locations of commercial importance plus a few others, and an automatic algorithm is used to match the sample analysis with the library content. However, if a sample match is not found, then a predictive approach must be used. Among the well-understood parameters are U–Pb ratio, Sm–Nd, and Rb–Sr isotopic ratios with REE abundance patterns as well as the $^{18}O/^{16}O$ ratio, in addition to the $^{238}U/^{235}U$ and $^{234}U/^{238}U$ isotopic ratios mentioned earlier.

The nonvolatile organic compounds that remain in the UOC after the milling process can help identify the type of purification and preconcentration methods that were used (Kennedy et al. 2012). A solution of 20:80 methanol:water with internal deuterated hydrocarbon standards was used to extract the organic compounds from a slurry of UOC and analysis was carried out with a GC–MS. As expected, di-octylamine, tri-isooctylamine, and alamine 336 were found in a UOC sample that was processed with amines.

The sulfur isotope ratio $^{34}S/^{32}S$ can also support the determination of the source of UOC *yellow cake* (Han et al. 2013). Method validation was based on IAEA standards that were weighed and dissolved in ultrapure nitric acid at 95°C on a hot plate for 6 h. After cooling, the sulfate content was determined by ion chromatography and the $^{34}S/^{32}S$ isotope ratio was measured by MC-ICPMS. Samples of UOC from 18 different origins were examined after leaching with ultrapure water for 24 h at room temperature. The samples were filtered and an anion exchange resin (AG 1-X4, Cl⁻ form) was used for separation the sulfate ions and isotopic ratio measurements. The variations of the different samples ranged from −15.4‰ to +18.3‰ relative to the assigned $\delta^{34}S$ *reference value* of 0.0441493 ± 0.0000080 for IAEA-S-1 standard (Han et al. 2013). These values differ from the $\delta^{34}S$ value for sulfuric acid by −5‰ to +15‰, suggesting that the sulfur content in the uranium ore contributes to the sulfur content in the final product and can thus indicate the origin of the UOC.

Laser-induced breakdown spectroscopy (LIBS) has been used to determine the uranium concentration in ore samples (Kim et al. 2012). A pulsed Q-switched laser at 532 nm was used as the light source and the emitted light from the uranium atoms at 356.659 nm was monitored. Although this method was not directly applied to fingerprinting uranium ores, the fact that multielement analysis is obtained in LIBS measurements implies that it has a potential to be used to distinguish between ores of different origins without the need for dissolution of the sample.

Highlights: The source of uranium ore concentrates at the front end of the NFC can be determined on the basis of chemical and isotopic fingerprints. These include the accurate isotopic composition of uranium ($^{238}U/^{235}U$ and $^{234}U/^{235}U$), the $^{87}Sr/^{86}Sr$ and $^{207}Pb/^{206}Pb$ ratios, the levels of elemental impurities (mainly REEs), the isotopic ratios of sulfur ($^{34}S/^{32}S$) and oxygen ($^{18}O/^{16}O$), as well as the presence of nonvolatile organic compounds. However, one must bear in mind that the final composition of the *yellow cake* depends both on the source of the raw material and on the processing procedure that may alter these fingerprints. The use of databases for matching interdicted samples with items in the library is helpful, although the availability of these databases is limited at present to some government institutions and perhaps to the IAEA.

5.2.3 URANIUM CONVERSION FACILITIES AND ENRICHMENT PLANTS

Nuclear forensics also plays a role with regard to the processing of uranium ore concentrates in the UCF and fabrication of uranium oxide for fueling nuclear reactors or of uranium hexafluoride for isotope enrichment facilities. Characterization of the nuclear materials can detect unauthorized operations and partially ascertain that no undeclared activities are taking place but one should always bear in mind that *absence of evidence is not evidence of absence.* In some cases, relevant information may be obtained from bulk samples but highly significant details may be found in analysis of single particles. Some examples will be presented here, but more details are discussed in Section 5.4.

5.2.4 FINGERPRINTING OPERATIONS WITH UF_6

The production of UF_6 at the UCF and, more importantly, the isotope enrichment process, are inevitably accompanied by some release of gaseous UF_6. Once UF_6 is released into the atmosphere, it will react with moisture to form aerosols of uranyl fluoride and HF (Equation 5.1):

$$UF_6 + 2H_2O \rightarrow UO_2F_2 + 4HF \qquad (5.1)$$

This would probably be a sequential reaction with one water molecule at a time, but for the present discussion this is not relevant. Formally, uranyl fluoride may be in a cluster with HF and water molecules $[UO_2F_2 \cdot HF_n \cdot H_2O_m]$ and the hydrogen fluoride may also be clustered with other HF and water molecules $[HF_n \cdot H_2O_m]$.

The larger aerosol particles settle down on surfaces at the plant and in its vicinity (and will be amenable for detection by swipe samples) while lighter particles could be carried to a considerable distance from the UCF (and environmental sampling may enable detection). Uranium of natural isotopic composition is usually the raw material in the UCF, so isotopic analysis is not sufficient to discriminate between uranium in the natural geochemical background and uranium-containing particles released from the UCF operations, unless the slight variations in the isotopic composition of natural uranium are distinguishable (see Section 5.2.2). In order to do this, the morphology of the particles must be studied by other means, like SEM or SIMS as discussed in the examples given later. On the other hand, uranyl fluoride particles released from the product side or the tails end of an enrichment plant would have a distinct isotopic signature. The deconversion activity of transforming enriched UF_6 to uranium oxide or metal for the production of nuclear fuel or fissionable material may also lead to release of UF_6 and formation of aerosol particles of unnatural isotopic composition. In the UCF, the final stages of the production process may cause emission of UF_6 from the chemical reactors, storage cylinders, or other operations with UF_6. In the enrichment plant, the chemical form of UF_6 is not normally altered or destroyed so leaks may occur mainly during operations involving the connection and removal of cylinders. In the deconversion operations, both these types of activity are expected with a release probability that depends on the skill of the operators and frequency of the actions.

For safeguards purposes, swipe samples are collected from different surfaces inside the enrichment facility. These could include operational equipment like pipes, cylinders, and machinery or even surfaces from the walls and floor. Bulk isotopic analysis of these swipe samples would give the average enrichment of the released particles. However, for a complete understanding of the operations carried out in the facility, single particle analysis is more revealing, as discussed and demonstrated here.

Detection and identification of undeclared activities using UF_6 and the UO_2F_2 aerosols as markers has received considerable attention from the nuclear forensics and safeguards aspects. A special chamber was constructed in which UF_6 can be hydrolyzed under controlled conditions of temperature and humidity (Kips et al. 2007). The morphology and isotopic composition of the aerosol particles formed in this chamber were investigated by SEM-EDX and SIMS. The relative humidity affected the morphology and size distribution of the particles as shown in Figures 5.5 and 5.6. When the relative humidity was 68%, the maximum in the particle size distribution was ~0.9 µm while at a higher relative humidity there was a shift toward smaller particles and the maximum was at about 0.75 µm. The size distribution of these particles was quite similar to that found in a real swipe sample where the diameter of the uranyl fluoride particles was 0.4–1.0 µm. The temperature also had an effect on the particle size and on the U:F ratio. It was found that heating the particles may result in a loss of fluorine and distortion of the U:F ratio (Kips et al. 2007). Ions containing uranium and fluorine jointly were not observed in the mass spectrum of samples that were heated to 350°C, but in the SIMS mass spectrum of a sample that was interrogated with a beam of Ga^+ ions secondary ions of UF^+, UOF^+,

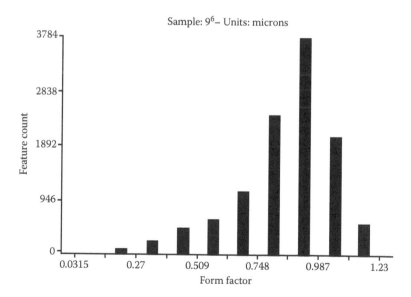

FIGURE 5.5 The particle size distribution of spherical uranium particles produced at a relative humidity of 68%. (From Kips, R. et al., *Microsc. Microanal.*, 13, 156, 2007. With permission.)

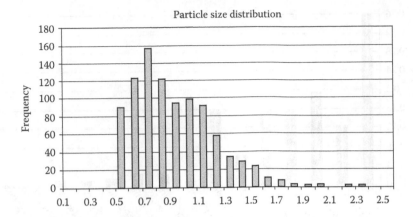

FIGURE 5.6 The particle size distribution of uranium particles produced at a high relative humidity. (From Kips, R. et al., *Microsc. Microanal.*, 13, 156, 2007. With permission.)

and UF_2^+ ions were found (note that there are isobaric overlaps between [235]UOF^+ and [238]UO_2^+ and between [238]UOF^+ and [235]UF_2^+). In a follow-up study by the same group, the effects of temperature, humidity, and UV-light on the loss of fluorine atoms from UO_2F_2 particles were examined, with the intention of using the F:U ratio as a means of determining the age of the particles (Kips et al. 2009a,b). As before, the UO_2F_2 particles were prepared in the reaction chamber by controlled hydrolysis of UF_6. The study involved one UF_4 particle that was aged for 29 months and 14 UO_2F_2 particles aged between one week and 28 months (and one that was aged for 6 h at 350°C). Three techniques were used to characterize the particles: scanning electron microscope-energy dispersive X-ray (SEM-EDX), ion microprobe-secondary ion mass spectrometry (IM-SIMS), and micro-Raman spectrometry (MRS). In general, the F:U ratio was reduced after 1–2 years, but exposure of the particles to UV radiation or to elevated temperatures increased the rate of fluorine loss. The MRS analysis showed a distinct peak at 865 cm^{-1} for UO_2F_2 particles but after heat treatment the Raman spectrum was similar to that of U_3O_8 particles indicating the fluorine was lost. Figure 5.7 shows the UF^+:U^+ (257/238 Da) ratio measured by IM-SIMS in the particles that were studied, and the loss of fluorine as a function of age, UV-radiation and temperature treatment is clearly seen. Table 5.6 summarizes the results of five representative particles. The conclusion was that combining these three analytical techniques for characterizing the particles can yield information on their age, although environmental conditions strongly affect the rate of fluorine loss.

As mentioned earlier, UCFs process mainly natural uranium so that the presence, or operation, of a clandestine UCF may be difficult to identify by uranium analysis alone or even by isotopic composition measurements. However, the co-existence of uranium and fluorine in micrometer-sized particles can help recognize such an operation (Faure et al. 2014). SIMS methodology was used for this purpose on two real-life UF_4 samples collected on filters and cotton cloth swabs in the purification and fluorination workshops and two CRM standards (CRM UF_4 and CRM UOC) were used for calibration. In addition, two swipe samples, to which a relatively large amount of

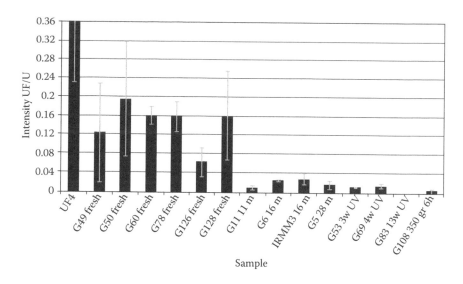

FIGURE 5.7 Average values for UF^+/U^+ (257/238) ratio obtained by IM-SIMS measurements on 14 UO_2F_2 particle samples and one UF_4 particle. (From Kips, R. et al., *Spectromchim. Acta B*, 64, 199, 2009a. With permission.)

TABLE 5.6
Summary of SEM-EDX, IM-SIMS, and MRS Results for Five Different Aged UO_2F_2 Samples

Sample Name	Aging	F/U SEM-EDX	UF/U IM-SIMS	MRS
G50	2 months	0.20–0.25	0.20 ± 0.12	Distinct peak at 865 cm^{-1}
G5	28 months	0.10–0.28	0.015 ± 0.008	Peak at 865 cm^{-1} in all 6 particle spectra
G108	350°C for 6 h	F peak not detected	0.004 ± 0.003	U_3O_8 features in 40 out of 50 particles
G83	3 months UV light exposure	F peak not detected	0.0008 ± 0.00001	Peak at 865 cm^{-1} in all 8 particle spectra
G69	4 weeks UV light exposure	0.04–0.14	0.014 ± 0.004	Peak at 865 cm^{-1} in all 6 particle spectra

Source: Adapted from Kips, R. et al., *Spectromchim. Acta B*, 64, 199, 2009a. With permission.

Notes: SEM-EDX F/U values are peak height ratios after subtraction of interpolated background levels; (ion-microprobe) IM-SIMS 1σ values quoted show the site-to-site variation; MRS, micro-Raman spectrometry.

sodium fluoride was added, were prepared in order to test the effect of nonuranium-bearing (but fluorine containing) particles on the SIMS analysis. First, an automatic scan to identify uranium-containing particles was carried out and this was followed by analysis of individual particles under microbeam conditions to determine the uranium content, its isotopic composition, and the amount of fluorine. As mentioned earlier, heating the sample should be avoided as this leads to loss of fluorine not only from UO_2F_2 but also from UF_4. Two ion sources were tested for the SIMS analysis: uranium peaks were not observed with the Cs^+ ion beam, while the use of O_2^+ as a primary beam yielded U^+, UC^+, UO^+, UF^+, UC_2^+, and UO_2^+ ions (the carbon containing ions come from the planchet). The F:U ratio in the samples that contained NaF was larger than in the unspiked samples; however, the authors concluded that the UF^+/U^+ ratio in the real UF_4 samples was larger than in the spiked samples. SIMS analysis fulfils three objectives: particle mapping, trace element measurements, and determination of the uranium isotopic composition. The fact that the process can be practically automated holds great potential for nuclear forensics applications (Faure et al. 2014).

A feasibility study for determining the $^{235}UF_6/^{238}UF_6$ isotope ratio by mid-range infrared spectrometry with a diode laser was carried out (Grigorev et al. 2008). The optical cell was 200 mm long and equipped with BaF_2 windows and a diode laser operated at 7.68 μm wavelength tuned to the $\upsilon_1 + \upsilon_3$ absorption band was used. The UF_6 samples were purified to minimize the HF level and the whole gas handling system was passivated by fluorine and by UF_6 to prevent interaction of the sample with the tubing and optical cell surfaces. System validation and calibration were carried out with specially prepared samples of UF_6 in which the degree of enrichment was between 3.5% and 12%. The authors concluded that the accuracy of the method was lower than that of mass spectrometric techniques but better than gamma spectrometric methods.

Environmental sampling nearby, or even at some distance from, uranium conversion (and deconversion) or enrichment facilities may yield evidence of the activity carried out. While particles of nonvolatile compounds like UF_4 or UO_2 will only be found inside the plant or at close proximity to it, release of gaseous UF_6 may travel considerable distances until aerosols of UO_2F_2 are deposited. The dispersion of UO_2F_2 aerosols formed by UF_6 released from uranium conversion plants was modeled (Kemp 2008). According to the assumptions about the source term and atmospheric conditions, the airborne concentration drops from 10^{-6} μg m^{-3} near the plant to 10^{-8} μg m^{-3} about 100 km away (the isopleths are asymmetric so these are just rough ballpark estimates). Wide area monitoring has been proposed but is of limited value when access to the facility is possible (like within a safeguards implementation regime). Obviously, the sensitivity required for environmental analysis is much higher than that needed for swipe samples collected at or near the plant.

Highlights: During the execution of industrial operations involving UF_6, release of some material is practically inevitable. The aerosols formed in the atmosphere initially contain particulate UO_2F_2 in which the F:U atom ratio should theoretically be 2:1. The size distribution of these particles depends on the relative humidity and the temperature, and experimentally the diameter of most particles was found to be between 0.4 and 1.0 μm. The aging of the particles would lead to gradual loss of fluorine and decrease in the F:U ratio. However, the presence of other fluorides, not associated with uranium, could change that and lead to an increase in the apparent F:U ratio, especially in bulk swipe samples. Therefore, the ability to interrogate the morphology, elemental impurities, and the uranium isotopic composition of single particles is important. The combination of several analytical techniques—optical microscopy, SEM, SIMS, laser ablation-ICPMS, and TIMS for accurate isotope analysis—can yield information on the activity of an undeclared nuclear facility (UCF or enrichment) or of unauthorized operations in a declared activity.

5.2.5 Nuclear Fuel Fabrication and Spent Fuel

5.2.5.1 Uranium Oxide

The use of near-infrared reflectance spectroscopy (NIR) for obtaining forensic signatures of the processes used in the manufacture of uranium oxide materials was described (Plaue et al. 2013). Laboratory-derived samples from five different processes commonly used to precipitate uranium from acidic solutions (with NH_4OH, H_2O_2, NaOH, Na_2CO_3, and magnesia) were examined by XRD and NIR. Uranium oxide samples prepared under a variety of temperature conditions were also tested in order to see whether the *precipitation history* was retained after heating the samples to temperatures ranging from 85°C to 750°C. In addition, some real-world UOC and oxide samples from the laboratory archives were analyzed. The XRD technique was used to determine the chemical phase while NIR with a fiber optic bundle was deployed for fingerprinting the samples. The results showed that the different precipitation processes led to distinguishable absorption peaks and several bands had been assigned to the unique reagents. At present, although NIR spectroscopy was unable to sort out the precipitation methods in the uranium oxide samples that had the same phases, there appears to be a potential for utilizing NIR in nuclear forensics.

In another study, samples of depleted uranium oxide manufactured through different processes were characterized (Hastings et al. 2008). Three types of uranium oxides were prepared from uranyl nitrate hexahydrate ($UO_2(NO_3)_2 \cdot 6H_2O$) at different temperatures: UO_2 at 500°C–700°C, U_3O_8 at 800°C–1100°C, and UO_3 at 350°C–450°C. Optical spectroscopy and particle fractionation were used to characterize the oxides. The color of the oxides, their density, and granular appearance (aerodynamic diameter and size distribution) were somewhat affected by the preparation conditions. The gamma spectrum of the samples was also recorded. The conclusion was that variations in the processing conditions of the uranium oxides were reflected in the density and particle size distribution and these characteristics could be used for nuclear forensics.

Micro-Raman spectroscopy combined with scanning electron microscopy (MRS-SEM) was applied to discriminate between micrometer-size particles of different uranium compounds (Pointurier and Marie 2010). The compounds included UO_2, U_3O_8, $UO_4 \cdot (4H_2O)$, UO_2F_2, and UF_4 and the particle sizes were from a few micrometers up to 30 μm. Particles were collected by dabbing sticky carbon disks on surfaces contaminated with particles. Then SEM was used to locate the uranium-containing particles and MRS to analyze them. Characteristic Raman bands were found for each compound when the sample was excited by lasers operated at 514 or 785 nm. Some reference Raman spectra obtained for micrometer size particles of UO_3, U_3O_8, and UO_2F_2 are shown in Figure 5.8a through c, respectively. One interesting result

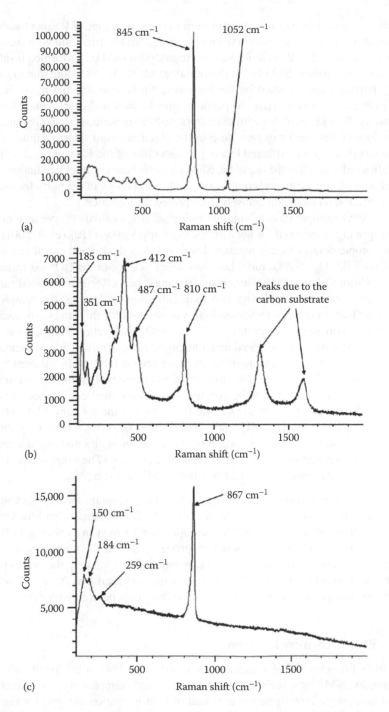

FIGURE 5.8 Reference Raman spectra obtained for micrometer size particles of UO_3 (a), U_3O_8 (b), and UO_2F_2 (c). (Adapted from Pointurier, F. and Marie, O., *Spectrochim. Acta B*, 65, 797, 2010.)

was that laser irradiation of metallic uranium led to appearance of Raman bands that are characteristic of U_3O_8 (indicating enhanced oxidation). In a follow-up study, the system was modified so that the Raman microanalysis could be performed inside the SEM vacuum chamber (SEM-SCA) (Stefaniak et al. 2014). As before, the uranium-bearing particles were located by the automatic SEM scan and the micro-Raman was then applied to characterize the particles thus located within the same chamber. Fresh uranyl fluoride particles with diameters <1000 nm were successfully analyzed without loss of fluorine or oxygen despite the electron-beam interrogation, but the particle morphology was affected leading to distortion of the Raman spectra. These two studies indicated that the use of the SEM electron beam or laser irradiation could cause changes in the elemental composition and morphology of the particles (mainly UO_2F_2 particles) so the results must be interpreted with caution.

The $^{18}O/^{16}O$ isotopic ratio in uranium oxide particles can also serve as a marker of the origin of the material for nuclear forensics applications (Pajo et al. 2001). The oxygen isotopic composition is measured with a TIMS device and calculated on the basis of the $^{238}U^{18}O_2/^{238}U^{16}O_2$ ratio. Due to the dependence of the UO^+ ion formation on the conditions of the TIMS filament and the mass bias effects (mentioned earlier), the total evaporation method with two filaments (one for loading the sample and evaporating it and the other for ionization) was selected for this study. The samples included uranium ore concentrates, U_3O_8 and UO_2 fuel pellets. The pellets were cleaned and all samples were stored under an argon atmosphere to prevent oxidation. Particles of several hundred nanograms were loaded on a Re or Ta filament by use of a suspension in dry benzene that also served to ensure that grains of similar sizes were selected (dissolution had to be avoided in order not to affect the isotopic composition). The temperatures of the sample filament and the ionizing filament were optimized to yield a high UO^+/U^+ ratio. The results showed that the $n(^{18}O)/n(^{16}O)$ ratio depended on the geographic origin of the uranium oxides and that the ratio in UO_2 pellets correlated with that in rainwater precipitation. The range of the $U^{18}O_2/U^{16}O_2$ ratios varied from 2.12‰ to 2.06‰ in the different samples.

Highlights: Uranium oxide particles that are the most common form of nuclear fuel can be characterized by spectroscopic methods (NIR and Raman) with or without a combination of SEM and the isotopic composition of oxygen is also an indicator of the geographic source. The production process used in the manufacture of the particles is reflected in the analysis and can thus help to identify the origin of the particles. It should be noted that the analytical method used may have an effect on the particle consistence and morphology and should thus be used with care.

5.2.6 POSTDETONATION FINGERPRINTS

The world's first ever site of a nuclear detonation (July 1945 at the Trinity site near Alamogordo, NM) was revisited decades later, and samples were collected for detailed analysis to investigate the potential of postdetonation analysis for accurate source attribution (Fahey et al. 2010; Bellucci et al. 2013). Although the Gadget (the code name given to the first test device) consisted of a plutonium core, detailed isotopic analysis of the uranium in trinitite (the glassy material formed by melting of the soil from the test) yielded information on the construction of the device. The uranium found in the samples came from two sources: from the uranium tamper used in the device and from natural uranium in the geological background. The age of the materials and the isotopic composition were determined. Higher than natural content of ^{234,235,236}U reflected the alpha decay of ^{238,239,240}Pu, respectively, and the ratios between them showed the isotopic composition of the plutonium core (concurrent with *supergrade* plutonium, i.e., low ^{240}Pu and ^{238}Pu contents). The isotopic composition of lead and the content of other impurities were also determined (but at the time of writing are *awaiting future publication*). The study included analyzing 75 points in 12 samples of trinitite glass by laser ablation-multicollector ICPMS (LA-MC-ICPMS). The glassy samples were cut into thin slices that were polished and the activity of the ^{152}Eu fission product (half-life ~13.5 years) that was measured by gamma spectrometry served to calculate the yield as a function of the distance from ground zero. Ablated sample introduction was by a 193 nm excimer laser and the ^{234}U, ^{235}U, and ^{236}U isotopes were recorded simultaneously by ion count detectors while ^{238}U was measured by a Faraday cup detector. Calibration was performed by bracketing with two NIST glass SRMs (610 and 612). The analysis time required for each sample was 2 min, demonstrating that this technique can be used to rapidly obtain analytical information from postdetonation debris and assist in nuclear forensics attribution (Bellucci et al. 2013).

In an earlier study, microanalysis of the trinitite glassy material and secondary materials yielded information that positively identified the nuclear components (Fahey et al. 2010). The study focused on a single piece of trinitite, weighing about 7.5 g, purchased from a mineral collector. A thin polished slice was mapped by SEM, and microfocusing XRF and SIMS were used to characterize the spots that showed enhanced alpha and beta activity in autoradiography. Gamma spectrometry identified several radionuclides (natural, fission, and decay products and activation products), including ^{137}Cs, ^{239}Pu, ^{241}Am, ^{60}Co, ^{133}Ba, and ^{152,154}Eu. SIMS was used to explore the isotopic composition of uranium, plutonium, and lead. The advantage of this microanalytical approach relative to bulk analysis of the nonhomogeneous samples was that the spatial correlation between uranium and plutonium found here would not be observed in the bulk analysis. Thus, the combination of spatially resolved radiological, elemental, and isotopic compositions provided detailed information about the construction of the device (Fahey et al. 2010).

Highlights: There are not many published examples of analysis of postdetonation debris. The reported results from the 1945 Trinity test show that information on the core and constituents of the device can be derived from advanced analytical techniques, with repercussions for nuclear forensics analysis.

5.3 BULK ANALYSIS AND SINGLE PARTICLE ANALYSIS

5.3.1 BULK ANALYSIS FOR SAFEGUARDS PURPOSES

The term *environmental swipe samples* used in many studies should be understood as referring to collection of samples for safeguards purposes in and around the facility where nuclear materials are processed and manufactured. This should not be confused with *wide area monitoring* that includes sample collection from localities that could be remote and not in the immediate vicinity of declared plants, like aquatic sampling (Wogman et al. 2001) or atmospheric air sampling (Krey and Nicholson 2001).

The collection of environmental swipe samples for safeguards purposes and the sample processing and analysis procedures practiced at the Safeguards Analytical Laboratory (SAL) of the IAEA in Seibersdorf were described in detail and can serve as a reference point for assessment of newer analytical methods (Vogt et al. 2001). Bulk swipe samples are collected with a 10 cm × 10 cm cotton cloth or round cellulose wipers for use in *hot cells*. After swiping the surface, the organic matrix is destroyed by dry-ashing in a furnace first at 450°C and then at 650°C. The residue is dissolved in nitric acid. Additionally, wet ashing can be used with a mixture of nitric acid followed by perchloric acid. Microwave-assisted digestion is not used routinely because of sample size limitations and cold-ashing requires meticulous cleaning between samples making it less convenient for routine work. In preparation for TIMS analysis, radiochemical separation is required and is performed by solid-phase extraction with tri-n-octylphosphine oxide (TOPO) and then separation and preconcentration with anion exchange columns (UTEVA chromatographic columns were still under investigation when this was published). The samples are loaded on a filament and analysis is carried out either by isotope-dilution (ID-TIMS) that gives quantitative and isotopic composition information or by mass spectrometry for isotopic data only. The detection limit of the uranium signal was 10^7 atoms (~$5 * 10^{-15}$ g); however, due to background and blank levels, the practical detection limit is higher by two orders of magnitude.

A simpler and faster method for analysis of swipe samples for safeguards purposes was developed (Pestana et al. 2013). Preparation of the swipe samples involved simply acid leaching assisted by sonication instead of total digestion and as the analysis was by ICP-MS then radiochemical separation was not required. The swipe samples were collected from a nuclear fuel production plant in Brazil with maximum uranium enrichment of 20%. The method described earlier (BA—dry-ashing in a furnace, dissolution of the residue, and chromatographic separation) was compared with ultrasonication for 15 min with 0.29 M HNO_3 followed by centrifugation for 3 min (designated by the authors as UAL) and direct analysis of an aliquot by ICP-MS for uranium isotopic composition.

A special case of interest in nuclear forensics is collection of samples from building materials that may have a very rugged surface in which trapped uranium bearing particles may be difficult to access. A simple, inexpensive method that can be deployed in the field for detecting and quantifying uranium in concrete, Plexiglas, glass, and steel was described (Greene et al. 2005). First, the sample was rinsed by a buffer solution (pH = 2.2) and the uranium in the solution was captured in an arsenazo-III complex with C18 crown-ether solid phase extraction (SPE) and could be determined by colorimetric spectroscopy at 654 nm. Although arsenazo-III is not specific for uranium, its combination in the C18 SPE complex separates U(VI) from interfering ions. The reported detection limit was 40 ng L^{-1} (or 5 ng cm^2 on the examined surface) and the extraction efficiency was estimated at >80%.

Method optimization and validation were performed with a standard (CRM NBS U2000) that contained 20% enriched uranium. The relative errors were below 1%

for $^{234}U/^{238}U$, $^{235}U/^{238}U$, and $^{236}U/^{238}U$ ratios, but the uncertainties were 7.73%, 0.94%, and 0.94%, respectively. For five real swipe samples that were processed by both methods (BA and UAL), the uncertainty was similar for the three isotopic ratios, although in some cases the values differed by 10% for the measured ratios. In conclusion, the likelihood of cross-contamination was reduced and the sample throughput was significantly increased, although the sensitivity, accuracy, and uncertainty were inferior to the TIMS method described earlier.

Methods for bulk analysis and particle analysis of nuclear materials for detection of undeclared activities were described in a review article (Piksaikin et al. 2006). The bulk detection methods included radiometry (alpha, beta, and gamma spectrometry) based on the natural decay of the radionuclides, including the use of $^{234}Th/^{230}Th$ gamma activity ratio for age determination (see detailed discussion later) and $^{226}Ra/^{235}U$ ratio that should be about 21 for undisturbed ores: higher in mining tails (after most of the uranium has been removed); and lower in mining products (after leaving most of the radium in the tails). On the whole, the passive methods are useful for screening bulk samples. Active detection methods based on neutron activation (NAA), excitation by x-rays (XRF), or proton induced x-ray emission (PIXE) can be used for the assessment of the total uranium content. Other bulk methods include extraction and chemical separation of the radionuclides and the use of mass spectrometric techniques (TIMS, ICPMS, and AMS) for the analytical measurement. The advantages and limitations of each method were described in the review (Piksaikin et al. 2006) and will not be discussed here.

Highlights: Analysis of bulk samples can yield invaluable information for nuclear forensics applications rapidly and relatively inexpensively. It is particularly useful for screening swipe samples and deciding which samples should be processed for single particle analysis. The total digestion of the organic matter of the swipe itself can be replaced by a simpler leaching procedure and separation of uranium that is required for TIMS analysis can be eliminated if ICPMS analysis is used. The results obtained from this relatively *quick and dirty* procedure are not as accurate as those that can be acquired from TIMS after uranium separation, but are helpful for preliminary assessment of the uranium content in the sample. As sample digestion is a destructive method the samples must be divided, if particle analysis is desired: one part for screening by bulk methods and another part preserved for particle analysis.

5.3.2 SINGLE PARTICLE ANALYSIS

The main difference between bulk analysis and single particle analysis is analogous to the difference between low-definition and high-definition photographs—you see the object in the photo in both cases, but in order to see fine details you need high resolution. Thus, the ultimate goal of a nuclear forensics investigation is to determine the physical, chemical, and isotopic characteristics of a nuclear material from analysis of a single particle or a small number of particles. This is particularly important for safeguards applications where bulk samples containing uranium show only the average degree of enrichment while the presence of individual particles with enrichment levels above the declared value would indicate illicit activities. Particle analysis requires extreme sensitivity of the analytical methods, and this has indeed been accomplished as demonstrated in the examples presented later. Some examples of the use of particle analysis for nuclear forensics applications were also presented earlier in the sections related to operations in the NFC.

In the late 1990s, methods for locating and analyzing individual particles containing plutonium or highly enriched uranium by use of SIMS were published, mainly by the group at the Institute for Trans Uranium Elements (ITU) in Karlsruhe, Germany, as seen for example (Betti et al. 1999). It was shown that the measurements of the isotopic composition of uranium obtained by SIMS compared favorably with TIMS results and the morphology (shape and dimensions) of the particles matched up with SEM analysis. In addition, SIMS provided information on the elemental composition of the particles. Since then, the methods have evolved and improved due to the availability of superior instrumentation and better understanding of the methodology.

The review of methods for the detection of undeclared activities mentioned earlier also lists the main analytical techniques that can be applied for characterization of single particles (Piksaikin et al. 2006). The list includes scanning electron microscopy (SEM) for the study of particle morphology combined with x-ray fluorescence for elemental analysis (SEM/XRF). Secondary ion mass spectrometry (SIMS) simultaneously provides morphological, elemental, and isotopic information on individual particles, as mentioned earlier. Other mass spectrometric techniques are used mainly for determining the isotopic composition of uranium after the particles have been located and transferred to the analytical device (placed directly on a filament for TIMS, designated for laser-ablation-ICPMS or dissolved for nebulizer-based ICPMS). Total reflection x-ray fluorescence (TXRF) spectrometry can be used to locate uranium-containing particles in samples that include very few target particles where SIMS is less effective. The importance of cleanliness and meticulous quality control are emphasized in this review that also presents the results of the Nusimep-3 exercise in which the uranium isotopic ratios determined by different methods were compared (Tresl et al. 2004). Several examples of actual studies in which particle analysis was carried out are shown later.

Thermal ionization mass spectrometry (TIMS) has been widely used for accurate determination of the isotopic composition of uranium in general, and particularly in single particles. Improved methods for the analysis of single uranium oxide particles of ~1 μm diameter (and as low as ~0.4 μm) have been developed for nuclear safeguards purposes (Kraiem et al. 2011a). SEM was used to locate the particles and morphologically characterize them and then they were transferred to the TIMS filament. The filament pretreatment and the technique used for placing the particle on the filament strongly affect the quality of the measurement due to their effect on ionization efficiency. The TIMS device was equipped with a multi-ion-counting (MIC) system for simultaneous measurement of ^{234}U, ^{235}U, ^{236}U, and ^{238}U adding to the accuracy and sensitivity. Some of the samples consisted originally of uranium fluoride particles that were produced from UF_6 in an aerosol deposition chamber and were subsequently heated to release the fluorine and form the U_3O_8 particles. Other particles were formed by dissolving uranium oxide from a CRM and deposition of 1 μL droplets with different uranium concentrations and enrichment levels on a pre-degassed and carburized zone-refined rhenium filament. The measurement

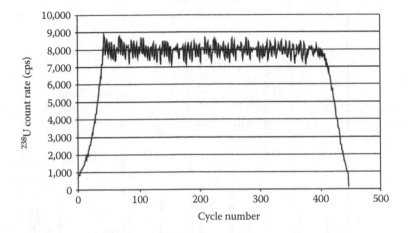

FIGURE 5.9 The ^{238}U count rate as a function of the cycle number obtained during the TIMS measurement of a single NBS-U200 particle. Each cycle represents an integration time of 1 s. (From Kraiem, M. et al., *Anal. Chim. Acta*, 688, 1, 2011a. With permission.)

with the MIC-ICPMS was carried out by the total evaporation procedure through which the particle is completely consumed avoiding mass bias often encountered in TIMS measurements. Special care was taken to diminish the background level (by a factor of 5–10) by applying a baking procedure of the sample filament (at ~1750°C for 1 h) before each measurement from a second filament positioned on the side of the sample filament. The filament carburization was carried out by exposing the heated Re filament to benzene vapors in a special chamber and optimal performance was found when the filament was heated to ~1750°C for 30 min with benzene pressure of $3 * 10^{-3}$ mbar (Kraiem et al. 2011a). Figure 5.9 shows the count rate of ^{238}U$^+$ from a single particle (NBS U-200) on a filament produced by this procedure as a function of the cycle number. Evidently, the reproducibility of measurement is very good. The ionization efficiency of uranium deposited on the carburized filament was calculated as 0.2%–0.3% as shown in Table 5.4.

An interesting example of the use of SIMS and SEM techniques for the characterization of *hot particles* found at the site of the 1968 crash of a nuclear weapon in Thule, Greenland, was presented (Ranebo et al. 2007). The samples were recovered from sediment core samples close to the crash site and isolated by sampling splitting methods based on the gamma activity of ^{241}Am. The target particles were manipulated and placed on adhesive carbon tape. Particles with high Z appeared brighter on the SEM image operated in backscattered mode and spot EDX measurements were made to obtain the U/Pu ratio. SIMS with a primary O$_2$$^+$ ion beam was used to obtain the mass spectrum and ion imaging patterns, as well as depth profiles. The SIMS mass spectrum (shown in Figure 5.10) covers the mass region of 234–242 Da and several isobaric interferences can be observed from hydride ions as well as from overlapping masses (Ranebo et al. 2007).

The development of a large geometry (LG-)SIMS with high transmission improved the performance of the technique by affording a high secondary ion yield (1.2%) at a resolution of 3000 (Ranebo et al. 2009). Other advantages were the reduction

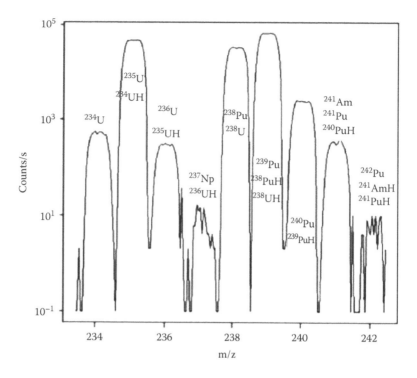

FIGURE 5.10 The SIMS mass spectrum from the *hot* particle found in a core sample at Thule, Greenland the crash site. Note isobaric interferences arising from hydride ions and overlapping isotopes in the 234–242 Da mass range. (From Ranebo, Y. et al., *Microsc. Microanal.*, 13, 179, 2007. With permission.)

of analysis time even for dust samples with few uranium-bearing particles and the decrease of interferences by isobaric ions that may be present in an environmental matrix. The standard (small geometry) SIMS instruments usually have limited sensitivity (due to lower ion yields) and lower resolution (R = 450) and hence suffer from background levels and isobaric interferences (as seen in Figure 5.10). Nevertheless, the standard SIMS is a powerful tool for single particle analysis as these particles are located and analyzed by one device without need for transfer between different instruments. The performance of the LG-SIMS also favorably compared with TIMS, where extensive sample preparation and separation steps are required prior to analysis, except for accuracy in measurement of the $^{236}U/^{238}U$ ratio due to the need for $^{235}UH^+$ isobaric correction in the LG-SIMS. System optimization and calibration were performed with mono-dispersed uranium particles generated by an aerosol generator from uranyl-nitrate solution (from CRM U020-A). The samples were then calcined in a furnace to form uranium-oxide particles. The size distribution of the particles was determined by SEM, and it was found that 90% of them were formed by a single aerosol droplet (larger particles were formed by clustering of two or more droplets). Two actual swipe samples collected from enrichment plants were measured by Small Geometry-(SG)SIMS and LG-SIMS. The SG-SIMS gave high

^{234}U and ^{236}U values with a large scatter that do not correspond to the degree of ^{235}U enrichment and this was attributed to isobaric interferences, while the LG-SIMS showed a linear correlation between ^{234}U and ^{235}U. As mentioned earlier, the ^{236}U measurements include contribution from the ^{235}UH$^+$ hydride ion that affects the accuracy and requires the use of a correction factor. This is a minor problem for depleted, natural and low-enriched uranium, but the higher the abundance of ^{235}U the larger the effect of the hydride ion on the ^{236}U measurement. For very accurate isotope ratio determination, the mass fractionation (also called mass bias or mass discrimination) should also be considered and corrected. Standard TIMS analysis may have a mass fractionation of 0.3%–0.4% Da^{-1}, but if total evaporation, multicollector method is used, this can be practically eliminated. SIMS analysis also has a mass bias that can reach almost 1% per Da for sputtering of large particles at high rates especially at the start of the analysis. Pre-sputtering of small particles has been shown to minimize this mass discrimination effect. Thus, compared to standard SIMS, the LG-SIMS can handle a larger number of particles per planchet, can give multi-isotope imaging, higher sensitivity, better accuracy, and provides a faster screening for swipe samples (Ranebo et al. 2009).

Laser ablation-ICPMS has been applied to measurement of the isotopic composition (^{234}U, ^{235}U, ^{236}U, and ^{238}U) of single uranium oxide particles with dimensions down to 10 μm, as shown in Figure 5.3 (Varga 2008). The uranium oxide samples that were studied included depleted uranium (DU), natural uranium (NU), low-enriched uranium (LEU), and high-enriched uranium (HEU). For solution-based analysis, the samples were dissolved in slightly heated 6 M HNO$_3$ and for the laser-ablation studies the particles were placed on a double-sided tape and covered with polyethylene transparent tape to prevent detachment during the analysis. A sector-field ICPMS was used for the analysis, and it was operated in the low-resolution setting (R = 300) that was sufficiently accurate for measurement of the ^{235}U/^{238}U ratio or the medium resolution setting (R = 4000) that was needed for the ^{234}U/^{238}U and ^{236}U/^{238}U. The results showed good agreement between the solution-based and the laser-ablation measurements.

Different strategies for isotope ratio measurements of single particles by laser-ablation-MC-ICPMS were discussed (Kappel et al. 2013). The samples included a series of U$_3$O$_8$ reference materials and two samples of particles from the interlaboratory comparison exercise with ^{235}U content ranging from 0.726% to 50%. Particles were manipulated and placed on silicon planchets that were preferred over carbon planchets because the dark particles could be more readily observed. The work was carried out in two different laboratories, each equipped with a multicollector ICPMS, and two types of lasers (a femtosecond [fs] laser operated at 795 nm and a nanosecond [ns] excimer laser at 193 nm) were used for this study. Four different data treatment strategies were deployed for calculation of the ^{235}U/^{238}U ratio from the transient signals acquired by laser ablation analysis of single particles. Some processing strategies give higher weights to particles that yielded higher signal intensities, and these are suitable in cases where the uranium-containing particles were previously located (by SEM, FTA, or SIMS). However, in cases where a *blind* scan is performed, some particles may be underrepresented because of the scan speed so other signal processing strategies are preferred.

A slightly different approach was used for precise and accurate isotopic analysis of microscopic uranium oxide grains from contaminated soil and dust using LA-MC-ICP-MS (Lloyd et al. 2009). The origin of the particles was consistent with depleted uranium from a factory that produced DU penetrators, counterweights, and radiation shielding from 1954 to 1984. The combustion of the scrap metal in the factory resulted in release into the environment of uranium oxide particles. An analytical procedure was used to differentiate these particles from the background natural and anthropogenic uranium. The soil and dust samples collected near the plant contained considerable levels of uranium that was evidently depleted (soil 90 ± 9 mg kg^{-1} with $^{235}U/^{238}U = 0.21\% \pm 0.1\%$; dust 385 ± 33 mg kg^{-1} with $^{235}U/^{238}U = 0.22\% \pm 0.1\%$). Before analysis, a simple treatment procedure was used involving drying at 60°C to remove moisture and sieving (<250 μm) to remove coarse grains. Then iron scrap was separated by a hand-magnet and floatation in a dense solvent (di-iodomethane, $d \sim 3.3$ g cm^{-3}) to remove silica, silicates, and fine particulate matter so that only large (>20 μm) dense ($d \sim 10.96$ g cm^{-3}) UO_2 grains sink while other particles float. These grains were mounted in epoxy resin and scanned by SEM-EDX to map the uraniferous grains that can then be easily located for ablation by the laser and analysis. Some grains were removed and dissolved for MC-ICPMS analysis that was compared with the laser ablation results. The $^{236}U/^{238}U$ measurements showed a large spread clustering around 27 ppm (maximum 50 ppm) with the $^{235}U/^{238}U$ ratio around 0.2% (maximum of 0.24%). This inferred that a variety of batches with different isotopic compositions were processed at the plant and indicated that environmental sampling of the soil could help determine what activities took place there.

A novel analytical strategy using laser ablation-sector field-ICP-MS for isotope ratio determination was proposed (Marin et al. 2013). The Nd:YAG laser (266 nm) was operated continuously at a low power setting and the beam was defocused so the sample was practically intact after the analysis—almost like nondestructive assay—and the $^{235}U/^{238}U$ ratio was found to be 16.36 ± 0.15 (good accuracy) for a high-enriched uranium sample. The system parameters were optimized with the glass SRM (NIST 610) and validated with a UO_2 pellet (NBL CRM 125-A). The HEU sample was obtained from the Round Robin exercise (RR3). Isotope ratios were corrected for mass fractionation and discrimination, but large deviations were found for the minor isotopes ($F_{234} = -18.11\%$ and $F_{236} = -54.66\%$) requiring large correction factors.

In a study of uranium-bearing particles in swipe samples, fission track (FT) analysis was used for locating the relevant particles and TIMS for the measurement of the isotopic composition (Shen et al. 2008). The particles were detached from the swipe sample by wiping on a thin film of a Lexan polycarbonate substrate (3×3 cm^2, 50 μm thick). A thin slice of freshly prepared mica served as a detector and was attached to the Lexan substrate by tape (Figure 5.11a). After neutron irradiation of the sample, craters were formed in the mica film where uranium-bearing particles were present (Figure 5.11b). The mica sheet was removed, etched with 40% HF to enlarge the craters so that they could be readily observed in an optical microscope and their coordinates recorded. The particles on the Lexan substrate that correspond to the craters on the mica film were then cut out of the Lexan substrate and manipulated for transfer to the TIMS filament. The efficiency for locating the relevant particles was

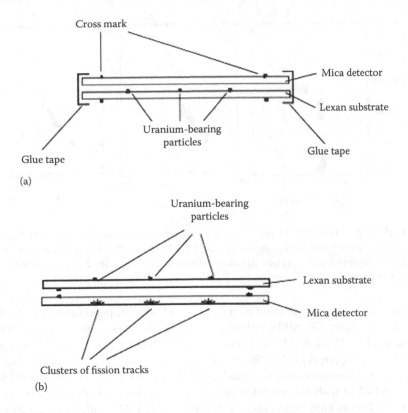

FIGURE 5.11   (a) A schematic diagram of a sandwich of a Lexan substrate with uranium bearing particles and mica track detector; (b) overlapping of the mica track detector and the Lexan substrate with uranium-bearing particles to make it easy to pick-up the particles. (From Shen, Y. et al., *Radiat. Meas.*, 43, S299, 2008. With permission.)

estimated as 90% for particles with diameters of 0.5–20 μm, and the accuracy of the location coordinates was within 5 μm. The success rate of transferring the particles to the filament of the TIMS instrument was size-dependent ~90%, ~73%, and ~43% for particles of 0.5–5, 5–10, and 10–20 μm, respectively. This approach is especially effective when the swipe sample contains a small number of uranium-bearing particles amidst many irrelevant ones.

The determination of the isotope ratios of uranium in individual particles by SIMS and ICPMS was compared (Esaka et al. 2009). Method development was first carried out with particles of U_3O_8 powder (CRM 050) that were smeared on a standard cotton cloth used for safeguards sampling. For the ICP-MS analysis the particles were transferred from the cloth to a silicon disk that was placed in a SEM chamber and EDX analysis was used to locate the uranium particles. Single particles were then placed in the center of a silicon chip and each chip was transferred to a clean Teflon vial where the particle was dissolved by addition of 40% HNO_3. Analysis of the solution was carried out with an Apex-Q desolvation system. In addition to the mono-atomic ions (^{234}U, ^{235}U, ^{236}U, and ^{238}U), the $^{238}U^{16}O^+$ ion was also measured to monitor the stability of the instrument. The contribution of $^{235}UH^+$ ions

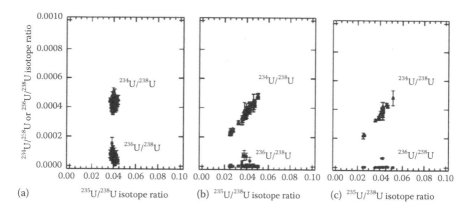

FIGURE 5.12 The $^{234}U/^{235}U$ and $^{236}U/^{235}U$ isotope ratios measured for individual particles in a swipe sample by (a) SIMS, (b) manipulated SIMS and (c) ICP-MS with desolation. The error bars represent one standard deviation. (From Esaka, F. et al., *Talanta*, 78, 290, 2009. With permission.)

to the $^{236}U^+$ peak was corrected by a factor of $5 * 10^{-5}$ (50 ppm) calculated from the $^{238}UH^+/^{238}U^+$ ratio. The SIMS instrument was used to measure the uranium isotopes at masses 234, 235, 236, 238, and mass 239 for hydride ion correction (that was about $4 * 10^{-4}$ or 400 ppm in the SIMS). A beam of O_2^+ ions was focused on individual particles for the measurements. Comparison of the methods for individual particles of the U050 CRM with diameters 0.6–4.2 μm showed that the ICP-MS measurements with desolation gave higher count rates than the ICP-MS without desolation and better relative standard deviation (RSD) values for isotope ratios. It was also found that the SIMS relative standard deviation that was based mainly on larger particles was slightly superior to the ICP-MS results. The results for the test swipe samples that spanned a $^{235}U/^{238}U$ enrichment range of ~2.5%–5.5% for SIMS manipulated-SIMS and ICP-MS with desolation are shown in Figure 5.12. Evidently the SIMS results show the effect of agglomeration—all samples fall within a narrow range (Figure 5.12a) while the manipulated-SIMS and ICP-MS clearly show the enrichment range of individual particles (Figure 5.12b and c, respectively).

In a subsequent publication by the same group, solution nebulization ICP-MS was combined with particle screening and microsampling for analysis of individual particles (Esaka et al. 2013). Two methods were used for locating the uranium-bearing particles: fission track analysis (FTA) and automated SIMS (described earlier). The particles were then manipulated inside a SEM and placed on silicon wafers and then dissolved for ICP-MS analysis. After method development and validation with particles from certified reference materials (NBL CRM U050, U100 and U500) that were smeared on a cotton swab, a real-life sample collected at a nuclear facility was examined by both techniques. The measured $^{235}U/^{238}U$ ratio of the particles that were located by SIMS varied from 2.59% to 4.32% while in those found by FTA the range was between 2.46% and 3.76%. The authors presented a plot of the ratio of the minor isotopes ($^{234}U/^{238}U$ and $^{236}U/^{238}U$) as a function of the enrichment of the particles ($^{235}U/^{238}U$) and clearly showed that while the $^{234}U/^{238}U$ ratio increased almost linearly with the degree of enrichment the $^{236}U/^{238}U$ ratio was unchanged and represented the background level. Finally, the time needed for analysis of 30 particles by SIMS alone (4 days), by automated SIMS for particle location combined with ICPMS isotopic analysis (8 days) and for FTA particle location with TIMS and ICPMS analysis (23 days for each method) was compared. The authors concluded that the proposed ICP-MS methods are time consuming compared to SIMS and TIMS, but state that microsampling of individual particles followed by dissolution and chemical separation enable the use of ICP-MS to collectively determine other actinides (e.g., plutonium and americium) on the same sample without separation (Esaka et al. 2013).

The three main analytical methods used for characterization of the isotopic composition of single uranium-containing particles—LA-ICPMS, FT-TIMS, and SIMS—were compared (Pointurier et al. 2013). The variability of the measurements of several particles from the Nusimep-7 intercomparison exercise by each method and the error bars for each individual particle measurement were presented. The results obtained by LA-ICP-QMS were inferior to the other two methods, but it was noted that the signal intensity obtained in the LA-ICPMS measurements with a quadrupole-based mass spectrometer was lower by a factor of 10–20 than the SIMS and FT-TIMS, which accounted for the relative uncertainties. In an earlier work by the same group, the isotopic composition of UO_2F_2 particles from the Nusimep-6 intercomparison exercise was analyzed by the three methods. The time needed for analysis of two samples, each with 30 uranium particles, was estimated for SIMS, SEM-LA-ICPMS, FT-TIMS, and FT-LA-ICPMS to be 4, 3.5, 21, and 13 days, respectively (Pointurier et al. 2011), in partial agreement with the estimates given above (Esaka et al. 2009).

The use of fission track (FT) analysis for determining the ^{235}U content in uranium oxide particles was described (Stetzer et al. 2004). Evidently, the amount of ^{235}U nuclides in a particle is the product of the uranium content and the enrichment level, so that in principle a particle containing 10 pg of 10% ^{235}U should yield the same signal (tracks radiating from crater) as a 100 pg particle with 1% enrichment. Thus, if the particle size is known (approximately), then the degree of enrichment can be approximately assessed from the FT analysis alone. Particles with a narrow size distribution were recovered from actual swipe samples, placed on a graphite substrate, heated to ~100°C and covered by a thin layer of graphite to fix them on the substrate. The size distribution was determined by SEM analysis and then a piece of a polycarbonate detector film was placed on the substrate that was inserted into a vial and irradiated by a thermal neutron flux of $1.7 * 10^{12}$ cm^{-1} s^{-1}. Standards of uranium oxide with enrichment between 0.5% and 90% and particle sizes between 0.8 and 1.3 μm were used for validation and calibration and some real swipe samples were also measured. The dark field image of an FTA detector in an optical microscope is shown in Figure 5.13a, where the number near each crater is the number of counted tracks. Figure 5.13b shows the histogram of counted fission tracks from the same sample of uranium oxide particles (10% ^{235}U, 0.8 μm diameter). The conclusion of this study was that FT can be used as an effective screening method for swipe samples and in some cases enrichment can be assessed to some degree. With some reservations, individual particles of highly enriched (HEU) can trigger a preset alarm even when they are present among several LEU particles (Stetzer et al. 2004).

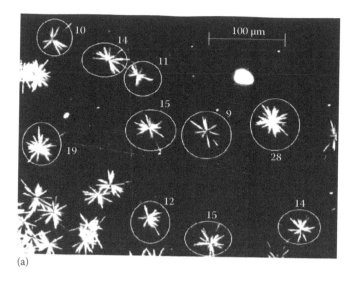

(a)

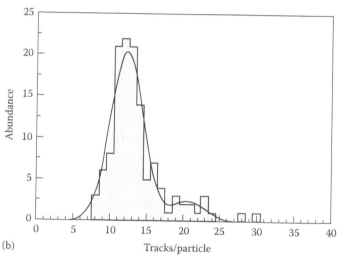

(b)

FIGURE 5.13   (a) The dark field image of an FTA detector in an optical microscope of uranium oxide particles (10% ^{235}U, 0.8 μm diameter) irradiated for 90 s. The number near each crater is the number of counted tracks. (b) The histogram of counted fission tracks from the same sample. (From Stetzer, O. et al., *Nucl. Instrum. Methods Phys. Res. A*, 525, 582, 2004. With permission.)

The ultimate performance of analyzing single particles for nuclear forensics applications is the derivation of information on the feed material, the degree of enrichment, and the process used (Kraiem et al. 2011b). The method used combined the SEM device for screening and derivation of elemental and morphological information, followed by relocation of the relevant particles to a carburized rhenium filament and TIMS multiple ion counters (MIC) isotopic composition analysis. The creation of scatter plots that depict the abundance of ^{234}U and ^{236}U as a function of the degree of ^{235}U enrichment can be used to investigate the possible enrichment process and the nature of the feed material as well as design features of the enrichment plant and operational parameters. Typical trend lines of the enrichment of reprocessed uranium (RU), natural uranium (NU), and depleted uranium (DU) as feed materials, based on measurement of individual particles, were presented in the article. The authors caution that the presence of ^{236}U indicates that RU was used at some time (not necessarily at the present time), and also that multiple recycling or different feed material origins may affect the abundance of ^{234}U and ^{236}U and the derived trend lines. Thus, interpretation of data obtained from analysis of single particles may be quite complex (Kraiem et al. 2011b).

Highlights: From the discussion presented in this section, the conclusion is that the analysis of single uranium-bearing particles can be divided into two stages: first, locating the target particles in the sample and then characterization of their physical, chemical, elemental, and isotopic properties. The highlights concerning these two stages are described later.

5.3.3 LOCATING URANIUM IN SINGLE PARTICLES

The main passive techniques for locating the particles are based on the natural radiation (alpha, beta, and gamma) emitted from the radioactive uranium nuclei and their progeny. Devices that are sensitive to radiation, like old-fashioned photographic films or their modern electronic equivalent (Fuji plate), have the advantage of size (large samples with many particles can be measured simultaneously) but their sensitivity is low so that long exposure times are required.

Therefore, several active methods were developed for locating uranium-containing particles. Scanning electron microscopy (SEM) is one of the most powerful techniques used for locating uranium-containing particles. The sample is placed on a stub that is inserted into the vacuum chamber of the SEM and is scanned by an electron beam. Typical peaks are obtained whenever uranium-bearing particles are encountered so that they can be mapped and their coordinates recorded. Characterization of elemental composition can also be carried out by SEM, but for isotopic measurements the particles must be transferred to a suitable device (usually TIMS or ICPMS).

Fission track (FT) is among the most sensitive active methods and is based on the fact that ^{235}U nuclides may undergo fission when bombarded by neutrons. The sample is placed between two thin films and is irradiated by neutrons (see Figure 5.11). The recoil of the fission products creates small deficiencies in the film that can be developed by alkaline treatment (or HF) to form craters that can be seen by an optical microscope (see Figure 5.13a). The FT method is useful for detection and location of the particles, and may also provide some semiquantitative information (the density of the tracks around the craters) but the particles then need to be transferred (relocated) for further characterization.

Another sensitive method is based on measurement of the secondary ions emanating from the sample when it is bombarded by a primary beam of energetic ions (SIMS). For particle location, only a rapid scan is carried out. A mass spectrometer detects the presence of uranium whenever signals at mass 238 Da are observed. The location of the relevant particles can be mapped (imaged) or characterized by the same device (requiring a longer residence time on each particle). Caution must be practiced not to obliterate the particle during the mapping and imaging stage.

Laser-ablation coupled to ICPMS can also be used for screening a sample and locating the uranium-containing particles. Here, too, a rapid scan is useful for mapping and imaging the distribution of relevant particles in the sample and a slower scan rate is required for characterization. As with SIMS analysis, the laser intensity should not be too strong in order to avoid destruction of the particle before it can be properly characterized.

5.3.4 CHARACTERIZATION OF URANIUM IN SINGLE PARTICLES

Some of the methods for locating uranium-containing particles, mentioned earlier, also provide morphological, elemental, or isotopic information. In principle, the whole sample or stub can be transferred from the screening apparatus to other analytical devices and the coordinates of the located particles can be used for further characterization or alternatively individual particles may be manipulated and placed on a TIMS filament, on a silicon planchet or dissolved for determination of the isotopic composition.

SEM is probably the optimal method for characterizing the particle morphology—shape, size, and porosity. With the addition of energy-dispersive spectroscopy (EDS) or wavelength-dispersive spectroscopy (WDS), the elemental composition of the particle can also be obtained in the SEM. The whole stub can be transferred to a SIMS chamber or LA-ICPMS cell and the coordinates of the particles, obtained from the SEM, can be used for direct measurement of the isotopic composition or, as mentioned earlier, individual particles can be manipulated and transferred.

Particles located by FT analysis can be treated in similar fashion, based on the tracks and craters found on the film they can either be manipulated individually or the whole film transferred to SIMS or LA-ICPMS analysis (Figure 5.11). The combination of FT-TIMS is considered one of the most sensitive (FT) and accurate (multicollector TIMS) techniques for detection and isotopic characterization of uranium particles for nuclear forensic applications.

SIMS, especially large-geometry-SIMS, can provide slightly less accurate results but in a much shorter time than FT-TIMS, as shown earlier (Ranebo et al. 2009).

Methods based on excitation by energetic x-ray or gamma radiation and emission or fluorescence of characteristic x-rays from excited uranium nuclei can also be deployed for screening samples.

5.4 TRACE ANALYSIS AND SAFEGUARDS

One of the main objectives of nuclear forensics is to substantiate compliance with declared activities within licensed nuclear facilities by applying safeguards. Verification requires a twofold approach: ascertaining that the nuclear materials produced in the plant are within the permitted range of enrichment for uranium and confirming that the quantity of the product is within the operational parameters of the facility. Nuclear forensics uses trace analysis techniques to characterize the nuclear materials but in its own is unable to monitor the exact quantities of the product. In addition, the declaration by the facility regarding the time period of given activities can be established or refuted by applying nuclear forensics techniques for age determination.

5.4.1 PRINCIPLES OF AGE DETERMINATION

After uranium is purified to nuclear grade, all natural decay products (see Chapter 1) should be absent. However, the gradual build-up of the decay products continues so that the ratio between a decay product and its parent uranium isotope can serve as a chronometer for *age dating* or *age determination* that is an indicator of the most recent time uranium was purified. Due to the relative long half-lives of the natural uranium isotopes, the amount of progeny nuclides is quite small and very sensitive analytical methods are needed to measure this ratio. In addition, enough time to achieve secular equilibrium must be assumed and an integral sample is needed (closed system), one in which the uranium and progeny nuclides are wholly retained because selective loss (e.g., by leaching or emanation of gases) would distort the calculated age even if the analysis itself is accurate. Cross-contamination must also be avoided for the same reasons. A *beginner's guide* provides details of the possible chronometers and the advantages and demerits of using each natural and common artificial isotope of uranium for age dating (Stanley 2012). Before going into a discussion on the different chronometers, two common features should be noted: as mentioned earlier, sensitive and accurate analytical techniques are required to measure the abundance ratio between the progeny and parent nuclide, and the daughter nuclide is of a different elements (first thorium that is produced by alpha decay from all uranium isotopes and thorium may decay further as seen in Figures 1.1 through 1.4) so efficient separation techniques are required to avoid interferences. As mass spectrometric techniques are commonly used, the main potential interference is from isobaric nuclides but alpha spectrometry is a viable alternative in some cases as shown later. The fundamental assumption for age dating of uranium-containing materials is that during the purification process all progeny radionuclides were removed. This may not always be the case as demonstrated for several samples in which the age was grossly overestimated (Stanley 2012). Age dating relies also on the accuracy of the half-life values of the parent and daughter nuclides so uncertainty in these values could increase the error in the calculated age.

The main relevant decay products of the three natural uranium isotopes (5.2) through (5.4) and the three most common anthropogenic isotopes (5.5) through (5.7) are shown in the following equations:

$$^{238}U \rightarrow {}^{234}Th \rightarrow {}^{234}Pa \rightarrow {}^{234}U \qquad (5.2)$$

$$^{235}U \rightarrow {}^{231}Th \rightarrow {}^{231}Pa \qquad (5.3)$$

$$^{234}U \rightarrow {}^{230}Th \rightarrow {}^{226}Ra \qquad (5.4)$$

$$^{236}U \rightarrow {}^{232}Th \tag{5.5}$$

$$^{233}U \rightarrow {}^{229}Th \tag{5.6}$$

$$^{232}U \rightarrow {}^{228}Th \tag{5.7}$$

In order to understand the methodology of uranium chronometry, we shall follow the example presented in the *beginner's guide* for ^{234}U that is the most useful uranium isotope for age determination (Stanley 2012). The alpha decay of ^{234}U with a half-life of ($\sim 2.46 \pm 0.006) * 10^5$ years leads to the build-up of ^{230}Th that in turns emits another alpha particle with a half-life of ($\sim 7.54 \pm 0.03) * 10^4$ years to form ^{226}Ra (Equation 5.4). Thus, the number of ^{234}U nuclides after time, T, $N_{U\text{-}234}^T$ is

$$N_{U\text{-}234}^T = N_{U\text{-}234}^0 * e^{-\lambda_{U\text{-}234}*T} \tag{5.8}$$

where

$N_{U\text{-}234}^0$ is the number at time zero (T^0) that is the time the uranium was last purified
$\lambda_{U\text{-}234}$ is the activity constant for ^{234}U equal to $\ln(2)/T_{\frac{1}{2}}^{U\text{-}234}$

The fact that ^{234}U nuclides may be produced by decay of ^{238}U (Equation 5.2) is neglected due the relative long half-life of ^{238}U.

The number of ^{230}Th nuclides at given time, T, $N_{Th\text{-}230}^T$ is a more complex expression because some of the nuclides produced are lost through their own alpha emission:

$$N_{Th\text{-}230}^T = \left[\frac{\lambda_{U\text{-}234}}{\lambda_{Th\text{-}230} - \lambda_{U\text{-}234}} \right] * N_{U\text{-}234}^0 * \left(e^{-\lambda_{U\text{-}234}*T} - e^{-\lambda_{Th\text{-}230}*T} \right) \tag{5.9}$$

where

$N_{Th\text{-}230}^0$ at time zero, there should be no ^{230}Th atoms
$\lambda_{Th\text{-}230}$ is the activity constant for ^{230}Th equal to $\ln(2)/T_{\frac{1}{2}}^{Th\text{-}230}$

Figure 5.14 depicts the ratio between the daughter and parent nuclides (calculated from Equations 5.8 and 5.9) as a function of the elapsed time since the last purification of uranium, assuming that at time zero no daughter nuclides were present. Figure 5.14a is for the $^{230}Th/^{234}U$ ratio and Figure 5.14b for the $^{231}Pa/^{235}U$ ratio discussed later. Note the different absolute scales showing that the build-up of ^{230}Th is three orders of magnitude faster due to the half-life ratio of the parent nuclides (^{234}U and ^{235}U).

Once the relative number (the ratio) of ^{234}U and ^{230}Th atoms is known from the measurement, the age of the material (time since last purification) can be determined (Equation 5.10):

$$T = \left[\frac{1}{\lambda_{U\text{-}234} - \lambda_{Th\text{-}230}} \right] * \ln \left\{ 1 - \left(\frac{N_{Th\text{-}230}^T}{N_{U\text{-}234}^T} \right) * \left(\frac{\lambda_{Th\text{-}230} - \lambda_{U\text{-}234}}{\lambda_{U\text{-}234}} \right) \right\} \tag{5.10}$$

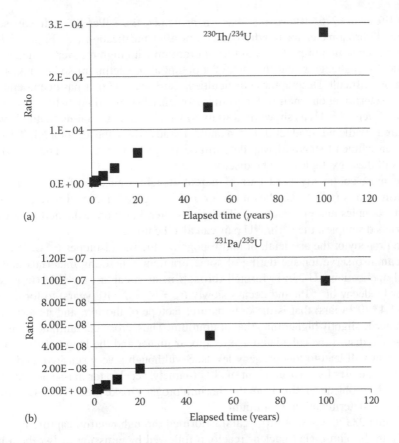

FIGURE 5.14 The daughter–parent ratio as a function of the elapsed time since the last purification of uranium (based on Equations 5.8 and 5.9): (a) $^{230}Th/^{234}U$ and (b) $^{231}Pa/^{235}U$. Note the difference in absolute ratios.

Although in principle this approach can be adopted for age determination for any uranium isotope (Equations 5.2 through 5.7), there are practical limitations arising from interferences, and exceptionally short or long lifetimes of the progeny or parent nuclei (Stanley 2012).

The discussion presented here on age dating with decay of the different uranium isotopes is based mainly on the beginner's guide (Stanley 2012) and this is followed by specific examples of actual investigations.

Uranium-238: The most abundant natural isotope, ^{238}U, has a low chronometric value for two main reasons: the background level of natural ^{238}U is high and interferes with measurement and in addition the $^{234}Th/^{238}U$ reaches a secular equilibrium after about 150 days (due to the short, ~24.1 day half-life of ^{234}Th) so practically no changes in the ratio are expected after that period. The next progeny, ^{234}U, is naturally abundant and greatly complicates the ratio determination to the point of being impractical.

Uranium-235 ($t_{1/2}$ = ~7.04 * 10^8 years) decays by alpha emission to form ^{231}Th that has a short lifetime (~25.5 h) and emits a beta-particle to produce ^{231}Pa that has a half-life of ~$3.276 * 10^4$ years. Thus, in principle, the ^{231}Pa/^{235}U pair can serve as a good chronometer for hundreds of years after purification (see Figure 5.14b). However, separation of protactinium from uranium with high recovery efficiency is analytically challenging and the lack of a potential protactinium spike makes this even more difficult. Despite these difficulties, the ^{231}Pa/^{235}U pair has been deployed for age determination, mainly for samples of enriched uranium (where there are elevated levels of ^{235}U) as shown in a study in which three age-dating methods were compared (Wallenius et al. 2002) or in another study where the age of a NIST U-100 (~10% enrichment) standard was determined (Eppich et al. 2013). The analytical aspects of these examples will be discussed later.

The most commonly used pair of parent–daughter nuclides for uranium age determination is ^{230}Th/^{234}U (shown earlier and in Figure 5.14a) and several experimental examples and procedures will be given later. For more extended periods or for enriched samples, the ^{214}Bi/^{234}U pair can also be used.

The progeny of the artificial, or anthropogenic, uranium isotopes (^{236}U, ^{233}U and ^{232}U) can also serve for age dating of some uranium-containing materials and for special applications. U-236 is formed through neutron capture by ^{235}U (in reactor fuel) or by decay of ^{240}Pu and decays slowly ($t_{1/2} = $ ~$2.342 * 10^7$ years) to form ^{232}Th ($t_{1/2} = $ ~$1.4 * 10^{10}$ years) that is the only natural isotope of thorium and its terrestrial abundance is slightly higher than that of uranium. Thus, cross-contamination of thorium would distort the calculated age to appear higher and the long half-life of the ^{236}U parent will lead to low progeny levels. So although a very sensitive and accurate method exists for measurement of ^{236}U (namely, accelerator mass spectrometry (AMS)), the problems arising from the thorium progeny make the ^{232}Th/^{236}U impractical for age determination of uranium.

Uranium-233 ($t_{1/2} = $ ~$1.59 * 10^5$ years) is formed through neutron capture by ^{232}Th (mainly in thorium-fueled nuclear reactors) followed by emission of two beta particles. Alternatively, ^{233}U may be formed by alpha decay of ^{237}Np that in turn is produced through a sequence of nuclear reactions in uranium-fueled reactors. The ^{229}Th decay product of ^{233}U is the relatively short-lived ($t_{1/2} = $ ~7880 years) so the ratio between these nuclides is linear for thousands of years after purifications. Alpha spectrometry and mass spectrometry can be used for measuring the ^{229}Th/^{233}U ratio. In special cases, this pair of nuclides can serve as a chronometer for nuclear proliferation, particularly if fissile ^{233}U is produced in reactors fueled with thorium.

Finally, ^{232}U that is produced by a sequence of nuclear reactions when thorium is irradiated or by decay of ^{236}U (and two beta emissions) has a very short half-life ($t_{1/2} = $ ~68.9 years) and its alpha decay product is ^{228}Th ($t_{1/2} = $ ~1.913 years). The short lifetimes of the parent and progeny imply that alpha spectrometry is very sensitive for their determination; however, this also limits the applicability for age dating to a few years.

5.4.2 ANALYTICAL PROCEDURES USED FOR URANIUM-BASED AGE DATING

Due to its importance in safeguards, several studies developed analytical procedures for age dating of uranium-containing materials for nuclear forensics applications. As mentioned earlier, the ^{230}Th/^{234}U pair is the most useful followed by the ^{231}Pa/^{235}U pair.

In one study, three pairs of radionuclides—^{230}Th/^{234}U, ^{231}Pa/^{235}U, and ^{232}Th/^{236}U— were used to determine the age of U_3O_8-enriched uranium standard reference materials (SRMs) with 50%, 80%, and 85% ^{235}U (Wallenius et al. 2002). Sample treatment started with dissolution of the uranium oxide samples in nitric acid. For multicollector ICPMS measurements, no further separation was required, but for isotope dilution-alpha spectrometry (ID-AS) or ID-TIMS the thorium and uranium had to be separated, so after dissolution a thorium isotopic spike was added (^{228}Th for alpha spectrometry or ^{232}Th for TIMS). Separation was carried out on a chromatographic column with TEVA resin according to the schematic diagram shown in Figure 5.15. The spiked dissolved sample was loaded on the TEVA column and the uranium was eluted with 2 M HNO_3. This was followed by elution of the thorium with 6 M HCl. The solution was evaporated to dryness and redissolved in nitric acid for isotope dilution-TIMS analysis. A second TEVA column was used for further purifying the thorium and after washing with 2 M HNO_3 the sample was loaded and the thorium eluted with 0.02 M HNO_3/HF and deposited on a disk for alpha spectrometry. For the ^{230}Th/^{234}U pair, the assigned age for the ID-TIMS and ID-alpha spectrometry measurements was within about −2% of the assumed age and with good agreement between these two methods, while for the MC-ICPMS analysis a larger bias ranging from −2% to −9% was found. The authors note that the MC-ICPMS results improved

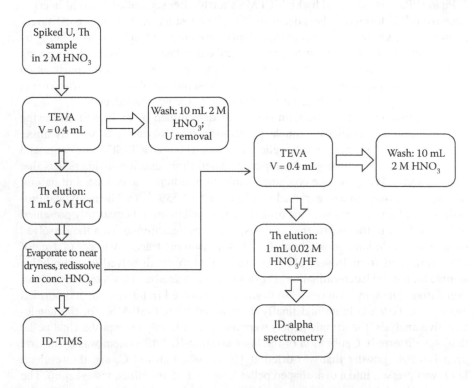

FIGURE 5.15 A schematic diagram of the experimental procedure for separation uranium and thorium for age determination. (Based on Wallenius, M. et al., *Anal. Bioanal. Chem.*, 374, 379, 2002.)

after correcting for the difference in ionization efficiency of thorium and uranium. The age could not be determined by MC-ICPMS for the ^{231}Pa/^{235}U pair in these samples due to low counting statistics and this was not even attempted by the other two methods (Wallenius et al. 2002).

A study that focused on the ^{231}Pa/^{235}U pair demonstrated that it is suitable for age determination of ^{235}U-enriched samples as well as for uranium-rich materials (Eppich et al. 2013). A MC-ICPMS instrument was used to determine the *age* or *date* (terms used by the authors) of a suite of CRMs varying from depleted to highly enriched uranium. Isotope dilution of protactinium was carried out with a home-made ^{233}Pa spike (t$_{1/2}$ = 26.967 days) that was freshly prepared from ^{237}Np (t$_{1/2}$ = ~2.14 * 10^6 years). The separation procedures of the protactinium spike from the neptunium parent and of the protactinium from the uranium matrix were described (Eppich et al. 2013). A ^{233}Pa spike was added to the sample and was equilibrated in a sealed Teflon vial on a hot plate for several hours. A drop of concentrated $HClO_4$ was added and the mixture was dried and then redissolved in 9 M HCl with a little HNO_3 and H_3BO_3. Samples were loaded on a column with AG-1 X8 anion exchange resin and washing with 9 M HCl removed the matrix elements while the protactinium was retained on the column and was eventually eluted with 9 M HCl + 0.05 M HF. After some further steps, the protactinium was concentrated on quartz wool and separated. The ratio of ^{231}Pa to ^{233}Pa was measured by MC-ICPMS shortly after separation to avoid interference from ^{233}U formed by beta decay of ^{233}Pa. The results were in quite good agreement with the known dates that the standards were prepared and also with dating by the ^{230}Th/^{234}U pair, but a consistent bias toward older dates was observed, indicating that some protactinium and thorium progenies were retained after the purification process. Once again this demonstrated that the assumption of *perfect purification* is not necessarily valid, and thus, a bias toward older age was introduced.

The production date of uranium oxide was determined by an ICP-SF-MS using the conventional destructive sample preparation by dissolution as well as direct measurements of uranium oxide pellets by laser ablation-ICP-SF-MS (Varga and Suranyi 2007). The investigated samples included three uranium-oxide pellets that were seized in Hungary: one contained depleted uranium, one consisted of natural uranium, and one had low-enriched uranium with ~2.55% ^{235}U. The analytical procedures used for separation of the uranium and thorium are schematically presented in Figure 5.16. In the destructive analysis, ^{234}U was determined from the dissolved sample after dilution and addition of ^{233}U as an isotopic tracer. A tracer of 1 pg of ^{229}Th (prepared from 18-year-old ^{233}U) was added to the dissolved uranium oxide sample for the ^{230}Th determination. Thorium was separated on TEVA resin by extraction chromatography, evaporated to dryness, redissolved in nitric acid, and purified again on a TEVA column and finally determined by ICP-SF-MS. For the nondestructive analysis, the sample pellets were analyzed directly (except the Unat pellet that was distorted). Calibration of the laser ablation-ICPMS system was by natural uranyl acetate powder that was dried at 110°C, ashed at 400°C, and the resultant U_3O_8 was pressed into a disk-shaped pellet (diameter ~5 mm thickness ~1 mm). The chromatogram of a laboratory prepared uranium oxide sample showing the intensity of the ^{230}Th and ^{234}U signals as a function of time, recorded at medium resolution (R = 4000) of the ICP-SF-MS, is shown in Figure 5.17 (Varga and Suranyi 2007).

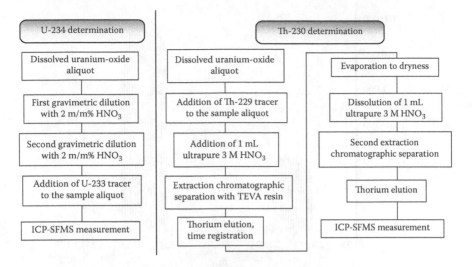

FIGURE 5.16 A schematic presentation of the chemical procedures used for separation of the uranium and thorium for ICP-SF-MS analysis. (From Varga, Z. and Suranyi, G., *Anal. Chim. Acta*, 599, 16, 2007. With permission.)

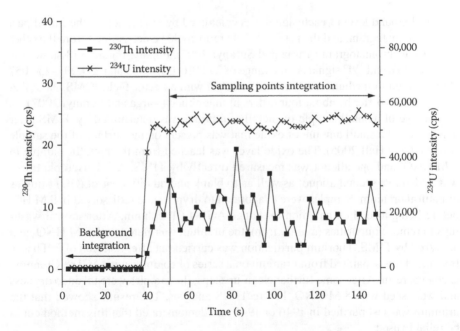

FIGURE 5.17 The chromatogram of a laboratory prepared uranium oxide sample showing the intensity of the ^{230}Th and ^{234}U signals as a function of time, recorded at medium resolution (R = 4000) of the ICP-SF-MS. (From Varga, Z. and Suranyi, G., *Anal. Chim. Acta*, 599, 16, 2007. With permission.)

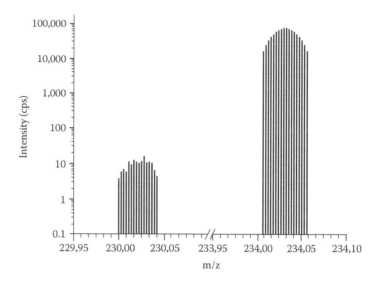

FIGURE 5.18 The average ^{230}Th and ^{234}U signals in the range of 230.016–230.0433 and 234.007–234.057 Da measured in medium resolution (R = 4000) with a Sector Field-ICMS. (From Varga, Z. and Suranyi, G., *Anal. Chim. Acta*, 599, 16, 2007. With permission.)

The background level of each signal was calculated by integration of the initial part of the chromatogram, and the net ^{230}Th/^{234}U ratio could be determined from the other part of the chromatogram (Varga and Suranyi 2007). Finally, Figure 5.18 shows the average ^{230}Th and ^{234}U signals in the range of 230.016–230.0433 and 234.007–234.057 Da measured in medium resolution (R = 4000) with a Sector Field-ICMS. Note that the intensities differ by about four orders of magnitude (Varga and Suranyi 2007).

The age of a small sample of metallic uranium was determined by TIMS after dissolution of a small amount of oxide that was present on the surface of the sample (LaMont and Hall 2005). The oxide layer was leached from the metallic sample in 8 M HNO$_3$ and one aliquot was measured directly by TIMS. A ^{233}U isotopic tracer was added to a second aliquot as well as to blank and quality control (QC) samples of natural uranium. Samples were evaporated to dryness and redissolved in 9 M HCl before being loaded on an anion-exchange AG1x4 Cl$^-$ column. After several washings to remove impurities (mainly iron), the uranium was eluted with 8 M HNO$_3$ and analyzed by TIMS. Thorium purification was carried out after addition of ^{229}Th as a tracer and was separated from uranium on a series of columns with AG1x8 Cl$^-$ anion exchange resin. After some further clean-up steps, the samples were taken to dryness and wet-ashed with 8 M HNO$_3$ before TIMS analysis. The results showed that the uranium was last purified in 1946 or 1947 and demonstrated that this method could be reliably used.

The isotope dilution-ICPMS method was used for age determination, based on the ^{230}Th/^{234}U ratio, of dissolved uranium compounds without radiochemical separation of thorium and uranium (Varga et al. 2010b). Spectral mathematical deconvolution was used to overcome the problem of *abundance sensitivity* (the tailing effect due to contribution of a very large peak to the baseline level of neighboring peaks) arising from the ^{238}U peak that was 10 orders of magnitude higher than the ^{230}Th peak. This was carried out by using the medium resolution (R = 4000) feature mentioned earlier of the sector field-ICPMS and made it possible to rapidly measure the ^{230}Th/^{234}U ratio directly in the samples for age dating purposes. The difference between the ^{230}Th peak shape in the low-resolution (R = 300) and medium-resolution setting (R = 4000) was clearly demonstrated in the study of two certified reference materials (CRM U010 and U030 with ~1.01% and ~3.14% ^{235}U, respectively) with known production dates. The samples were dissolved in 8 M HNO$_3$ and diluted to 100 µg g^{-1}. For isotope dilution, 200 pg g^{-1} of ^{229}Th was added as a tracer and ^{232}Th (10 ng g^{-1}) was introduced as a carrier. The content of ^{234}U was measured in the diluted samples at the low-resolution setting. The ^{230}Th was first measured with the low-resolution setting and after smoothing and baseline subtraction and correcting for tailing a flat-top peak was obtained. The authors used a different deconvolution procedure for the medium-resolution spectrum by superimposing a Gaussian function. The results for the age dating of the two CRMs determined by the low- and medium-resolution measurements, with and without the tracers, were in good agreement with the known production dates (Varga et al. 2010b).

Highlights: From the nuclear forensics point of view, the materials of interest are only those that were produced since the 1940s, that is, those of anthropogenic origin. Although similar age dating methods can serve in archeological and geological research, they are not relevant for nuclear forensics. Age dating of short-term chronometers provides a powerful tool for nuclear forensics and safeguards. However, this tool should be wielded judiciously and with care because several complications may lead to erroneous results. The following underlying assumptions must be verified and whenever possible additional measures should be taken before drawing conclusions:

1. The last uranium purification left no traces of radioactive progeny. Studies have demonstrated that this is not always true.
2. The sample was contained in a *closed system* to prevent selective loss of parent or progeny nuclides. This is not a problem for CRMs, SRMs, or seized materials, but for environmental samples (including swipe samples) this could be a major concern.
3. Sample preparation must be carried out meticulously to avoid cross-contamination.
4. The analytical procedure (radiochemical separation and measurement) must be robust and accurate. If possible, orthogonal analytical devices (e.g., mass spectrometry and alpha spectrometry) should be used.
5. If possible, age dating should be validated by at least two pairs of radionuclides.
6. Isotopic tracers to determine recovery should be used whenever available to ascertain correction for unequal recovery of parent and daughter nuclides.
7. Age dating should be performed on several subsamples to ensure representative results for inhomogeneous samples.

5.5 NUCLEAR FORENSICS

5.5.1 Databases for Nuclear Forensic Investigations

The characterization of seized illegal nuclear materials according to chemical, physical, isotopic, and nuclear attributes is very important, but without a reference database (library) is little more than an analytical exercise. However, if the characteristic features of the contraband sample can be compared to a database, then the origin of the material may be determined, the culprits identified, and properly prosecuted. Such databases exist but are not routinely shared among all interested parties although the owners of these *libraries* may provide the relevant information on a case-to-case basis. The need for such a nuclear forensics data bank was clearly stated in view of the increasing risk in the twenty-first century from terrorists, shadowy nonstate actors and nontransparent regimes and the fact that special nuclear materials are available in more states than previously and on the *black market* (Garvey 2010).

The existence of some of these databases is acknowledged publicly. For example, as mentioned earlier, LLNL has a database that includes 1800 samples of *yellow cake* (Kristo and Turnet 2013), and the Nuclear Forensics Analysis Center (NFAC) in Savannah River National Laboratory (SRNL) provides support for the FBI's Radiological Evidence Examination Facility (REEF) (Nichols 2011). The latter contains a database of spent nuclear fuel from several reactors in the United States and other countries. An example of the processing of interdicted nuclear material at REEF uses traditional forensics combined with nuclear forensics to determine the origin and make attribution. The results of the isotopic measurements are compared to known compositions in the database based on reactor physics models (see flowchart in Figure 5.19).

The IAEA has a database on illicit trafficking of nuclear materials and incidents involving these materials (IAEA 2013). This database includes statistics on seized materials and is available to all member states but does not contain analytical procedures.

5.5.2 Case Studies

The scientific and technical nuclear forensics literature contains several studies of characterizing seized nuclear materials (e.g., Moody et al. 2005; Wallenius et al. 2006; IAEA 2013). A detailed description of the characterization of two types of samples (uranium nuclear fuel pellets and uranium oxide powder) that were seized in the mid-1990s can serve as an example of the application of the methodology developed at the ITU (Wallenius et al. 2006). The seized nuclear materials were characterized according to the parameters shown in Table 5.2. A nondestructive gamma spectroscopy test of the powder was carried out and ^{235}U enrichment level of 89.59% ± 0.43% was found. Then samples were dissolved and the isotopic composition was determined by TIMS and MC-ICPMS. The values by both methods for ^{234}U, ^{235}U, ^{236}U, and ^{238}U were around 1.077%, 87.78%, 0.21%, and 10.93%, respectively. SIMS was used to directly determine the isotopic composition of

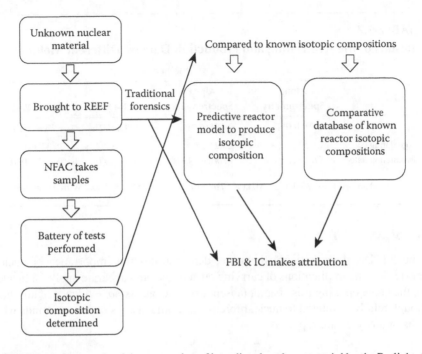

FIGURE 5.19 Example of the processing of interdicted nuclear material by the Radiological Evidence Examination Facility (REEF) operated by the FBI at Savannah River National Laboratory (SRNL). Traditional forensics is combined with nuclear forensics to determine the origin and make attribution. (Adapted from Nichols, T.F., Nuclear Forensic Analysis Center: Forensic analysis to data interpretation, DA-AC09-08SR22470, Savannah River National Laboratory US Department of Energy, Savannah River, NC, 2011.)

smaller and larger particles and no significant variations of the composition were observed. The uranium content was determined by potentiometric titration and isotope dilution mass spectrometry (ISMS) and the range of the uranium content was from 82.26% to 86.09% depending to some extent on the particle size (probably due to the oxide stoichiometry and humidity). The impurity content of 15 elements was measured by sector field-ICPMS with rhodium as an internal standard. The results varied from 1.25 µg g^{-1} for manganese to 69.5 µg g^{-1} for iron. The determination of the age of the material by four methods (alpha spectrometry for the $^{234}U/^{230}Th$ and $^{235}U/^{231}Pa$ pairs and ICPMS with a quadrupole mass spectrometer and multicollector-ICPMS for the $^{234}U/^{230}Th$ pair) was compared as shown in Table 5.7 (Wallenius et al. 2006). The agreement between three of the results was good, while the ICP-QMS deviated by about 10%. Finally, a sample was fixed on an aluminum stub for SEM/EDX measurements and the morphology of the larger particles (0.1–1 mm in diameter) of the powder was found to be an agglomerate of very fine particles (~0.25 µg). The conclusion drawn from all the data acquired by the nuclear forensics examination was that the origin of the seized materials (pellets and powder) was consistent with the fuel made in 1975 for a fast-breeder reactor in Russia (Wallenius et al. 2006).

TABLE 5.7
Results of the Age Dating and Production Date by Different Methods

	Method			
	Alpha Spectrometry	Alpha Spectrometry	ICP-QMS	MC-ICPMS
	$^{234}U/^{230}Th$	$^{235}U/^{231}Pa$	$^{234}U/^{230}Th$	$^{234}U/^{230}Th$
Age dating	26.52 ± 0.7	25.21 ± 0.8	29.9 ± 5.0	25.32 ± 3.5
Production date	October 1976	February 1978	June 1973	January 1978

Source: Adapted from Wallenius, M. et al., *Forensic Sci. Int.*, 156, 55, 2006. With permission.

5.6   SUMMARY

Frame 5.1, based on fictional literature, demonstrates the importance of nuclear forensics and the implications of carrying out a timely and accurate analysis. In addition, the quote opening this chapter (absence of evidence is *not* evidence of absence), although totally unrelated to nuclear forensics, should always be on our minds when results of analyses are reported.

FRAME 5.1   NUCLEAR FORENSICS TO THE RESCUE

"Was this the face that launch'd a thousand ships", Christopher Marlowe, in Doctor Faustus

"Will this be the nuclear forensics analysis that will launch a thousand missiles", ZK, 2014

The most impressive application of nuclear forensics to date, fortunately only in fictional literature, is in the Tom Clancy novel *The Sum of All Fears* (Berkley Books, New York, 1991). The climax of the novel (and this is a real spoiler) is when an improvised nuclear device (IND) is exploded by terrorists at a football stadium in Denver in the hope of provoking a nuclear war between the United States and Russia. Fortunately the device fizzles out and fails to perform according to plan. A Rocky Flats technician carries out gamma spectrometry measurements of a sample from the debris and detects the presence of tritium, plutonium, neptunium, americium, gadolinium, curium, promethium, and uranium. The dataset (the book) shows that the plutonium 239/240 mix originated from Savannah River where they *had that gadolinium problem* and the source of the material was tracked to the K Reactor, dated February 1968. The fictional hero, Dr. Jack Ryan, manages to convince the President that the nuclear material was not of Russian origin and a full-scale nuclear war is averted...

In this fictional case, nuclear forensics came to the rescue and prevented the launching of a thousand missiles, but this story serves as a demonstration of the heavy responsibility that rests on the shoulders of the scientists in charge of the analysis.

The best estimate of the extent of the problem of illicit trafficking of nuclear and radioactive materials can be obtained from IAEA periodic reports on the incident and trafficking database (ITDB) of nuclear and other radioactive materials out of regulatory control (IAEA 2013). The report includes a summary of the confirmed incidents involving unauthorized possession and related criminal activities. Between 1993 and 2012, about 400 such incidents were reported (of course, the number of actual, unreported cases, is unknown) and in 16 of those unauthorized possession of highly enriched uranium or plutonium was confirmed. In most cases, gram quantities were seized and it is suspected that these may have been samples from larger stockpiles. Another compilation summarized the confirmed incidents involving theft or loss and shows a sharp increase from less than 10 incidents in the early 1990s to a record high of ~130 incidents in 2006 followed by a sharp decrease to about 20 incidents per year in 2012. The number of confirmed incidents involving other unauthorized activities and events increased from 10 to 15 between 1993 and 1998 to a record high of ~150 in 2007 and gradually decreasing to ~80 in 2012. It should be noted that as yet there were no incidents on public record in which the quantity and quality of the seized nuclear material were sufficient for producing an improvised nuclear device (IND) or nuclear weapon.

Application of nuclear forensics provides a powerful tool for safeguards purposes and attribution of illicit activities involving nuclear materials. A generic flowchart of the processes involved in nuclear forensics analysis is shown in Figure 5.20.

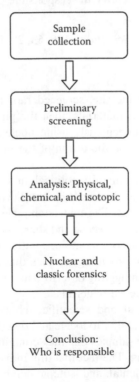

FIGURE 5.20   A generic flowchart of the processes involved in nuclear forensics analysis.

FRAME 5.2   NUCLEAR FORENSICS: FALSE
HOPES AND PRACTICAL REALITIES

This article cautions that blindly relying on nuclear forensics as a means of attributing acts of nuclear terrorism should be avoided, yet assigns a limited role to the technology that may serve as a deterrent. "Nuclear forensics can contribute somewhat to reducing the risks of nuclear proliferation and nuclear terrorism, but its limitations need to be fully understood." These limitations are technical as well as political. The author states that the fact that data concerning the composition of nuclear materials provided by participating countries could be used against them is especially disturbing, and would prohibit full disclosure. In addition, the data may contain information that would contribute to proliferation that is counterproductive. Furthermore, he raises an *authenticity problem* that intentionally falsified data may be given in order to remove suspicion from the culprit and point at an adversary. Nuclear forensics "is unlikely to reduce the range of possibilities to a single or even a few states with sufficient certainty as to make evident the deliberate responsibility of their governments for the incident."

It remains an open question whether governments that were involved in the past in illicit nuclear activities would cooperate with an international regime that's sole purpose is to identify the responsible parties and perhaps punish them.

Source: Weitz, R., *Polit. Sci. Quat.*, 126, 53, 2011.

Samples are collected from the site, screened rapidly to determine whether they contain radioactive or nuclear materials, then the physical, chemical, and isotopic features are characterized meticulously using nuclear and classical forensics and finally conclusions are drawn on the origin of the material and for identifying the responsible parties.

However, the limitations—mainly political and not scientific or technical—should also be kept in mind, as discussed in detail (Weitz 2011). Some of the points raised are quoted in Frame 5.2. The understandable reluctance of countries with advanced nuclear technologies to expose and share the details of the composition of these materials with other countries, combined with the mutual suspicions that some governments may have about providing samples that may later be used to incriminate them in acts of nuclear smuggling or terrorism, impose severe limitations on the usefulness of databases of nuclear materials.

Nevertheless, the technical and scientific achievements of nuclear forensics analysis are impressive. The ability to locate and detect minute amounts of uranium in a single particle in a swipe samples that may contain copious amounts of dust and soil particles is quite amazing. The fact that this individual particle may be singled out and its morphology, elemental, and isotopic compositions determined is indicative of the progress of analytical techniques.

REFERENCES

Anderson, P. (2004). Absence of evidence is not evidence of absence: We need to report uncertain results and do it clearly. *Br. J. Med.* 328, 476–477.

Aregbe, Y., Truytens, J., Kips, R. et al. (2008). NUSIMEP, P-6: Uranium isotope amount ratios in uranium particles. Geel, Belgium: Interlaboratory Exercise, Joint Research Commission.

Bellucci, J.J., Simonetti, A., Wallace, C. et al. (2013). Isotopic fingerprinting of the world's first nuclear device using post-detonation materials, *Anal. Chem.* 85, 4195–4198.

Bentridi, S.E., Gall, B., Gauthier-Lafaye, F. et al. (2011). Inception and evolution of Okla natural reactors, *Comptes Rendus Geosci.* 343, 738–748.

Betti, M. (2005). Review: Isotope ratio measurements by secondary ion mass spectrometry (SIMS) and glow discharge mass spectrometry (GDMS), *Int. J. Mass Spectrom.* 242, 169–182.

Betti, M., Tamborini, G., and Koch, L. (1999). Use of secondary ion mass spectrometry in nuclear forensics analysis for the characterization of plutonium and highly enriched uranium particles, *Anal. Chem.* 71, 2616–2622.

Brennecka, G.A., Borg, L.E., Hutcheon, I.D. et al. (2010). Natural variations in uranium isotope ratios of uranium ore concentrates: Understanding the ^{238}U–^{235}U fractionation mechanism, *Earth Planet. Sci. Lett.* 291, 228–233.

Burger, S., Riciputi, L.R., Bostick, D.A. et al. (2009). Isotope ratio analysis of actinides, fission products, and geolocators by high-efficiency multi-collector thermal ionization mass spectrometry, *Int. J. Mass Spectrom.* 286, 70–82.

Donohue, D.L. (2002). Peer reviewed: Strengthened nuclear safeguards, *Anal. Chem.* 74, 28A–35A.

Donohue D.L. and Zeisler, R. (1993). Analytical chemistry in the aftermath of the Gulf war, *Anal. Chem.* 75, 359A–368A.

Eppich, G.R., Williams, R.W., Gaffiney, A.M. et al. (2013). ^{235}U–^{231}Pa age dating of uranium materials for nuclear forensic investigations, *J. Anal. At. Spectrom.* 28, 666–674.

Esaka, F., Magara, M., and Kimura, T. (2013). The use of solution nebulization ICP-MS combined with particle screening and micro-sampling for analysis of individual uranium-bearing particles, *J. Anal. At. Spectrom.* 28, 682–688.

Esaka, F., Magara, M., Lee, C.G. et al. (2009). Comparison of ICP-MS and SIMS techniques for determining uranium isotope ratios in individual particles, *Talanta* 78, 290–294.

Fahey, A.J., Zeissler, C.J., Newbury, D.E. et al. (2010). Postdetonation nuclear debris for attribution, *PNAS* 107, 20207–20212.

Faure, A.L., Rodriguez, C., Marie, O. et al. (2014). Detection of traces of fluorine in micrometer sized uranium bearing particles using SIMS, *J. Anal. At. Spectrom.* 29, 145–151.

Garvey, J.L. (2010). Nuclear containment for the twenty-first century: A mandatory international nuclear forensics data bank, *J. Conflict Security Law* 15, 301–346.

Glaser, A. and Bielefeld, T. (2008). Nuclear forensics: Capabilities, limits and the "CSI effect." http://www.princeton.edu/~aglaser/talk2008_forensics.pdf (accessed November 27, 2013).

Greene, P.A., Copper, C.L., Berv, D.E. et al. (2005). Colorometric detection of uranium(VI) on building surfaces after enrichment by solid phase extraction, *Talanta* 66, 961–965.

Grigorev, G.Y., Lebedeva, A.S., Malyugin, S.T. et al. (2008). Investigation of ^{235}UF$_6$ and ^{238}UF$_6$ spectra in mid-IR range, *Atom. Energy* 104, 398–403.

Han, S.H., Varga, Z., Krajko, J. et al. (2013). Measurement of the sulphur isotope ratio 34S/32S in uranium ore concentrates (yellow cakes) for origin assessment, *J. Anal. At. Spectrom.* 28, 1919–1925.

Hastings, E.P., Lewis, C., FitzPatrick, J. et al. (2008). Characterization of depleted uranium oxides fabricated using different processing methods, *J. Radioanal. Nucl. Chem.* 276, 475–481.

Hiess, J., Condon, D.J., McLean, N. et al. (2012). $^{238}U/^{235}U$ systematics in terrestrial uranium-bearing minerals, *Science* 335, 1610–1614.

Hinrichsen, Y. (2010). *Fingerprinting of Nuclear Material for Nuclear Forensics.* Hamburg, Germany: University of Hamburg.

Hocking, H.E., Burggraf, L.W., Duan, X.F. et al. (2012). Composition of uranium oxide particles related to TOF-SIMS ion distribution, *Surf. Interface Anal.* 45, 545–548.

IAEA. (2002). Advances in destructive and non-destructive analysis for environmental monitoring and nuclear forensics. Vienna, Austria: IAEA.

IAEA. (2006). Nuclear forensics support: Technical guidance reference manual, IAEA nuclear security series 2. Vienna, Austria: IAEA.

IAEA. (2013). IAEA incident and trafficking database (ITDB) (Incidents of nuclear and other radioactive materials out of regulatory control). 2013 Fact sheet. Vienna, Austria: IAEA. http://www-ns.iaea.org/downloads/security/itdb-fact-sheet.pdf (accessed December 4, 2013).

IAEA. (2014). Application of nuclear forensics in combating illicit trafficking of nuclear and other radiological material, *IAEA TECDOC Series 1730*, Vienna, 2014.

Inn, K.G.W., Johnson, C.M., Oldham, W. et al. (2013). The urgent requirement for new radiological certified reference materials for nuclear safeguards, forensics and consequence management, *J. Radioanal. Nucl. Chem.* 296, 5–22.

Jakopic, R., Sturm, M., Kraiem, M. et al. (2012). Certified reference materials and reference methods for nuclear safeguards and security. *INSINUME Sixth International Symposium on In Situ Nuclear Metrology as a Tool for Radioecology.* Brussels, Belgium: Joint Research Centre, Presentation.

Kabir, A., Holness, H., Furton, K. et al. (2013). Recent advance in micro-sample preparation with forensic applications, *Trends Anal. Chem.* 45, 464–479.

Kappel, S., Boulyga, S.F., Dorta, L. et al. (2013). Evaluation strategies for isotope ratio measurements of single particles by LA-MC-ICPMS, *Anal. Bioanal. Chem.* 405, 2943–2955.

Keegan, E., Wallenius, M., Mayer, K. et al. (2012). Attribution of uranium ore concentrates using elemental and anionic data, *Appl. Geochem.* 27, 1600–1609.

Kemp, R.S. (2008). Initial analysis of the detectability of UO_2F_2 aerosols produced by UF_6 released from uranium conversion facilities, *Sci. Global Security* 15, 115–125.

Kennedy, A.K., Bostick, D.A., Hexel, C.R. et al. (2012). Non-volatile organic analysis of uranium ore concentrates, *J. Radioanal. Nucl. Chem.* 296, 817–821.

Kim, Y.S., Han, B.Y., Shin, H.S. et al. (2012). Determination of uranium concentration in an ore sample using laser-induced breakdown spectroscopy, *Spectrochim. Acta B* 74–75, 190–193.

Kips, R., Kristo, M.J., Hutcheon, I.D. et al. (2009b). Determination of the relative amounts of fluorine in uranium oxyfluoride particles using secondary ion mass spectrometry and optical spectroscopy. *INMM Annual Meeting.* Tucson, AZ: LLNL.

Kips, R., Leenaers, A., Tamborini, G. et al. (2007). Characterization of uranium particles produced by hydrolysis of UF6 using SEM and SIMS, *Microsc. Microanal.* 13, 156–164.

Kips, R., Pidduck, A.J., Houlton, M.R. et al. (2009a). Determination of fluorine in uranium oxyfluoride particles as an indicator of particle age, *Spectromchim. Acta B* 64, 199–207.

Kraiem, M., Richter, S., Kuhn, H. et al. (2011a). Development of an improved method to perform single particle analysis by TIMS for nuclear safeguards, *Anal. Chim. Acta* 688, 1–7.

Kraiem, M., Richter, S., Kuhn, H. et al. (2011b). Investigation of uranium isotopic signatures in real-life particles from a nuclear facility by thermal ionization mass spectrometry, *Anal. Chem.* 83, 3011–3016.

Krey, P.W. and Nicholson, K.W. (2001). Atmospheric sampling and analysis for detection of nuclear proliferation, *J. Radioanal. Nucl. Chem.* 248, 605–610.

Kristo, M.J. and Turnet, S.J. (2013). The state of nuclear forensics, *Nucl. Instrum. Methods Phys. Res. B* 294, 656–661.

LaMont, S.P. and Hall, G. (2005). Uranium age determination by measuring the ^{230}Th/^{234}U ratio, *J. Radioanal. Nucl. Chem.* 264, 423–427.

Lloyd, N.S., Parrish, R.R., Horstwood, M.S.A. et al. (2009). Precise and accurate isotopic analysis microscopic uranium-oxide grains using LA-MC-ICP-MS, *J. Anal. At. Spectrom.* 24, 752–758.

Marin, R.C., Sarkis, J.E.S., and Nascimento, M.R.L. (2013). The use of LA-SF-ICP-MS for nuclear forensics purposes: Uranium isotope ratio analysis, *J. Radioanal. Nucl. Chem.* 295, 99–104.

Mathew, K.J., O'Connor, G., Hasozbek, A. et al. (2013). Total evaporation method for uranium isotope-amount ratio measurements, *J. Anal. At. Spectrom.* 28, 866–876.

Mayer, K. (2013). Expand nuclear forensics, *Nature* 503, 28, 461–462.

Mayer, K., Wallenius, M., and Fanhanel, T. (2007). Nuclear forensic analysis—From cradle to maturity, *J. Alloys Compd.* 444–445, 50–56.

Mayer, K., Wallenius, M., and Ray, I. (2005). Nuclear forensics—A methodology providing clues on the origin of illicitly trafficked nuclear materials, *Analyst*, 130, 433–441.

Mayer, K., Wallenius, M., and Varga, Z. (2013). Nuclear forensic science: Correlating measurable material parameters to the history of nuclear material, *Chem. Rev.* 113, 884–900.

Mayer, K. , Wallenius, M., Lutzenkirchen, K. et al. (2011). Nuclear forensics: A methodology applicable to nuclear security and to non-proliferation, *J. Phys.: Conf. Ser.* 312, 062003 (1–9).

Mercadier, J., Cuney, M., Lach, P. et al. (2011). Origin of uranium deposits revealed by their rare earth element signature, *Terra Nova* 23, 264–269.

Moody, K., Grant, P., and Hatcheon, I. (2005). *Nuclear Forensic Analysis.* Boca Raton, FL: Taylor & Francis.

Nichols, T.F. (2011). Nuclear Forensic Analysis Center: Forensic analysis to data interpretation, DA-AC09-08SR22470. Savannah River, NC: Savannah River National Laboratory US Department of Energy.

Pajo, L., Mayer, K., and Koch, L. (2001). Investigation of the oxygen isotopic composition in oxidic uranium compounds as a new property in nuclear forensics, *Fresenius J. Anal. Chem.* 371, 348–352.

Pestana, R.C.B., Sarkis, J.E.S., Marin, R.C. et al. (2013). New methodology for uranium analysis in swipe samples for nuclear safeguards purposes, *J. Radioanal. Nucl. Chem.* 298, 621–625.

Piksaikin, V.M., Pshakin, G.M., Roshchenko, V.A. (2006). Review of methods and instruments for determining undeclared nuclear materials and activities, *Sci. Global Security* 14, 49–72.

Plaue, J.W., Klunder, J.W., Hutcheon, I.D. et al. (2013). Near infrared reflectance spectroscopy as a process signature in uranium oxides, *J. Radioanal. Nucl. Chem.* 296, 551–555.

Pointurier, F., Hubert, A., and Pottin, A.C. (2013). Performance of laser ablation: Quadrupole-based ICP-MS coupling for analysis of single micrometric uranium particles, *J. Radioanal. Nucl. Chem.* 296, 609–616.

Pointurier, F. and Marie, O. (2010). Identification of the chemical forms of uranium compounds in micrometer-size particles by micro-Raman spectrometry and scanning electron microscope, *Spectrochim. Acta B* 65, 797–804.

Pointurier, F., Pottin, A.C., and Hubert, A. (2011). Application of nanosecond-UV laser ablation—Inductively coupled plasma mass spectrometry for the isotopic analysis of single submicrometer size uranium particles, *Anal. Chem.* 83, 7841–7848.

Ranebo, Y., Eriksson, M., Tamborini, G. et al. (2007). The use of SIMS and SEM for the characterization of individual particles with a matrix originating from a nuclear weapon, *Microsc. Microanal.* 13, 179–190.

Ranebo, Y., Hedberg, P.M.L., Whitehouse, M.J. et al. (2009). Improved isotopic SIMS measurements of uranium particles for nuclear safeguards purposes, *Anal. Chem.* 24, 277–287.

Richter, S., Alonso, A., DeBolle, W. et al. (1999). Isotopic "fingerprints" for natural uranium ore samples, *Int. J. Mass Spectrom.* 193, 9–14.

Richter, S., Eykens, R., Kuhn, H. et al. (2010). New average values for the n(238U)/n(235U) isotope ratios of natural uranium standards, *Int. J. Mass Spectrom.* 295, 94–97.

Roudil, D., Rigaux, C., Rivier, C. et al. (2012). CETAMA contribution to safeguards and nuclear forensic analysis based on nuclear reference materials, *Procedia Chem.* 7, 709–715.

Sela, H., Karpas, Z., Zoriy, M. et al. (2007). Biomonitoring of hair samples by laser ablation inductively coupled plasma mass spectrometry (LA-ICP-MS), *Int. J. Mass Spectrom.* 261, 199–207.

Shen, Y., Zhao, Y., Guo, S.L. et al. (2008). Study on analysis of isotopic ratio of uranium-bearing particle in swipe samples by FT-TIMS, *Radiat. Meas.* 43, S299–S302.

Sohier, A. and Hardeman, F. (2006). Radiological dispersion devices: Are we prepared? *J. Environ. Radioact.* 85, 171–181.

Stanley, F.E. (2012). A beginner's guide to uranium chronometry in nuclear forensics and safeguards, *J. Anal. At. Spectrom.* 27, 1821–1830.

Stanley, F.E., Stalcup, A.M., and Spitz, H.B. (2013). A brief introduction to analytical methods in nuclear forensics, *J. Radioanal. Nucl. Chem.* 295, 1385–1393.

Stefaniak, E.A., Pointurier, F., Marie, O. et al. (2014). In-SEM Raman microspectroscopy coupled with EDX—A case study of uranium reference particles, *Analyst* 139(3), 668–675. doi: 10.1039/c3an01872e.

Stetzer, O., Betti, M., van Geel, J. et al. (2004). Determination of the ^{235}U content in uranium oxide particles by fission track analysis, *Nucl. Instrum. Methods Phys. Res. A* 525, 582–592.

Svedkauskaite-Legore, J., Mayer, K., Millet, S. et al. (2007). Investigation of the isotopic composition of lead and of trace elements concentrations in natural uranium materials as a signature in nuclear forensics, *Radiochim. Acta* 95, 601–605.

Tamborini, G., Donohue, D.L., Rudenauer, F.G. et al. (2004). Evaluation of practical sensitivity and useful ion yield for uranium detection by secondary ion mass spectrometry, *J. Anal. At. Spectrom.* 19, 203–208.

Tresl, I., De Wannemacker, G., Quetel, C.R. et al. (2004). Validated measurements of the uranium isotopic signature in human urine using magnetic sector-field inductively coupled plasma mass spectrometry, *Environ. Sci. Technol.* 38, 581–586.

Truyens, J., Stefaniak, E.A., and Aregbe, Y. (2013). NUSIMEP-7: Uranium isotope amount ratios in uranium particles, *J. Environ. Radioact.* 125, 50–55.

Varga, Z. (2008). Application of laser ablation inductively coupled plasma mass spectrometry for the isotopic analysis of single uranium particles, *Anal. Chim. Acta* 625, 1–7.

Varga, Z., Katona, R., Stefanka, Z. et al. (2010a). Determination of rare earth elements in uranium bearing materials by inductively couple plasma mass spectrometry, *Talanta* 80, 1744–1749.

Varga, Z. and Suranyi, G. (2007). Production date determination of uranium-oxide materials by inductively coupled plasma mass spectrometry, *Anal. Chim. Acta* 599, 16–23.

Varga, Z., Wallenius, M., and Mayer, K. (2010b). Age determination of uranium samples by inductively coupled plasma mass spectrometry using direct measurement and spectral deconvolution, *J. Anal. At. Spectrom.* 25, 1958–1962.

Varga, Z., Wallenius, M., Mayer, K. et al. (2009). Application of lead and strontium isotope ratio measurements for the origin assessment of uranium ore concentrates, *Anal. Chem.* 81, 8327–8334.

Varga, Z., Wallenius, M., Mayer, K. et al. (2011). Analysis of uranium ore concentrates for origin assessment, *Proc. Radiochim. Acta* 1, 1–4.

Vogt, S., Zahradnik, P., Klose, D. et al. (2010). Bulk analysis of environmental swipe samples. *International Safeguards Symposium.* Vienna, Austria: IAEA, IAEA-SM-367/10/06, pp. 1–7.

Wallenius, M., Mayer, K., and Ray, I. (2006). Nuclear forensic investigations: Two cases studies, *Forensic Sci. Int.* 156, 55–62.

Wallenius, M., Morgenstern, A., Apostolidis, C. et al. (2002). Determination of the age of highly enriched uranium, *Anal. Bioanal. Chem.* 374, 379–384.

Weitz, R. (2011). Nuclear forensics: False hopes and practical realities, *Polit. Sci. Quat.* 126, 53–75.

Wogman, N.A., Wigmosta, M.S., Swindle, D.W. et al. (2001). Wide area aquatic sampling and analysis for detection of nuclear proliferation, *J. Radioanal. Nucl. Chem.* 248, 611–615.

6 Summary

And he shall judge among the nations, and shall rebuke many people: and they shall beat their swords into plowshares, and their spears into pruning hooks: nation shall not lift up sword against nation, neither shall they learn war any more.

Isaiah 2:4, King James Bible "Authorized Version," Cambridge Edition

After the general introduction chapter, which covered the properties of uranium and its main compounds and an overview of analytical techniques, a myriad of analytical procedures used for characterization of uranium were presented in Chapters 2 through 5. This large variety reflects the diversity of samples in which uranium has to be determined. On the one hand, there are the samples in which uranium is the major constituent, as expressed in Chapter 2, which described the nuclear fuel cycle, and the focus is on determining the level of impurities in the nuclear grade materials. A unique analytical challenge involves the characterization of highly radioactive samples like those encountered in spent nuclear fuel (after irradiation in a reactor). On the other hand, there are environmental samples that include soil, sediments, plants, water bodies, and air samples, in which uranium is present at low levels, usually in the parts-per-million to parts-per-billion range. In addition, there are biological samples, mainly of urine, feces, hair, and nails, in which uranium is present at trace level concentrations, usually in the parts-per-billion to parts-per-trillion range. For nuclear forensics and safeguards applications, uranium-bearing particles need to be detected and located prior to characterization according to their physical, chemical, and elemental properties. The isotopic composition of uranium is an essential feature of uranium characterization throughout this monograph.

The diversity of samples, the specifications, and analytical requirements have led to the development of procedures for sample preparation that often involve separation, purification, and preconcentration of the uranium. In some cases, where uranium is the major component, the uranium matrix is removed in order to improve the detection limits of trace impurities, while in other cases separation of uranium is needed for accurate isotope composition measurements.

The nuclear fuel cycle that follows the fate of uranium from the cradle (prospecting for deposits and mining operations) to the grave (sometimes literally the tomb and burial of uranium and other radioactive waste products) is schematically depicted in Figure 6.1. The rigid and strict specifications of nuclear materials and components of nuclear power plants are required to ensure proper operation, on the one hand, and to prevent the formation of highly radioactive products, on the other hand. The pathways through which the general public may be exposed to uranium compounds impose severe limits on the permissible concentration of uranium in food and drinking water, but it should always be kept in mind that uranium is a naturally occurring element and is ever present in our biosphere. Sensitive analytical methods

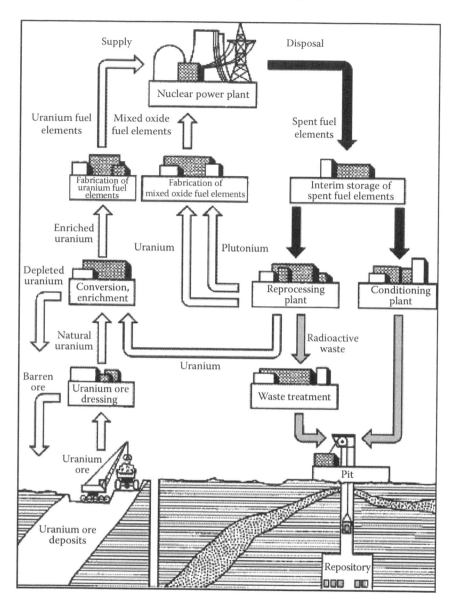

FIGURE 6.1 Schematic of the nuclear fuel cycle from the uranium deposits to the repository of uranium and other radioactive waste products. (From Informationskreis KernEnergie, Berlin, Germany; http://www.euronuclear.org/info/encyclopedia/images/nuc_fuel.gif, accessed August 5, 2014. With permission.)

were therefore developed for the assessment of exposure and the body burden on the basis of bioassays. The importance of strengthening the safeguards regime that is employed to prevent the proliferation of nuclear weapons has led to the development of extremely sensitive analytical instrumentation and procedures. The accurate determination of the full isotopic composition of uranium in swipe samples collected from declared and suspect nuclear facilities plays an important role in deploying the nonproliferation treaty.

Finally, highly enriched uranium could serve as the forerunner of the Biblical prophecy of beating *swords into plowshares* as more HEU from nuclear warheads is down blended into nuclear fuel containing low-enriched uranium. It is estimated that about 10% of the electric power produced by nuclear power plants in the United States originates from down-blending of HEU from ex-USSR sources.

Index